313

Topics in Current Chemistry

Topics in Current Chemistry
Recently Published and Forthcoming Volumes

Unimolecular and Supramolecular Electronics II
Volume Editor: Robert M. Metzger
Vol. 313, 2012

Unimolecular and Supramolecular Electronics I
Volume Editor: Robert M. Metzger
Vol. 312, 2012

Bismuth-Mediated Organic Reactions
Volume Editor: Thierry Ollevier
Vol. 311, 2012

Peptide-Based Materials
Volume Editor: Timothy Deming
Vol. 310, 2012

Alkaloid Synthesis
Volume Editor: Hans-Joachim Knölker
Vol. 309, 2012

Fluorous Chemistry
Volume Editor: István T. Horváth
Vol. 308, 2012

Multiscale Molecular Methods in Applied Chemistry
Volume Editors: Barbara Kirchner, Jadran Vrabec
Vol. 307, 2012

Solid State NMR
Volume Editor: Jerry C. C. Chan
Vol. 306, 2012

Prion Proteins
Volume Editor: Jörg Tatzelt
Vol. 305, 2011

Microfluidics: Technologies and Applications
Volume Editor: Bingcheng Lin
Vol. 304, 2011

Photocatalysis
Volume Editor: Carlo Alberto Bignozzi
Vol. 303, 2011

Computational Mechanisms of Au and Pt Catalyzed Reactions
Volume Editors: Elena Soriano, José Marco-Contelles
Vol. 302, 2011

Reactivity Tuning in Oligosaccharide Assembly
Volume Editors: Bert Fraser-Reid, J. Cristóbal López
Vol. 301, 2011

Luminescence Applied in Sensor Science
Volume Editors: Luca Prodi, Marco Montalti, Nelsi Zaccheroni
Vol. 300, 2011

Chemistry of Opioids
Volume Editor: Hiroshi Nagase
Vol. 299, 2011

Electronic and Magnetic Properties of Chiral Molecules and Supramolecular Architectures
Volume Editors: Ron Naaman, David N. Beratan, David H. Waldeck
Vol. 298, 2011

Natural Products via Enzymatic Reactions
Volume Editor: Jörn Piel
Vol. 297, 2010

Nucleic Acid Transfection
Volume Editors: Wolfgang Bielke, Christoph Erbacher
Vol. 296, 2010

Carbohydrates in Sustainable Development II
Volume Editors: Amélia P. Rauter, Pierre Vogel, Yves Queneau
Vol. 295, 2010

Unimolecular and Supramolecular Electronics II

Chemistry and Physics Meet at Metal-Molecule Interfaces

Volume Editor: Robert M. Metzger

With Contributions by

B. Branchi · C. Herrmann · K.W. Hipps · M. Hliwa · C. Joachim · C. Li · D.L. Mattern · R.M. Metzger · A. Mishchenko · M.A. Rampi · M.A. Ratner · N. Renaud · F.C. Simeone · G.C. Solomon · T. Wandlowski

Springer

Editor
Prof. Robert M. Metzger
Department of Chemistry
The University of Alabama
Room 1088B, Shelby Hall
Tuscaloosa, AL 35487-0336
USA
rmetzger@ua.edu

ISSN 0340-1022 e-ISSN 1436-5049
ISBN 978-3-642-43324-5 ISBN 978-3-642-27398-8 (eBook)
DOI 10.1007/978-3-642-27398-8
Springer Heidelberg Dordrecht London New York

Printed on acid-free paper

Springer is part of Springer Science+Business Media (www.springer.com)

Volume Editor

Prof. Robert M. Metzger

Department of Chemistry
The University of Alabama
Room 1088B, Shelby Hall
Tuscaloosa, AL 35487-0336
USA
rmetzger@ua.edu

Editorial Board

Topics in Current Chemistry Also Available Electronically

Topics in Current Chemistry is included in Springer's eBook package *Chemistry and Materials Science*. If a library does not opt for the whole package the book series may be bought on a subscription basis. Also, all back volumes are available electronically.

For all customers with a print standing order we offer free access to the electronic volumes of the series published in the current year.

If you do not have access, you can still view the table of contents of each volume and the abstract of each article by going to the SpringerLink homepage, clicking on "Chemistry and Materials Science," under Subject Collection, then "Book Series," under Content Type and finally by selecting *Topics in Current Chemistry*.

You will find information about the

- Editorial Board
- Aims and Scope
- Instructions for Authors
- Sample Contribution

at springer.com using the search function by typing in *Topics in Current Chemistry*.

Color figures are published in full color in the electronic version on SpringerLink.

Aims and Scope

The series *Topics in Current Chemistry* presents critical reviews of the present and future trends in modern chemical research. The scope includes all areas of chemical science, including the interfaces with related disciplines such as biology, medicine, and materials science.

The objective of each thematic volume is to give the non-specialist reader, whether at the university or in industry, a comprehensive overview of an area where new insights of interest to a larger scientific audience are emerging.

Thus each review within the volume critically surveys one aspect of that topic and places it within the context of the volume as a whole. The most significant developments of the last 5–10 years are presented, using selected examples to illustrate the principles discussed. A description of the laboratory procedures involved is often useful to the reader. The coverage is not exhaustive in data, but rather conceptual, concentrating on the methodological thinking that will allow the non-specialist reader to understand the information presented.

Discussion of possible future research directions in the area is welcome.

Review articles for the individual volumes are invited by the volume editors.

In references *Topics in Current Chemistry* is abbreviated *Top Curr Chem* and is cited as a journal.

Impact Factor 2010: 2.067; Section "Chemistry, Multidisciplinary": Rank 44 of 144

Preface

For these volumes in the Springer book review series *Topics in Current Chemistry*, it seemed natural to blend a mix of theory and experiment in chemistry, materials science, and physics. The content of this volume ranges from conducting polymers and charge-transfer conductors and superconductors, to single-molecule behavior and the more recent understanding in single-molecule electronic properties at the metal–molecule interface.

Molecule-based electronics evolved from several research areas:

1. A long Japanese tradition of studying the organic solid state (since the 1940s: school of Akamatsu).
2. Cyanocarbon syntheses by the E. I. Dupont de Nemours Co. (1950–1964), which yielded several interesting electrical semiconductors based on the electron acceptor 7,7,8,8-tetracyanoquinodimethan (TCNQ).
3. Little's proposal of excitonic superconductivity (1964).
4. The erroneous yet over-publicized claim of "almost superconductivity" in the salt TTF TCNQ (Heeger, 1973).
5. The first organic superconductor (Bechgard and Jérôme, 1980) with a critical temperature $T_c = 0.9$ K; other organic superconductors later reached T_c 13 K.
6. Electrically insulating films of polyacetylene, "doped" with iodine and sodium, became semiconductive (Shirakawa, MacDiarmid, Heeger, 1976).
7. The interest in TTF and TCNQ begat a seminal theoretical proposal on one-molecule rectification (Aviram and Ratner, 1974) which started unimolecular, or molecular-scale electronics.
8. The discovery of scanning tunneling microscopy (Binnig and Rohrer, 1982).
9. The vast improvement of electron-beam lithography.
10. The discovery of buckminsterfullerene (Kroto, Smalley, and Curl, 1985).
11. Improved chemisorption methods ("self-assembled monolayers") and physisorption methods (Langmuir–Blodgett films).
12. The growth of various nanoparticles, nanotubes, and nanorods, and most recently graphene.

All these advances have helped illuminate, inspire, and develop the world of single-molecule electronic behavior, and its extension into supramolecular assemblies.

These volumes bring together many of the leading practitioners of the art (in each case I mention only the main author). Bässler sets in order the theoretical understanding of electron transport in disordered (semi)-conducting polymers. Saito summarizes in fantastic detail the progress in understanding charge-transfer crystals and organic superconductivity. Echegoyen reviews the chemistry and electrochemistry of fullerenes and their chemical derivatives. Thompson reviews the progress made in organic photovoltaics, both polymeric and charge-transfer based. Ratner updates the current status of electron transfer theory, as is applies to measurements of currents through single molecules. Metzger summarizes unimolecular rectification and interfacial issues. Kagan discusses field-effect transistors with molecular films as the active semiconductor layer. Allara reminds us that making a "sandwich" of an organic monolayer between two metal electrodes often involves creep of metal atoms into the monolayer. Rampi shows how mercury drops and other techniques from solution electrochemistry can be used to fabricate these sandwiches. Wandlowski discusses how electrochemical measurements in solution can help enhance our understanding of metal–molecule interfaces. Hipps reviews inelastic electron tunneling spectroscopy and orbital-mediated tunneling. Joachim addresses fundamental issues for future molecular devices, and proposes that, in the best of possible worlds, all active electronic and logical functions must be predesigned into a single if vast molecular assembly. Szulczewski discusses the spin aspects of tunneling through molecules: this is the emerging area of molecular spintronics.

Many more areas could have been discussed and will undoubtedly evolve in the coming years. It is hoped that this volume will help foster new science and even new technology. I am grateful to all the coauthors for their diligence and Springer-Verlag for their hosting our efforts.

Tuscaloosa, Alabama, USA
Delft, The Netherlands
Dresden, Germany

Robert Melville Metzger

Contents

Top Curr Chem (2012) 313: 1–38
DOI: 10.1007/128_2011_227

Published online: 14 September 2011

Molecular Electronic Junction Transport: Some Pathways and Some Ideas

Gemma C. Solomon, Carmen Herrmann, and Mark A. Ratner

Abstract When a single molecule, or a collection of molecules, is placed between two electrodes and voltage is applied, one has a molecular transport junction. We discuss such junctions, their properties, their description, and some of their applications. The discussion is qualitative rather than quantitative, and focuses on mechanism, structure/function relations, regimes and mechanisms of transport, some molecular regularities, and some substantial challenges facing the field. Because there are many regimes and mechanisms in transport junctions, we will discuss time scales, geometries, and inelastic scattering methods for trying to determine the properties of molecules within these junctions. Finally, we discuss some device applications, some outstanding problems, and some future directions.

Keywords Conduction · Electron transfer · Electron transport · Molecular electronics

Contents

G.C. Solomon
Nano-Science Center and Department of Chemistry, University of Copenhagen, Universitetsparken 5, 2100 Copenhagen, Østerbro, Denmark

C. Herrmann
Institute for Inorganic and Applied Chemistry, University of Hamburg, Martin-Luther-King-Platz 6, 20146 Hamburg, Germany

M.A. Ratner (✉)
Department of Chemistry, Northwestern University, 2145 Sheridan Road, Evanston, IL, 60208 USA
e-mail: ratner@northwestern.edu; ratner@chem.northwestern.edu

1 Introduction

To a scientist, the fundamental properties of the real world break down into two broad categories, structure and dynamics. The two are often commingled – the baseball gives the home run, the planet gives its orbit, the muscle fiber gives contraction and expansion, and the donor/bridge/acceptor molecule gives phosphorescence, fluorescence, nonradiative decay, photovoltaic behavior, and electron transfer [1]. A molecular transport junction, which is the structure of most interest in this chapter, provides current flows as a function of voltage, temperature, geometrical arrangement, chemical composition, and density of environment.

Electron transfer in donor/bridge/acceptor molecules and currents in molecular transport junctions are closely related by the Born–Oppenheimer separation [2] that uses the mass difference between electrons and nuclei to permit isolated discussion of electron dynamics that almost always occur far faster than those of nuclei. The understanding that electron tunneling is a common feature between intramolecular electron transfer and transport in molecular junction structures was used by Nitzan to produce an approximate linear relation between the measurable quantities (rate constants for electron transfer, and conductances for molecular junctions) [3].

While chemists love structure (as a glance through any chemical journal will show), they are generally fascinated with mechanism. This short chapter is about some of the models, ideas, and understandings that have occurred in electron transport and molecular junctions [4–6]. The field is large, the problems are hard, the processes could be important both for our understanding and for many commercial applications, and finally the issues are fascinating for the chemical imagination.

The remainder of the chapter is structured as follows. Section 2 discusses the physical description of transport junctions, dealing with length scales, categorization, and the particular measurements that can be made. Section 3 is devoted to models – the general nature of models, and then the geometric, molecular, Hamiltonian, and transport models that are associated with molecular transport junctions and their interpretation.

Section 4 is entitled "Ideas" (for mechanisms and models). It deals with how we can interpret/calculate the behavior of molecular transport junctions utilizing particular model approaches and chemical mechanisms. It also discusses time parameters, and coherence/decoherence as well as pathways and structure/function relationships.

Section 5 is on one particular molecule, *p*-benzene dithiol. This is one of the most commonly studied molecules in molecular electronic transport junctions [7] (although it is also one of the most problematic). Section 6 discusses a separate measurement, inelastic electron tunneling spectroscopy [8, 9] (IETS). This can be quite accurate because it can be done on single molecules at low temperatures. It occurs because of small perturbations on the coherent transport, but it can be very indicative of such issues as the geometrical arrangement in the molecular transport junction, and pathways for electron transport through the molecular structure.

Finally, some remarks on the different subfields of the larger topic of molecular electronics are found in Sect. 7.

2 Physical Description of Molecular Transport Junctions

By definition, a molecular transport junction consists of a molecule extended between two macroscopic electrodes. The nature of the molecule, the environment (whether it is solvated or not), the electrode's shape and composition, the temperature, the binding of the molecule to the electrodes, and the applied field are all variables that are relevant to the measurement, which is usually one of differential conductance, defined as the derivative of the current with respect to voltage.

Figure 1 shows two things: a number of sketches of possible geometries for solid-state molecular transport junctions, and some electron microscopy images of actual functional transport junctions. There are two striking features to note: first, the

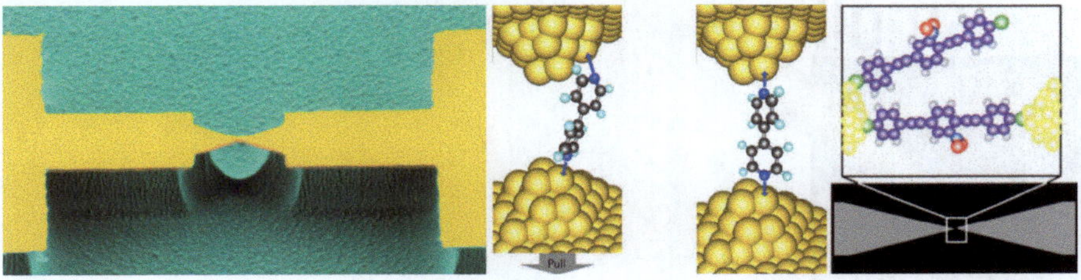

Fig. 1 Sketches of break junction-type test beds for molecular transport. On the far left is a tunneling electron microscopy (TEM) image of the actual metallic structure in (mechanical) break junctions from the nanoelectronics group at University of Basel. The sketches in the middle (Reprinted by permission from Macmillan Publishers Ltd: Nature Nanotechnology 4, 230–234 (2009), copyright 2009) and right (reproduced from Molecular Devices, A.M. Moore, D.L. Allara, and P.S. Weiss, in NNIN Nanotechnology Open Textbook (2007) with permission from the authors) show possible geometries for molecules between two gold electrodes, and (on the upper right) a molecule that has only one end attached across the junction

sketches are suggestive, but this is not evidence for their precision – in particular, we know essentially nothing about the coordination of molecules in transport junctions, nothing about the actual geometry (whether the molecule is standing, lying down, at a tilt angle). We do not know how many molecules are in the junction, and, if there is more than one, we certainly do not know their relative geometries.

On the other hand, the images show fairly definitively the structure of the metallic electrodes. We see that they are often regular at first sight (but often irregular at the atomic scale) and we know that when they are made of soft metals like gold or silver they can distort as the measurement is made. Therefore, our lack of understanding of length scales and geometries is one of the crucial aspects in molecular transport junctions that we will refer to time after time.

The structure of the molecular transport junction is reminiscent of the transport junctions used in a fascinating and important subarea of condensed matter physics: mesoscopic physics. In these (Fig. 2 shows an analogous chemical system), current, conductance, and higher derivatives are normally measured for systems containing a quantum dot or several quantum dots between two electrodes, usually in an environment in which gating can be applied. The two-dimensional electron gas is one of the standard systems in mesoscopic physics, one in which exquisite control can be achieved (Fig. 3). The striking difference between molecular and mesoscopic transport junctions is that the controls on geometry are very weak in the molecular situation – the fact that molecules are all the same as each other (one naphthalene is the same as every other naphthalene) does not help, because the length scale on which the system operates is so much larger.

Mesoscopic physics has defined many of the issues (Landauer limit transport [10, 11], Coulomb blockade regime [12], Kondo resonance regime [13–15]. . .) that will occur later in this chapter describing molecular transport junctions. These concepts are relevant, but must be reinterpreted to understand the molecular case.

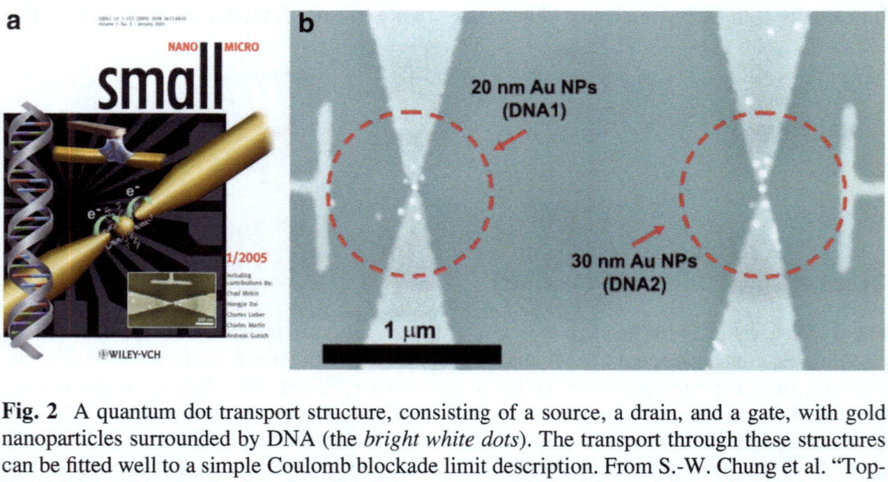

Fig. 2 A quantum dot transport structure, consisting of a source, a drain, and a gate, with gold nanoparticles surrounded by DNA (the *bright white dots*). The transport through these structures can be fitted well to a simple Coulomb blockade limit description. From S.-W. Chung et al. "Top-Down Meets Bottom-Up: Dip-Pen Nanolithography and DNA-Directed Assembly of Nanoscale Electrical Circuits" Small (2005) 1, 64–69. Copyright Wiley-VCH Verlag GmbH & Co. KGaA. Reproduced with permission

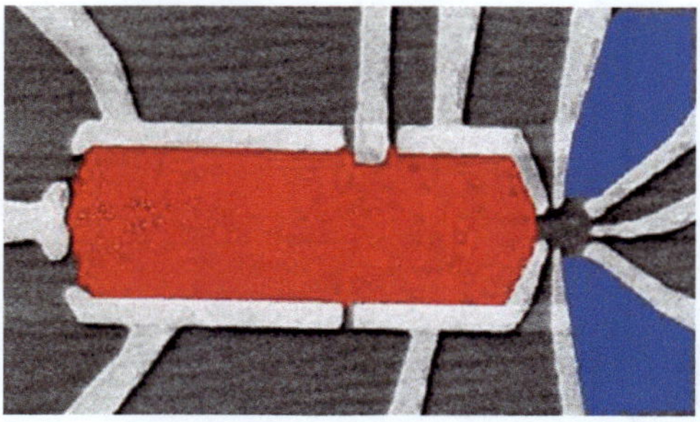

Fig. 3 A two-dimensional electron gas fabricated in the lab of David Goldhaber-Gordon by Ron Potok. These structures, from the realm of mesoscopic physics, can be tuned to provide many different sorts of transport structures, and their geometry is entirely controlled by fabrication. The *red region* is 3 μm long

2.1 Categories, Break Junctions, and Structure

While many different molecular junction structures have been developed and utilized, they fall into three large categories. The first are measurements that by their nature observe ensembles of molecules. These include a range of systems comprising self-assembled monolayers (SAMs) measured in various ways, from molecular chemistry to the use of nanodot collectors to mesa-type structures supporting a small number of molecules [16]. Ensemble measurements are also made using conducting atomic force microscopy [17–19] with or without quantum dots as collectors. Other approaches that measure ensembles of molecules and their transport include the approach of using a liquid drop as one of the electrodes [20, 21] (Fig. 4). Finally, the nanopore structure [22] developed by Reed and his colleagues is a more elegant, and smaller, ensemble sort of measurement.

In sharp contrast to these are single molecule measurements, shown schematically in Fig. 5. These are normally done utilizing a break junction technique – a mechanical break junction [23–27] is one in which a thin region in a single metal is broken either by bending or stretching; the molecule is often trapped between the broken structures (Fig. 1). The electrochemical break junction [28] is one in which a metallic strand is stretched to breaking in a solution containing a molecule that can then bind to both broken ends of the strand. The difference is that the mechanical break junction is almost always used in vacuum, whereas the electrochemical break junction is almost always used in solution. Both can be gated, but the gating is very different – the mechanical break junction is gated by a third planar electrode reminiscent of a traditional semiconductor structure [15], while the electrochemical break junction is gated by a reference electrode, so that the measurements [29, 30] resemble single molecule electrochemistry.

A group in England has developed a very nice idea based on fluctuations [31]: here a molecule is chemisorbed on one end to a surface, and a conductive scanning tip is brought to within about a molecule length from the supporting metal. Thermal excitation then permits molecules to form instantaneous transport bridge structures between the planar support and the conductive electrode – one observes fluctuations

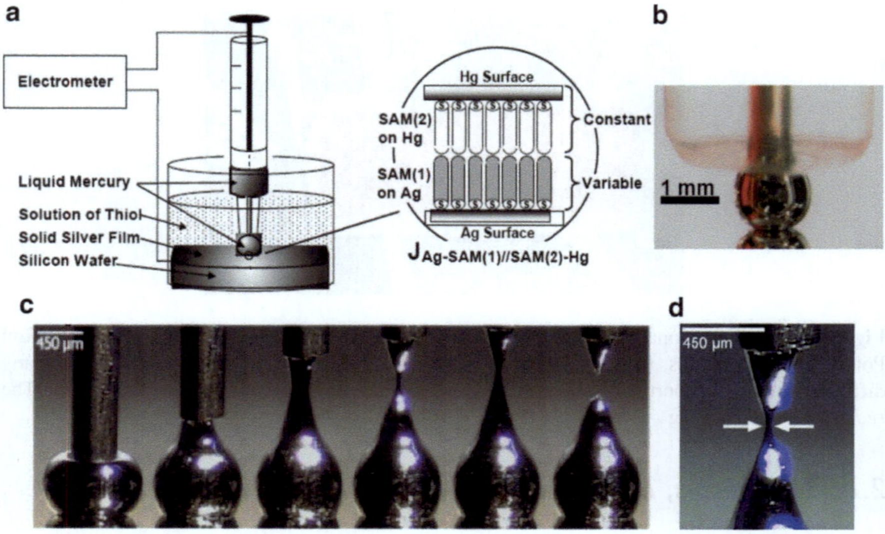

Fig. 4 The liquid metal droplet test bed for molecular conductance. As the drop comes into contact with the surface, the molecules contained on the surface of one can form a bridge to the other, resulting in an inexpensive, quite generally useful test bed for molecular transport (in this case it is a multimolecule transport situation). The setup is shown schematically in (**a**) and the liquid mercury drop on a surface in (**b**). Reprinted with permission from Michael L. Chabinyc et al. J. Am. Chem. Soc. (2002) 124, 11730–11736. Copyright 2011 American Chemical Society. An alternative liquid electrode is eutectic gallium indium (EGaIn) shown in (**c**, **d**); a protective oxide layer forms on the EGaIn surface making a second monolayer of molecules unnecessary. EGaIn has very different rheology from Hg making it possible to prepare narrower liquid tips. From R. C. Chiechi et al. "Eutectic Gallium–Indium (EGaIn): A Moldable Liquid Metal for Electrical Characterization of Self-Assembled Monolayers" Angew Chem Int Ed (2007) 120, 148–150. Copyright Wiley-VCH Verlag GmbH & Co. KGaA. Reproduced with permission

here, fluctuations that arise from the motion of the molecules between different bending geometries, as well as breaking the interaction with the tip altogether.

The categories just described compromise the majority of the measurements on molecular transport junctions.

The lack of information about the molecular geometry within the junction raises a crucial issue. It is one that we will continue to return to, because it is the most vexatious issue – especially in contrast to vapor phase measurements, crystal structures, and even NMR structures, where one can place very tight metric constraints on bond lengths (certainly 0.01 Å accuracy can be obtained even by crude scattering methods). This is emphatically not true in these measurements – while techniques such as IETS and simultaneous measurement of conductance and Raman spectra [32] may give indirect information on molecular bonding in the junction, no instruments exist to measure the geometries of a transport junction directly, even in the absence of current flow, and it is even more difficult in the nonequilibrium situation when current is flowing.

It is possible to use electronic structure calculations combined with measurements in which the geometry is purposely varied to make some elegant deductions about the adsorption of molecules on the electrodes. A beautiful example is provided by work

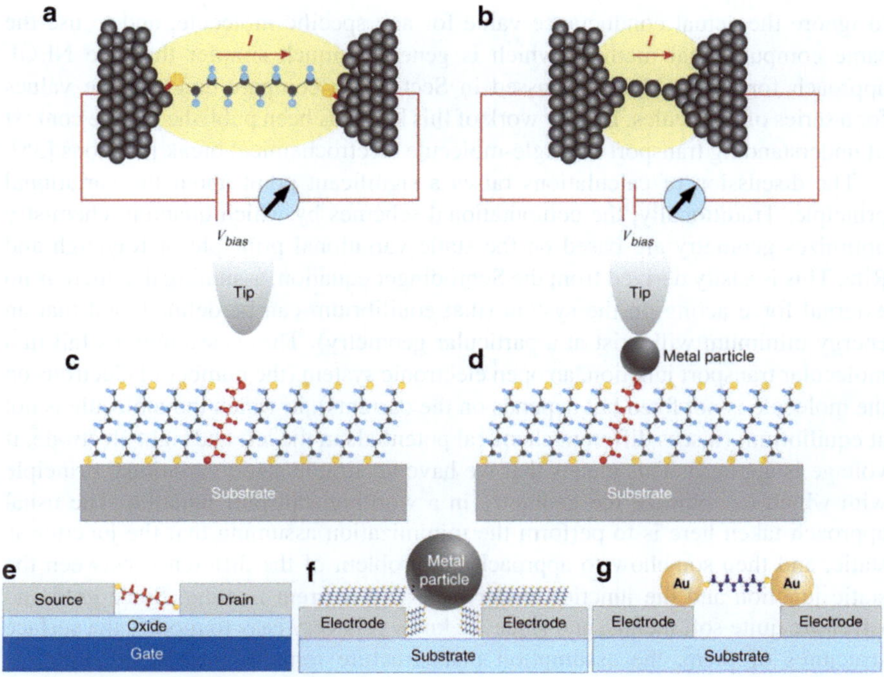

Fig. 5 (**a**) Current through a molecule covalently bound to two electrodes. (**b**) Current through a metal atom attached to two electrodes made of the same metal. (**c**) Scanning tunneling microscopy (STM) study of electron transport through a target molecule inserted into an ordered array of reference molecules. (**d**) STM or conducting atomic force microscopy (AFM) measurement of conductance of a molecule with one end attached to a substrate and the other end bound to a metal nanoparticle. Schematic illustrations of single-molecule conductance studies using different methods. (**e**) A single molecule bridged between two electrodes with a molecular-scale separation prepared by electromigration, electrochemical etching or deposition, and other approaches. (**f**) Formation of molecular junctions by bridging a relatively large gap between two electrodes using a metal particle. (**g**) A dimer structure, consisting of two Au particles bridged with a molecule, assembled across two electrodes (Reprinted with permission from Ann. Rev. Phys. Chem. (2007) 58, 535–564)

from the Columbia/Brookhaven group [33] employing electrochemical break junctions under extension, and using a combination of calculation and observation to suggest that the amine groups with which these molecules are capped select a single unsaturated gold atom to bind to – this is quite surprising in terms of the more standard sulfur terminations, and represents a real triumph of analysis. Similarly, beautiful measurements on gold wires [34] (not really a topic in molecular electronics, but one of great relevance, especially considering the role of the gold wires in electrochemical junctions) showed that there was a sharp correlation between the transport measurements and the electron microscopy measurements of geometric reorganization in the metal as current was passing through it.

In general, however, many relevant geometric parameters are unknown in molecular transport junctions, and therefore it is necessary to make assumptions, and calculations, to help in understanding the geometry. One interesting approach is to ignore the actual conductance value for any specific molecule, and to use the same computational method (which is generally much simpler than the NEGF approach for conductance discussed in Sect. 4) to compare conductance values for a series of molecules. Lovely work of this kind has been published in the context of understanding transport in single-molecule electrochemical break junctions [35].

The discussion of calculations raises a significant point about the variational principle. Traditionally, the computational schemes by which quantum chemistry optimizes geometry are based on the static variational principle of Rayleigh and Ritz. This is easily derived from the Schrödinger equation, assuming that there is no external force acting on the system (that equilibrium can be defined, and that an energy minimum will exist at a particular geometry). These assumptions fail in a molecular transport junction, an open electronic system (the number of electrons on the molecule is not fixed but depends on the currents), in which the molecule is not at equilibrium (it sees different chemical potentials in the left and right electrode, if voltage is applied). This means that we have no simple static variational principle with which to optimize the geometry in a working transport junction. The usual approach taken here is to perform the minimization assuming that the junction is static, and then somehow to approach the problem of the difference between the static junction and the junction under bias, with current flowing. Since gold and silver are quite soft metals, and since we know it is very easy to modify the surface structures of them, the assumption that structure remains unchanged during a current/voltage experiment seems dubious. Therefore, there is no good theoretical method to calculate the molecular geometry – this is one of the major open challenges in molecular transport junctions.

2.2 Measurements

The quantities to be measured in transport junctions are current, voltage, conductance, inelastic electron tunneling spectroscopy (essentially the derivative of the conductance with respect to voltage), and the conductance as the molecular structure is distorted, generally by stretching [33, 36–38]. Additional measurements are sometimes made, including optical spectroscopy, vibrational spectroscopy (in particular Raman spectroscopy) [32, 39] and using particular applications such as the MOCSER entity [40, 41] (essentially a molecular transistor developed by the Weizmann group).

3 A Bit on Models

Science is largely about the world around us, about reality insofar as we can grasp it. But since the days of Euclid, and particularly since Lucretius, scientists have constructed models – that is, scientists have made simulacra, either conceptual or physical, in an attempt to mimic aspects of what they perceive to be reality, but to do so in a more comprehensible or revelatory way. This tradition, now more than two millennia old, was reinvented by Newton, who modeled the universe in terms of particles with mass but no physical extension – Einstein followed with models for relativity, and modern physical science is probably most familiar with models used in dealing with the nature of quantum mechanics – that is, the nature of matter as we perceive it.

Several categories of models appear as the basis for the study of molecular electronics in general, and molecular transport junctions in particular. These are the geometrical (or molecular), Hamiltonian, and transport analysis models.

The geometrical models have been mentioned already, but must be referred to again. In building an understanding of transport junctions, we need to know the geometry at least at some level. The geometrical models are almost always simply atom placement, sometimes static and sometimes not. Since there is no legitimate way to compute the optimal geometry, it is simply assumed for some (possibly arbitrary) reason – this represents the geometric model, upon whose statics the dynamics of electron transport is pursued.

The molecular models are in a sense a subset of the geometrical ones – we assume that we know which molecules are present and we assume that we know their geometries (indeed sometimes we assume more than that, such as the usual assumption that thiol end groups lose their protons when forming their asymmetric bond with gold). In this we also necessarily assume that there are no other species, either on the electrode surface or in the surrounding media, that influence the current flow through the system.

Then, there are model Hamiltonians. Effectively a model Hamiltonian includes only some effects, in order to focus on those effects. It is generally simpler than the true full Coulomb Hamiltonian, but is made that way to focus on a particular aspect, be it magnetization, Coulomb interaction, diffusion, phase transitions, etc. A good example is the set of model Hamiltonians used to describe the IETS experiment and (more generally) vibronic and vibrational effects in transport junctions. Special models are also used to deal with chirality in molecular transport junctions [42, 43], as well as optical excitation, Raman excitation [44], spin dynamics, and other aspects that go well beyond the simple transport phenomena associated with these systems.

The Hamiltonian models are broadly variable. Even for an isolated molecule, it is necessary to make models for the Hamiltonian – the Hamiltonian is the operator whose solutions give both the static energy and the dynamical behavior of quantum mechanical systems. In the simplest form of quantum mechanics, the Hamiltonian is the sum of kinetic and potential energies, and, in the Cartesian coordinates that are used, the Hamiltonian form is written as

$$H = \sum_i P_i^2/2m + V(\vec{X}). \tag{1}$$

Here the electron mass is m, P_i is the momentum of the given particle i, and $\vec{x}$ represents the vector of all displacements, both electronic and nuclear. We have assumed that, following the Born–Oppenheimer approximation, electronic and nuclear motions are decoupled, and a purely electronic Hamiltonian can be defined as in (1) (with the nuclear coordinates entering only as parameters). For very simple systems like the hydrogen atom, quantum mechanics is solved in exactly this form by choosing the Coulomb potential for V and then finding the eigenvalues and eigenfunctions analytically.

For anything bigger than the hydrogen atom, however, solving directly in terms of the coordinates and momenta becomes extremely difficult. Far more common is to express the wave function in terms of basis functions, introducing the idea of second quantization [45]. A simple way to think of second quantization is that it describes the quantum mechanics, from the beginning, in terms of a set of basis functions.

As a simple example, if we choose to work on the problem of the spectroscopy of the benzene molecule, we might make a model in which we ignore all repulsions among the electrons, we ignore the σ electrons, and we take the π electron wave function to be represented in terms of six sites each containing a single $p\pi$ orbital and centered at a carbon nucleus. We then restrict the electronic interactions to exist only between neighboring carbons. Still retaining the assumption that these $p\pi$ orbitals are orthogonal and form a complete basis set for our model, the model becomes the standard Huckel model, that can be written as

$$H_{\text{Huc}} = 1/2 \sum_i \sum_j \beta_{i,j} (a_i^+ a_j + a_j^+ a_i). \tag{2}$$

Here the operator a_i^+ creates (and the operator a_i removes) an electron at site i; the nn denotes near-neighbors only, and $\beta_{i,j} = \int \mathrm{d}r \phi_i H \phi_j$ denotes a Coulomb integral if $i = j$ and a resonance integral otherwise. The second quantization form of this equation clearly requires a basis set. It is a model for the behavior of benzene – not a terribly accurate one, but one that helps us understand many things about its spectroscopy, its stability, its binding patterns, and other physical and chemical properties.

If the basis set is restricted to one $p\pi$ basis function on each sp^2 carbon, if the two-electron integrals ignore all three-center or four-center ones, and if we exclude exchange components, one has the Pariser–Parr–Pople model. If, further, all two-electron integrals are set to zero except for the repulsion between opposite spins on the same site and the one-electron tunneling terms are restricted to nearest neighbors, the result is the Hubbard Hamiltonian

$$H_{\text{Hub}} = H_{\text{Huc}} + U \sum_i n_{i,\uparrow} n_{i,\downarrow} \tag{3}$$

with β, U the parameters of the model and $n_{i\sigma} = a_{i\sigma}^+ a_{i\sigma}$ the number operator for an electron of spin σ on site i.

In molecular transport junctions, the Hamiltonian models are usually based on Kohn–Sham density functional theory [46–48]. They use relatively small basis sets because the calculations are sufficiently complicated, they take a number of empirical steps for dealing with the basis sets and their potential integrals, and they assume a static basis (that is, the ground and excited states are described in the same basis). The more complicated the model, the more complicated the calculation.

The tradition of model building only works when the right model is chosen for the right problem. For qualitative understanding of molecular charge transport, extended Huckel models can actually be useful [49] – to get quantitative information, one requires either a high level ab initio approach (going well beyond Hartree–Fock) or (much more commonly) a density functional theory with a fairly sophisticated functional, and with corrections to get the one-electron levels at roughly the right energy [50].

A great deal more could be said about models – to understand behavior like strong correlation, Coulomb blockade, and actual line shapes, it is necessary to use a number of empirical parameters, and a quite sophisticated form of density functional theory that deals with both static and dynamic correlation at a high level. Often this can be done only within a very simple representation of the electrons – something like the Hubbard model [51–53], which is very common in this situation.

General issues with models are discussed elsewhere. For our purposes here it is important to remember that model Hamiltonians are the only way in which any molecule larger than diatomic is ever described – in a sense, the science resides in using the right model for the right system, and solving it appropriately.

Models are also required for analysis of the transport. For calculations of current/voltage curves, current density, inelastic electron scattering, response to external electromagnetic fields, and control of transport by changes in geometry, one builds transport models. These are generally conceptual – more will be said below on the current density models and IETS models that are used to interpret those experiments within molecular transport junctions.

In mesoscopic physics, because the geometries can be controlled so well, and because the measurements are very accurate, current under different conditions can be appropriately measured and calculated. The models used for mesoscopic transport are the so-called Landauer/Imry/Buttiker elastic scattering model for current, correlated electronic structure schemes to deal with Coulomb blockade limit and Kondo regime transport, and charging algorithms to characterize the effects of electron populations on the quantum dots. These are often based on capacitance analyses (this is a matter of thinking style – most chemists do not consider capacitances when discussing molecular transport junctions).

Another set of models involves molecular mechanisms – how does current pass through molecules? We know that coherent transport (tunneling through the molecule) could occur in short molecules, and that the transition to hopping transport (electrons localized for long time scales compared to the scales on which they move between these localization sites) is common in electron transfer systems; by the Nitzan analogy we would expect the same to be seen in conductance junctions, and indeed this has been observed [54]. The mechanistic transition from tunneling to hopping is a fascinating one, with many areas still uncertain, particularly for ionic molecules like DNA.

The third set of models is for understanding the actual currents, and the pathways that the currents follow through molecular transport junctions. This is to some extent a matter of visualization and categorization, but it is very helpful in understanding the mechanism of molecular transport.

Occasionally terms from models can be misused badly. For example, the standard, nonequilibrium Green's function/density functional theory approach to transport (the most common one for general calculations on molecular junctions) [55–66] uses concepts like frontier orbitals [67] (homo/lumo) that come from a different part of chemistry. These are almost always used incorrectly – in frontier molecular orbital theory, the homo and lumo are well defined – one is the highest occupied molecular orbital, the other the lowest unoccupied molecular orbital. They are orbitals, they have shapes, and they have orbital energy levels. But they are one-electron constructs – for example, the lumo for naphthalene and for its cation, its anion, and its doubly charged dication are completely different. So that when, in a description of transport, we talk about electrons moving through the lumo, it is not the same lumo that is defined for the isolated molecule! The proper term would be "affinity level," but that proper term is hardly used. This is important, because the changes in energy between the lumo of a closed-shell molecule and the lumo of its anion or cation can be very large (electron volts), so that the nomenclature is wrong, in a serious way.

The thicket of models is complicated, and with misunderstood notation (including homo/lumo), the careful user or reader of models has to be aware of exactly what is being done in any given analysis. While it is possible to decry the use of (in particular) the homo/lumo language, that language is universal. This can be avoided simply by thinking of them as affinity levels and detachment levels, as they really are.

Given the understanding that our description of molecular transport junctions is based on a description of the model that we build, we can proceed to some of the concepts that characterize the mechanistic behaviors.

4 Ideas and Concepts (from Mechanisms and Models)

Molecular transport junctions differ from traditional chemical kinetics in that they are fundamentally electronic rather than nuclear – in chemical kinetics one talks about nucleophilic substitution reactions, isomerization processes, catalytic insertions, crystal forming, lattice changes – nearly always these are describing nuclear motion (although the electronic behavior underlies it). In general the areas of both electron transfer and electron transport focus directly on the charge motion arising from electrons, and are therefore intrinsically quantum mechanical.

4.1 Coherence and Decoherence, Tunneling and Hopping

The simplest and most significant new idea in trying to understand molecular transport junctions comes from mesoscopic physics, and in particular from the work of Landauer with Imry and Buttiker [10, 11]. This in turn is based on a simple observation – in mesoscopic physics transport junctions, or in molecular transport junctions, there is a disparity of size scales: the molecule or quantum dot is very small compared with the electrodes. These macroscopic electrodes, then, set the chemical potentials, and once an electron enters one of them, it can be thought of as losing its phase immediately, and simply becoming part of the electronic sea in that metal. This is the fundamental Landauer idea: when voltage is placed across a transport junction, electrons travel from one electrode to the other. They travel through the molecule or quantum dot, on which they may reside for a long time or a short time. But once they enter the downstream electrode, that acts as a perfect sink –all phase coherence is immediately lost, and the electron has disappeared into the Fermi sea. This fundamental idea is crucially different from understanding a classical wire, and thinking of conduction in terms of Ohm's law; in that situation there is no size separation, and the electrons are thought of as a current that generates heat and undergoes resistance as it moves–the description is initially classical, although it can easily be made quantal. In the Landauer/Imry/Buttiker approach, the transport is quite different – it is scattering (indeed it is elastic scattering in the simplest picture).

This approach to understanding transport leads to the Landauer/Imry/Buttiker formula for conductance which is

$$g = g_0 \sum_i T_{ii} = 2\frac{e^2}{h} \sum_i T_{ii}. \tag{4}$$

Here g, g_0, T_{ii}, e, and h are respectively the conductance, the quantum of conductance equal to 77.48 microsiemens, the transmission through channel i, the electronic charge, and Planck's constant. The idea that conductance can be quantized is a remarkably new one compared with ohmic behavior – Fig. 6 shows experiments that directly demonstrate quantization of transport in atomic gold wires.

The sum in (4) runs over all the transverse channels of the system – that is, the channels that extend from the upstream to the downstream electrode. They normally are thought of (qualitatively) in terms of the molecular orbitals on the molecule, with appropriate modifications for mesoscopic systems.

While quantum chemistry cannot be used to work with the Landauer/Buttiker/Imry formula as it stands, a very different approach based on nonequilibrium Green's functions yields a different formula (sometimes called the Caroli formula [68], or the NEGF formula in the Landauer/Imry/Buttiker limit). It is, for the current I:

$$I = 2e/h \int \mathrm{d}E Tr\{\Gamma(E,V)G^r(E,V)\Gamma^+(E,V)G^a(E,V)\}(f_\mathrm{L}(E,V) - f_\mathrm{R}(E,V)) \tag{5}$$

where G^r is the retarded Green's function for electrons, Γ is the spectral density (twice the imaginary part of the self energy), and f is the Fermi distribution function. This equation can be rewritten, for clarity, as

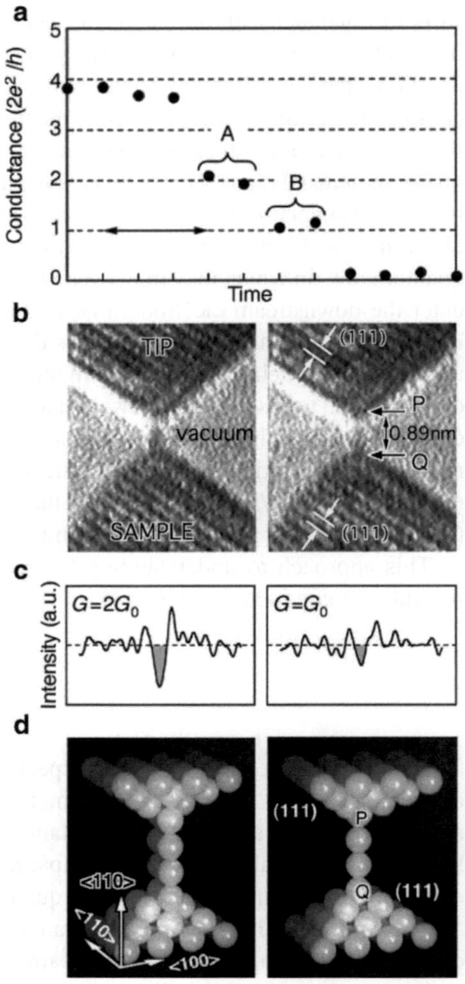

Fig. 6 (**a**) Conductance steps in a Au wire as an STM tip was retracted. (**b**) Electron microscope images of gold bridges obtained simultaneously with the conductance measurements in (a). Left, bridge at step A; right, bridge at step B. (**c**) Intensity profiles of the left and right bridges shown in (b). The shaded area is the intensity from the bridge after subtraction of the background noise. (**d**) Models of the left and right bridges. The bridge at step A has two rows of atoms; the bridge at step B has only one row of atoms. The distance from P to Q (see b) is about 0.89 nm, wide enough to have two gold atoms in a bridge if the gold atoms have the nearest-neighbor spacing of the bulk crystal (0.288 nm) (Reprinted by permission from Macmillan Publishers Ltd: Nature (1998) Nature 395, 780–783, copyright (1998))

$$I = 2e/h \int \mathrm{d}ET(E,V)\,(f_{\mathrm{L}}(E,V) - f_{\mathrm{R}}(E,V)), \tag{6a}$$

$$T(E,V) = Tr\{\Gamma(E,V)G^{r}(E,V)\Gamma^{+}(E,V)G^{a}(E,V)\}. \tag{6b}$$

Here, the transmission, T, is expressed as (6b). The Landauer/Imry/Buttiker formula (almost always called the Landauer formula) then says that the left-to-right electronic current through a molecular transport junction is the integral of the transmission through the molecule, weighted by the statistical requirements that the electrons begin in an occupied level of one electrode and finish in an unoccupied level of the other electrode. This form is quite general, and it is the one on which almost all of the quantum calculations of simple transport are based. It does need to be generalized to deal with issues like electron correlation, photonic excitation, thermal processes, decoherence and dephasing, very strong correlation, magnetic effects, and other aspects of molecular transport junctions – but it is the basis from which most of that work is done.

One way to think about the Landauer formula is to say "conductance is scattering" [69]. In fact, conductance is elastic scattering, because in the original Landauer approach, all scattering is considered to be elastic – particles leave the electrode and are scattered elastically until they make it into the other electrode (or not). Inelastic events are not included, at least conceptually.

This language is a bit different from our ordinary understanding of conduction and resistance, but it is the right approach for systems that are by their nature quantal, and that have the length scale separation characteristic of transport junctions.

Mechanistically, it is a bit hard to swallow the idea that conduction through a molecule must go by elastic scattering. For example, suppose the molecule in question were really long – something like a DNA double helix with a hundred base pairs. Elastic scattering through such a structure would fall off exponentially with length, and therefore any transport that was seen could not be explained. The model that is used to derive the Landauer equation – that is, the model that assumes the space scale separation quoted above, and the elasticity of all collisions, can begin to fail. This brings in a series of chemical mechanisms that occur because of the nature of the molecules. These chemical mechanisms are well understood from problems like conductive polymers and electron transfer in molecular systems – they might be expected to occur in molecular transport junctions, and indeed they do.

One way to think about mechanistic change is in terms of time scales. This is familiar from classical kinetics where (for example) the steady state assumption assumes that the reactive intermediate is made and destroyed on exactly the same time scale, so that (after the induction period of the chemical reaction) the rate of the overall reaction could be found by assuming that the reactive intermediates exist at steady state. This leads to the idea of chemical mechanisms for dynamical processes, and to the question of time scales. The time scale problem in molecular transport junctions is complicated, but extremely important. One time scale that is unfamiliar to most chemists is the so-called Landauer/Buttiker time or contact time [70]. This is conceptualized as the time that the electron actually spends in contact with the molecule. This is not the same as the inverse of the rate, which describes how long it takes for an electron to go from one end to the other, but rather tells about how much time the electron is actually "on" the molecule – when it can contact other molecular degrees of freedom such as the vibrations through the electron/vibration interaction [71, 72]. A simple argument based on the uncertainty principle (that can be supported by scattering theory analysis) is that this Landauer/Buttiker contact time is given approximately by

$$\tau_{\mathrm{LB}} = \frac{n\hbar}{\Delta E_{\mathrm{g}}}. \tag{7}$$

Here the variables are n, which is the dimensionless length of the system in terms of subunits and ΔE_{g}, which is the gap energy between the Fermi level of the electrode and the relevant molecular energy level. This formula looks like the uncertainty principle multiplied by a length, which seems reasonable. The uncertainty principle part is slightly counterintuitive: it says that the higher the injection barrier, the smaller the contact time. This is only unexpected because, if one were to talk about rates, the higher the barrier, the slower the rate, and therefore the longer the rate time. Conceptually, one gets around this by thinking of the Landauer/ Buttiker contact time as describing how long the electron is under the barrier – in the original analysis this could be tested by looking at a spin flip within the barrier, as modulated by the presence of the tunneling electron.

Qualitatively, for a characteristic transport molecule like an alkane thiol or a small ring system, the gap is more than 1 V, the contact time is less than 1 fs, and there is simply not enough time for strong interaction between the electrons and the vibrations. But as resonance is approached, the time τ_{LB} can approach the period of molecular vibrations or motions, which can then enter into resonance. This mechanistic change is important – once the resonance regime is approached, the scattering is certainly not elastic, the behavior does not occur simply by tunneling, thermalization is possible, vibrational subpeaks should be seen in the transport, and the mode of transport is closer to the hopping mechanism seen in conductive polymers than to the tunneling mechanism also seen in conductive polymers [71].

Many other time parameters actually enter – if the molecule is conducting through a polaron type mechanism (that is, if the gap has become small enough that polarization changes in geometry actually occur as the electron is transmitted), then one worries about the time associated with polaron formation and polaron transport. Other times that could enter would include frequencies of excitation, if photo processes are being thought of, and various times associated with polaron theory. This is a poorly developed part of the area of molecular transport, but one that is conceptually important.

The Landauer formula assumes elastic processes. If the electrons move coherently (that is without any loss of energy or of phase) they will tunnel; if the energy gap through which they must tunnel becomes relatively small, they can tunnel a long way. Generally, the conduction in the tunneling regime is written as

$$g = k_0 \mathrm{e}^{-\beta x} \tag{8}$$

where k_0 is a constant depending on the system, x is the distance between the electrodes, and β is the decay parameter corresponding to tunneling through a given molecular system.

4.2 Pathways and Analysis

The orbital description of electrons in molecules suggests that it should be possible to map the actual physical pathways by which electrons transfer through a molecule between two electrodes, or at least identify the parts of a molecule responsible for mediating the electronic interaction between the two electrodes. Some of these pathways have been roughly described on the basis of inelastic electron tunneling spectroscopy – this is discussed in Sect. 6. However, a more general and useful analysis (this time based on theory rather than experiment) has been developed in terms of channels [73–80]. The most recent extension of the channels idea is based on continuity: if one imagines planes perpendicular to the line between the two electrode tips, then the current through all such planes must be identical at steady

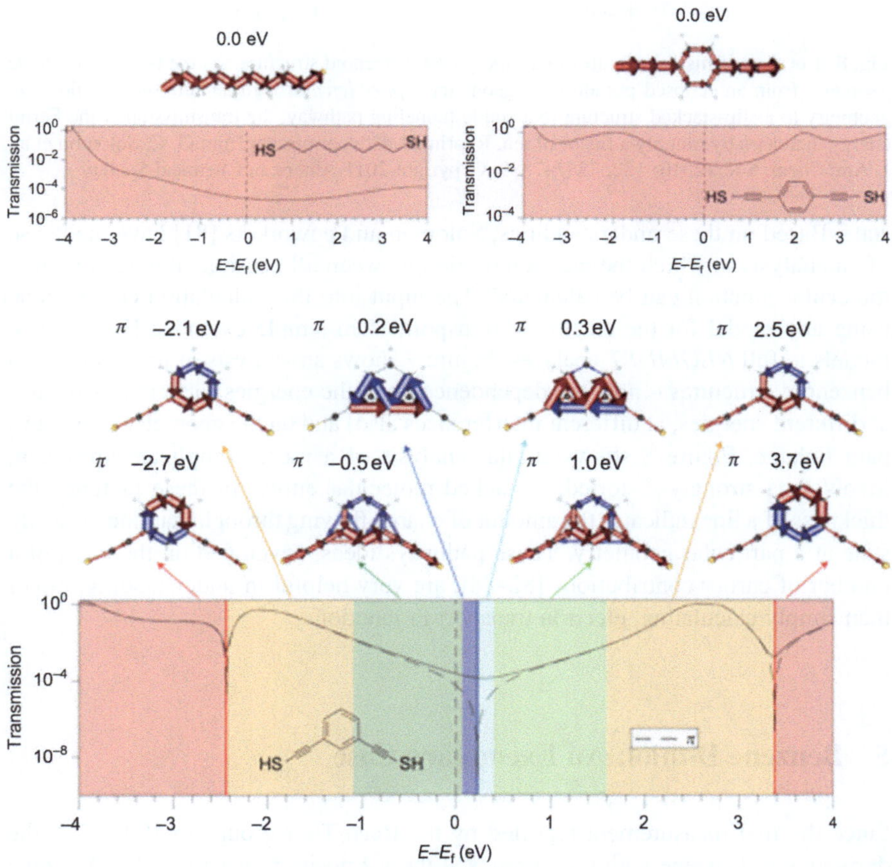

Fig. 7 Local transmission description of transport through an extended alkane (*top left*), a para linked di(thioethyne) benzene species (*top right*), and a meta-linked benzene species (*lower figures* and *panels*). In the two upper cases, transport goes through a single simple pathway in the alkane, and through two symmetrically disposed pathways in the para-benzene – this gives a relatively flat conductance or transmission spectrum as a function of voltage or energy. In the meta-benzene, different interference features occur (at roughly −2.5, 0.2, and 3.4 eV). The interference patterns shown near these features are characterized by ring-current reversal moving from one side of the interference feature to the other. Reproduced from [81]

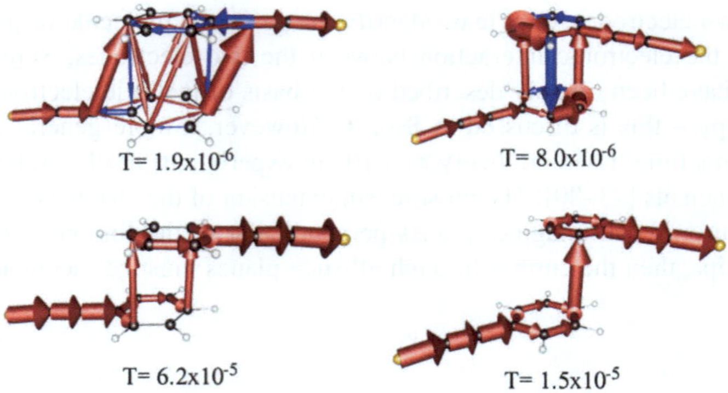

Fig. 8 Local transmission pictures in a superposed benzenoid structure. As the two rings change geometry from an eclipsed pseudo *para* geometry (*upper left*) through an eclipsed pseudo *meta* geometry to a slip-stacked structure to a single tunneling pathway, the transmission at the Fermi energy increases by roughly a factor of ten. Reprinted with permission from G. C. Solomon et al. J. Am. Chem. Soc. (2010) 132, 7887–7889. Copyright 2011 American Chemical Society

state. Based on these understandings, Solomon and coworkers [81] have made use of an analysis in which the electron motion between all possible atomic pairs in a molecular junction can be calculated. The input into this calculation can be done using any model for the electronic transport, from simple extended-Huckel type models to full *NEGF*/*DFT* analyses. Figure 7 shows an analysis of the transport in benzenoid structures – note the dependence upon the energies (different pathways at different energies, at different interferences also) and on the geometry of meta vs para linkage. Figure 8 shows similar analysis of a more complicated problem, involving a strongly distorted, π-stacked molecular entity. In these pictures, the thickness of a line indicates the amount of charge flowing through that line in steady state at a particular geometry. These pathways ideas, developed on the basis of a number of earlier contributions [82–88], are very helpful in understanding, rather than simply calculating, electron transport in junctions.

5 Benzene Dithiol: An Exemplary Case

Since the first measurement reported by the Reed/Tour groups in 1997 [23], the derivative of benzene with thiol groups at the 1,4 position (usually called benzene dithiol) has become the standard case for the discussion of molecular transport junctions. That measurement by the Reed group was made with a mechanical break junction, and reported both the zero-voltage and the voltage-dependent conductance. Specific values were given for both, and the cartoons in the paper suggested that the thiol group lost its hydrogens, and that the sulfur atoms were uniquely coordinated to the gold electrodes. Since it was entirely a measurement paper, there was no discussion of possible binding geometries. This important paper was one of the first reported single-molecule transport measurements, and therefore has been instrumental in the entire area.

Questions about what was being measured, and the geometries of what was being measured, began immediately. It was suggested that perhaps the junction contained two molecules, one bound to one electrode, and the other to the counterelectrode, with a sort of π-type stacking in between them [89]. A large number of calculations using different methods were published. These modeling activities suggested that different interactions of the molecule with the gold could produce substantially different transport. Since the measurement from the Reed lab was the standard, that value has been enshrined.

There are almost 100 papers that discuss benzene dithiol's conductance. As the point about geometric distributions became well understood, it was realized that statistical analysis was extremely useful. Accordingly, electrochemical break junction techniques, both in their original form of crashed electrodes being separated to form the gap or in the newer electrochemistry form, in which a gap is created and then electrochemically modified, have proliferated. The important thing is that statistical measurements can be made [24, 90], with hundreds or thousands of data points. Not surprisingly, distributions are observed (as the earlier computations had suggested).

The closest thing to a unique measurement was reported by the group at Columbia University/Brookhaven National Laboratory [91]. They used amine rather than thiol end groups. Both the narrowness of the experimental distributions and very nice theoretical work integrating molecular dynamics and transport calculations [33] suggested that the amine likes to bind to a coordinatively unsaturated site on a single gold atom, so the narrowness of the distribution here is greater than is typical for thiols.

Some measurements showing high conduction for benzene itself and some benzene derivatives are best explained by a geometry very different from the extended one first suggested by Reed, and serve as the basis for much calculation. In the measurements from Ruitenbeek's laboratory, the conductance is close to the atomic unit of conductance [92]. The simplest way to explain this phenomenon is that the molecule is oriented perpendicular to the interelectrode coordinate, and electrodes are very near one another. So the molecule really does not assist substantially in the transport, although it can be seen in the IETS spectra.

In a 2007 overview [7], simple NEGF/DFT calculations were compared with reported experiments, and the outlier was benzene dithiol. It is now clear that (particularly with small molecules) geometry dependence can (indeed must) give distributed values for the conductance. This is entirely in keeping with the understanding of single molecule spectroscopy [93, 94] that is demarked by such phenomena as blinking (in many cases) and spectral wandering (in essentially all cases). These arise from fluctuations, be they fluctuations of charge density or fluctuations in geometry of the environment in which the molecule is measured. From the viewpoint of fundamental understanding, these fluctuational quantities are well described by simple statistical mechanics – fluctuations scale as the inverse square root of the sample number, so that with millions of samples, an average number can be readily agreed upon. A small number of measurements would be expected to give a fairly wide distribution of observed behaviors – this is indeed seen [28] in benzene dithiol, and probably should be seen (and has been) in many other molecules.

6 Inelastic Electron Tunneling Spectroscopy

In the Landauer/Imry limit, the transport through the junction is due to elastic scattering. If the gap between the injection energy and the frontier orbital resonance is large, the Landauer/Buttiker contact time is very small, so that the charge is present on the molecule for a very short time. This means that its interaction with any vibration will be weak, because there just is not time to complete a full vibrational period before the charge has gone into the electrode sink.

There will be vibronic interactions in any molecular system, because the charged states will always have a different geometry from the uncharged ones. This means that the charge on the molecule will cause the geometry of the molecule to change, and that will be reflected in a vibrational side peak in the transport spectra. The simplest and most useful measurement to make on such systems is inelastic electron tunneling spectroscopy [8] (IETS), in which one measures the second derivative of the current with respect to the voltage, and plots that (divided by its value at a reference voltage) as a function of voltage. Figure 9 shows both the schematic behavior. This experiment, first reported in molecular junctions by Reed and coworkers [95] and by Kushmerick and coworkers [96] in 2004, is a significant way to investigate molecules in junctions.

When the gap is large, the sketch in Fig. 9 shows that a second channel will open when there is a vibrational resonance – that is, when $eV = \hbar\omega$, with ω one of the vibrational frequencies of the molecule. This is vibronic resonance, and energy will transfer from the momentum of the tunneling electrons into the vibrations of the molecule. The interaction is quite weak (because the tunneling time is so short);

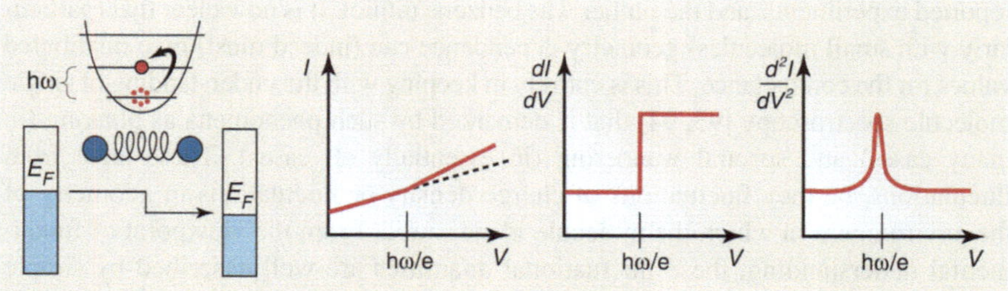

Fig. 9 Schematic of the inelastic electron tunneling phenomenon. From M. Galperin et al. Science (2008), 319, 1056–1060. Reprinted with permission from AAAS

IETS spectra are usually reported at very low temperatures, and careful data management is required to see the IETS features.

The interpretation of IETS is helpful in understanding molecular junctions. Several workers have developed techniques for doing so [97–102], some based on quite complex analyses of the full Green's function [99–101], others based on a much simpler analysis in which the fact that the response is so weak is used as the basis for perturbative expansion[98]. The results of these analyses fit the spectra well.

From these analyses, a number of major advances have followed. First, the presence of the molecular vibrations indicates that the molecules are indeed in the junction, and that the transport is passing through them. Second, the pathway of the current through the molecule can to some extent be determined based on which vibrational modes are enhanced – as is not surprising, if the electron density between atoms on which a particular normal mode exhibits large amplitude is not substantially modified upon charging, then that molecular mode will be silent in the IETS spectra. This leads to a set of propensity rules [103–106] that have helped substantially in interpreting both IETS spectra and (more interestingly) the actual geometries of the junction.

As has been stated several times, the geometry problem in junctions is difficult. Several papers have utilized the differences in the IETS calculated spectrum at different trial geometries to compare with the experimental spectrum, and thereby to deduce the true geometry of the structure. Figure 10 shows some results by Troisi [107], in which he was able to deduce the angle between the molecular backbone and the electrode, based on agreement with the IETS spectrum.

It is also possible to deduce pathways in a more adventurous way by noting which modes are enhanced, doing the normal coordinate analysis to find out where those modes have their maximum amplitudes, and arguing that this describes the pathway for the electron going through the molecule. An example is shown in Fig. 11, also from Troisi's work [108].

It was noted early by Reed and others that the IETS spectrum could exhibit both absorption and emission peaks – that is, the plots of Fig. 9 could have positive excursions and negative excursions called peaks and dips. The simple analysis suggested in Fig. 9 implies that it should always be absorptive behavior, and therefore that there should always be a peak (a maximum, an enhancement) in the IETS spectrum at the vibrational resonances. It has been observed, however, that dips sometimes occur in these spectra. These have been particularly visible in small molecules in junctions, such as in the work of van Ruitenbeek [92, 109] (Fig. 12). Here, formal analysis indicates that, as the injection gap gets smaller, the existence of an inelastic vibrational channel does not contribute a second independent channel to the transport, but rather opens up an interference [100]. This interference can actually impede transport, resulting in a dip in the spectrum. Qualitatively, this occurs because the system is close to an electronic resonance; without the vibrational coupling the conductance is close to g_0, and the interference subtracts from the current.

These IETS features have been observed. The technique is a very good one for addressing certain aspects of a molecular structure in the junction, and the molecular pathways. Of all the areas of molecular transport, this one is probably the most quantitatively accurate for comparison with experiment.

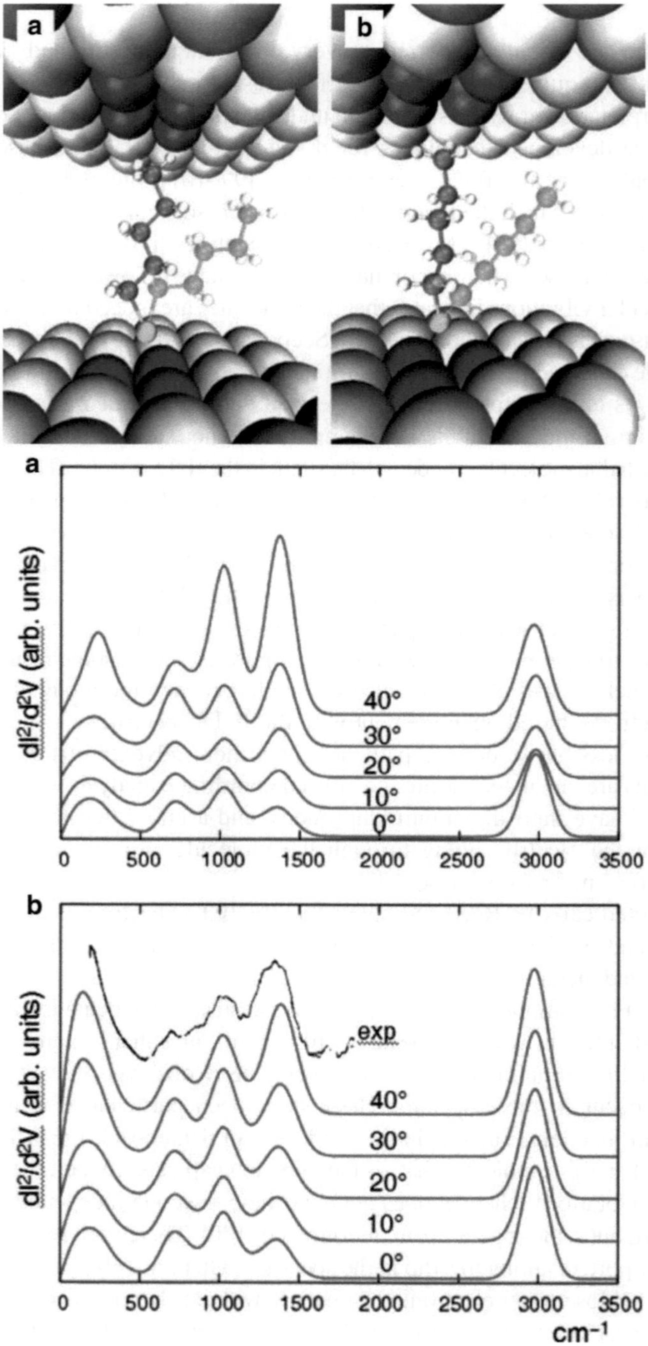

Fig. 10 Estimation of the tilt angle for an alkane between gold electrodes, determined by fitting the computed IETS spectrum with the experiment (panel b below). Result is a 40 degree tilt angle perpendicular to the plane of the carbon chain, as illustrated in the lighter shade structure in the sketch (b) above. Sketch (a) above and panel (a) below refer to the alkane tilted in the plane of the carbon chain. The structures in sketch (a) do not fit so well an those in (b), suggesting the methyl group position shown in (b) above. From [107]. Reproduced by permission of the PCCP Owner Societies

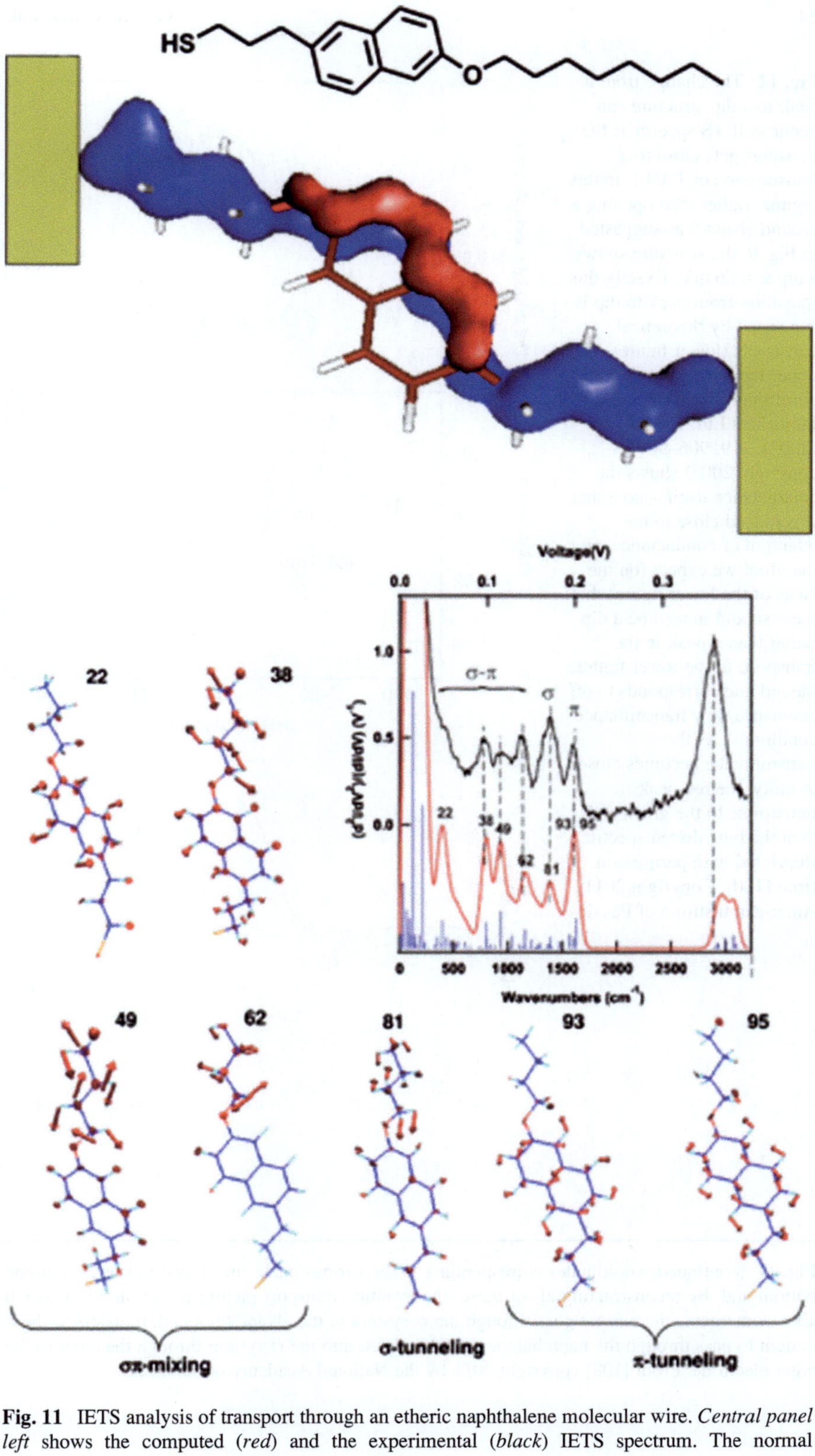

Fig. 11 IETS analysis of transport through an etheric naphthalene molecular wire. *Central panel left* shows the computed (*red*) and the experimental (*black*) IETS spectrum. The normal

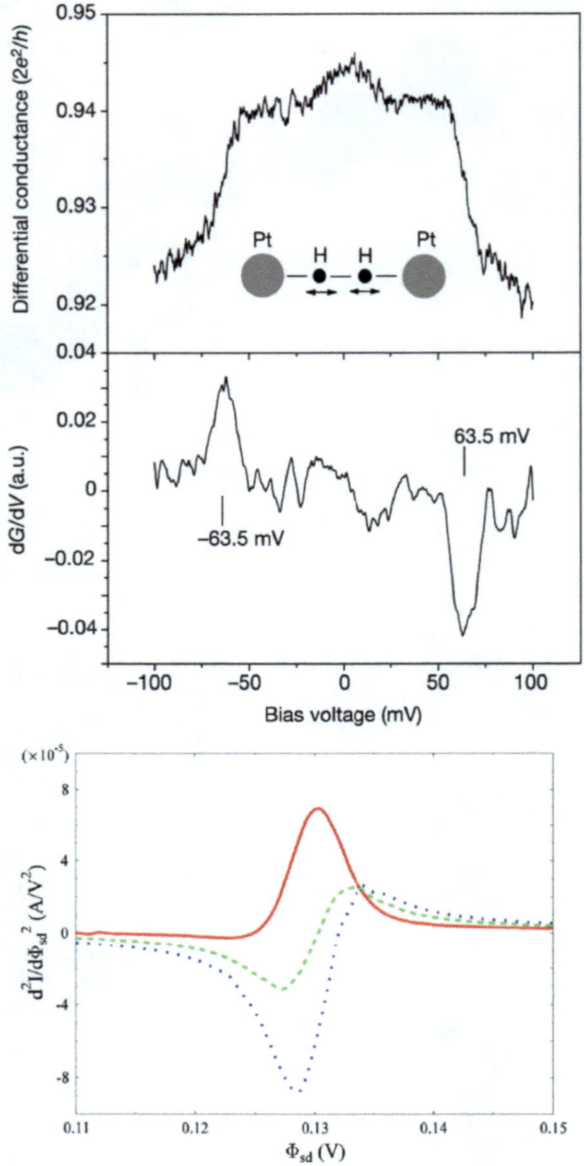

Fig. 12 The change from a peak to a dip structure can occur in IETS spectra as the transport gets close to a transmission of T 1/4 1. In this regime, rather than opening a second channel as suggested in Fig. 9, the structure shows a dip at 63.6 mV. Exactly this transition from peak to dip is suggested by theoretical constructs (lower figure). The upper figure (Reprinted by permission from Macmillan Publishers Ltd: Nature (2002), 419, 906–909, copyright 2002) shows the conductance itself – note that it is indeed close to the quantum of conductance, and therefore we expect (on the basis of the lower figure) that there should indeed be a dip rather than a peak in the transport. In the lower figure, the red line corresponds to off resonance, low transmittance conditions. As the transmittance becomes closer to unity, the red peak transforms to the green and then the blue dotted spectra. Reprinted with permission from [100]. Copyright 2011, American Institute of Physics

Fig. 11 (continued) coordinates corresponding to the various peaks are shown in sketches on the bottom and, by reconstructing all of these, the eventual transport picture in the blobbish sketch above emerges – the current goes through the σ system in the alkane thiol end, transfers to the π system to pass through the naphthalene, and then back into the σ system through the ether to the other electrode. From [108] copyright 2011 by the National Academy of Sciences

7 Challenges

The field sometimes called molecular electronics actually should extend well beyond simple measurement of current/voltage characteristics of single molecules. The latter topic, single molecule transport, has comprised by far the dominant reported molecular electronics measurement and modeling, and, as has been discussed above, the community is reaching some agreement in this area.

The original vision of molecular electronics was not only single transport, but actual applications of that single molecule transport, and more complicated mechanisms, towards electronic phenomena (and perhaps electronic devices) based on the functional use of molecules [110, 111]. In its most visionary manifestation, these would be single molecules; there have been a number of important efforts in the single-molecule device area that extend beyond simple transport [112–116]. Issues such as decoherence and fabrication complexity have plagued this area, but it remains as an important intellectual challenge. The original idea of molecular rectification has been experimentally established in important, extensive work by Metzger and colleagues [117, 118].

Since the notion of single molecule devices was put forward in the 1970s, the field of electronics has moved on in a very elegant (and profitable) fashion. Several areas have been developed, areas in which molecules might well provide advantages. We will complete this overview with a very brief description of some of these situations.

In this chapter, we will deal with the challenges involved in single molecule electronics, or possibly the electronics of systems with a few molecules. This is to be distinguished from multimolecule effects in electronics, where a good deal of progress has been made. Organic electronics [119] now comprises an entire field, as does organic optoelectronics [120]. In these areas, molecules act as functional devices, but not on a single molecule basis. Areas such as organic photovoltaics, organic transistors, and organic light emitting diodes are already the stuff of commerce, and represent challenging fields for the chemical sciences and the materials and electronic engineering fields.

Sometimes molecules are used as layers in other devices. For example, molecules can act as capacitors, and this relatively new field is promising for applications in energy storage as well as providing typical capacitance behavior in thin film devices [121, 122].

SAMs containing molecules (originally based on thiol/gold interactions, but now extended to many different molecular terminal groups and multiple solids including metals, semiconductors, and some insulators) represent another significant area of application. One obvious application is the control of the affinity and ionization potentials of a given macroscopic material that can be provided by dipolar layers of SAMs on their surfaces [123, 124]. One can also make mixed SAMs, in particular SAMs of molecules that should be pretty good insulators (such as alkane thiols) and unsaturated molecules that should be good transporters [125, 126]. Many groups have examined statistics, sometimes asking about whether the current through *n* conducting wires is actually *n* times the current through a given wire (the answer is "sometimes") [127–133] and often using the nonconductive host material to stabilize the guest [96, 132, 134]. Once again, switching and geometric change, both static and dynamic, have been observed in such systems.

7.1 Strong Correlations

The two most common manifestations of the so-called strong correlation effects are the Kondo peaks that arise with an odd spin on the molecule [13–15] and the Coulomb blockade phenomena that arise when the molecular coupling to the electrodes is weak compared to intramolecular energies such as the Fermi gap or averaged electron repulsions [14]. It is then necessary to deal with a strongly correlated molecular Hamiltonian, but the effects of mixing with the electrodes can normally be handled according to simple master equation kinetics. There have been both extensive measurements and extensive modeling [135] to approach these limits, largely in imitation of the situation found in mesoscopic physics.

Because vibronic coupling is strong in most molecules (unlike silicon itself or even silanes), most molecules exhibit relatively large geometry changes between molecules and their ions – this requires relatively simple extensions of the Born–Oppenheimer approximation, so that the individual ionic states are well defined. But when current passes through a molecule, the change in electron density can cause a geometric change, and that will change the spectroscopy, the transport mechanism, and the nature of the correlations.

7.2 Spintronics

Spintronics [136–138] is a word used to describe transport in a mesoscopic junction in which the transport medium (or the electrodes) contain unpaired electron spins. There are different aspects of spintronics, but the simplest idea is that one can transport spin without necessarily transporting charge. This leads to the idea of a molecular spin transistor, and other spin phenomena such as spin valves and spin gates.

One difficulty with the spintronics area using molecules [139–141] has been that, like simple transport, it will change with the geometry of the interface. Nevertheless, spintronic applications are intriguing, and this has become a new focus area for molecular electronics.

7.3 *Optoelectronics*

The simplest structural entity in optical electronics would be a molecule whose geometry can be made to change by optical excitation. Systems such as azobenzenes and dithienylethenes immediately come to mind, and indeed such switching has been seen [142–146] and computationally explained [147–151]. Photoswitching single molecules bound in conducting junctions presents an interesting problem for system design: on the one hand strong coupling to the electrodes is desirable to maximize current flow, on the other hand this strong coupling can quench the molecular excited states thereby inhibiting switching. More recently, optoelectronic switching in a distributed set of molecules binding together quantum dots has been explored [145] – such systems might have important applications in photovoltaics, but also represent an interesting way of constructing controlled networks. Thus both single molecule optoelectronics and optoelectronics in molecule-based extended nanosystems are of substantial current interest.

7.4 *Dynamical Control of Transport Properties*

The first five sections of this chapter discussed the relationship between the geometry of molecular transport and the magnitude of the conductance. But once a transport junction is assembled, its function can be changed by controlling the potentials that the molecule feels. Three outstanding examples are gating of transport junctions either by means of a third gate electrode (very similar to traditional mesoscopic transistors) or by the ionic environment, which also [152] provides a gate (and relates to dynamical control of processes in electrochemistry). A third, very new approach is based on coherent control – that is, modifications of the structure of the bridge caused by incident laser fields. This is still mostly a theoretical endeavor [153], but clearly follows directly from the geometry changes involved in photoexcitation.

Representing the actual optical field in a coherent control scheme for a single molecule transport junction is complicated by the optical inhomogeneity of this space (with vacuum, metal, and molecular components). Nevertheless, coherent control is scientifically very intriguing, and such processes might well be useful for trapping and storing charge and energy.

7.5 *Chirality and Broken Symmetry*

Many molecules are chiral – that is, these molecules are not superimposable on their mirror images. One way for molecules to obtain chirality is to have so-called asymmetric carbons (that is, a carbon atom with four covalent bonds, none of which are equivalent). But there are many other structures that are also chiral, ranging from helicenes through simple twisted molecules such as biphenyl.

It seems clear based on symmetry arguments (and on more formal considerations) that a chiral molecule symmetrically placed in a symmetric transport junction should have a symmetric current/voltage characteristic – that is, it should not exhibit rectification. However, it also seems clear that adding a second symmetry-breaking effect (such as a magnetic field, polarized light, or an odd spin) should change the overall transport symmetry, allowing rectification. Extensive work on this subject has been reported by the Weizmann Group [154, 155], and these fascinating experiments provoke a series of questions. Very simple models [42] have been used to show different transport for different interfaces between the molecule and the metal, interfaces whose difference might well arise from chirality.

If one were interested in making devices, these chiral effects might be quite significant. Magnetic sensors, molecular sensors, and strong rectification structures might well be made by using the phenomenon of chirally-induced transport. Even very simple questions remain to be answered: are two different molecules that have different chiral strengths (that is, rotate the vector of polarized light to different extents) expected to show stronger effects of gating or switching or rectification? These are very new experiments, which should be quite challenging, and perhaps significant for applications in the life sciences.

7.6 Crosstalk, Interference, and Decoherence

Crosstalk between two molecular wires occurs when the current through the pair of wires differs from twice the current through a single wire [156–161]. As indicated above, these issues were first explored in mixed SAM films; several measurements have been made in which crosstalk was discussed as part of the explanation for the observations. Classically, since two parallel pipes of the same diameter will carry twice as much water as a single pipe, one would not expect to see crosstalk. It therefore corresponds to typical quantum behavior (in which the probability for two parallel channels might be as large as four times the transport in one, or might even be smaller than the transport in one, depending on the nature of the couplings between the wires [128, 162]).

Crosstalk has been discussed fairly extensively, as one of a series of interference phenomena that can lead to a different kind of control of molecular transport than has been discussed in Sect. 7.4. It is also possible to observe intramolecular interference effects. For example, with cross-conjugated molecules [163] or benzene dithiol linked in the 1,3 (or meta) configurations [164–171], both are expected to show substantially reduced transport.

These interference patterns are wonderful manifestations of wave function behavior, and are not found in classical electronics or electrodynamics. Since the correspondence principle tells us that quantum and classical systems should behave similarly in the limit of Planck's constant vanishing, we suspect that adequate decoherence effects will change the quantum equation into classical kinetics equations, and so issues of crosstalk and interference would vanish. This has been suggested theoretically in a number of cases, and the disappearance of quantum mechanical effects upon increasing temperature is crucial to modern magnetic resonance and coherent optical spectroscopies. We should see the same behavior in transport, as more sophisticated measurements are made on appropriate systems.

One classic example is the turnover from coherent tunneling at low temperatures and large gaps to hopping conduction at higher temperatures and smaller gaps. This has been suggested theoretically in a number of situations [54, 172, 173], and there are some good experimental examples in systems like DNA and oligomers [174]. Understanding how decoherence occurs, and how to control it, is one of the major problems in contemporary chemical physics [175], and there have been specific applications of the general theory to molecular transport [176, 177]. Indeed, the theoretical difficulty of how one describes a dynamical subsystem interacting with a very large extended host is a frontier area of science in 2011, and molecular junctions are an area in which its understanding and mastery will be helpful. Indeed, much of the initial concern about the practicality of single molecule and few-molecule electronics had to do with decoherence – the argument being that the environment would destroy some of the delocalization properties of wave functions, and that the localized results would have lost the typical behaviors of coherence interactions, upon which quantum computing (and other quantum processes) are based. Therefore, understanding of these decoherence phenomena, and how they relate to the structure of the molecule and its environment, is one of the most challenging problems in the entire area.

7.7 *Quantum Cellular Automata and Cascade Devices*

Very thought-provoking extensions of the idea of molecular electronics have been published by Eigler (quantum cascade logic [178]) and by the Notre Dame group (cellular automata based on molecular subunits [179]). In Eigler's example, lines of CO molecules adsorbed onto a single crystal can be caused to change their angles with respect to the plane (almost like dominoes falling down in a chain). STM images of such logic are quite beautiful, and these elegant pictures demonstrate that, with sufficient attention paid to the details, such molecular dominoes can indeed perform Boolean logic. It is not clear how such structures could be made to scale with size, such that many (as opposed to a few) interactions and subunits could be used to form devices, memory, or logic.

Quantum-dot cellular automata (QCA) provide an alternative approach for the design of molecular electronics. In the QCA scheme, binary information is stored in the charge configuration of single cells and transferred via Coulomb coupling between neighboring cells. Decreased resistive heating makes possible extremely high device densities without dissipating catastrophic amounts of energy [179].

7.8 True Devices

At one level, a simple metal/molecule/metal or metal/molecule/semiconductor junction is a device. From an applications point of view, devices should have functions that are useful for information storage, logic, energy transfer, energy storage, polarization control, thermal switching, or some other behavior that could become, through appropriate engineering, an entity in the marketplace of devices, as well as ideas. The field of organic electronics, as mentioned at the beginning of this section, has already done that for systems based on many molecules. For systems based on a single molecule, the usual arguments (fragility, reproducibility, difficulties with fabrication, chemical reactivity) have been invoked to suggest that it would be difficult to make true single-molecule or few-molecule devices that would be active as technological systems. While some of these arguments are quite persuasive, even thinking about molecular electronics is only three and a half decades or so old, so that true devices may still be built, based on the use of the intrinsic degrees of freedom (including chirality, isomerization, switching, and binding) that characterize molecular systems.

The discussion in this contribution has been largely qualitative, and impressionistic. This seems in keeping with a volume of this kind – most of the topics discussed here are still very much alive, and it seems that molecular electronics, defined as the understanding and technological application of electronic properties of single molecule systems or few-molecule systems, remains as a challenge to the molecular sciences of the twenty-first century.

Acknowledgments We are grateful to Robert Metzger for the opportunity to contribute to this volume, and to the very large number of wonderful colleagues and coworkers who contributed to our understanding in this general area, starting with Dr. Ari Aviram and extending to our current research groups. We are also grateful to the MRSEC program of the NSF for support of this research. C.H. would like to thank the Landesexzellenzinitiative Hamburg (Nanospintronics) for funding. G.C.S. acknowledges funding from The Danish Council for Independent Research/ Natural Sciences.

References

1. Wasielewski MR (2006) Energy, charge, and spin transport in molecules and self-assembled nanostructures inspired by photosynthesis. J Org Chem 71(14):5051–5066
2. Born M, Oppenheimer R (1927) Zur Quantentheorie der Molekeln. Annalen der Physik 389 (20):457–484
3. Nitzan A (2001) A relationship between electron-transfer rates and molecular conduction. J Phys Chem A 105(12):2677–2679
4. Cuevas JC, Scheer E (2010) Molecular electronics: an introduction to theory and experiment. World Scientific Publishing, Singapore
5. Datta S (2005) Quantum transport: atom to transistor. Cambridge University Press, Cambridge
6. Ventra MD (2008) Electrical transport in nanoscale systems. Cambridge University Press, Cambridge

7. Lindsay SM, Ratner MA (2007) Molecular transport junctions: clearing mists. Adv Mater 19(1):23–31
8. Jaklevic RC, Lambe J (1966) Molecular vibration spectra by electron tunneling. Phys Rev Lett 17(22):1139
9. Stipe BC, Rezaei MA, Ho W (1998) Single-molecule vibrational spectroscopy and microscopy. Science 280(5370):1732–1735
10. Landauer R (1957) IBM J Res Dev 1:223
11. Büttiker M, Imry Y, Landauer R, Pinhas S (1985) Generalized many-channel conductance formula with application to small rings. Phys Rev B 31(10):6207
12. Grabert H, Devoret MH (eds) (1992) Single charge tunneling. Coulomb blockade phenomena in nanostructures. Plenum, New York
13. Kondo J (1964) Resistance minimum in dilute magnetic alloys. Prog Theor Phys 32:37
14. Park J, Pasupathy AN, Goldsmith JI, Chang C, Yaish Y, Petta JR, Rinkoski M, Sethna JP, Abruna HD, McEuen PL, Ralph DC (2002) Coulomb blockade and the Kondo effect in single-atom transistors. Nature 417(6890):722–725
15. Liang W, Shores MP, Bockrath M, Long JR, Park H (2002) Kondo resonance in a single-molecule transistor. Nature 417(6890):725–729
16. Reinerth WA, Jones L II, Burgin TP, Zhou C-w, Muller CJ, Deshpande MR, Reed MA, Tour JM (1998) Molecular scale electronics: syntheses and testing. Nanotechnology 9(3):246
17. Wold DJ, Frisbie CD (2000) Formation of metal-molecule-metal tunnel junctions: microcontacts to alkanethiol monolayers with a conducting AFM tip. J Am Chem Soc 122 (12):2970–2971
18. Cui XD, Primak A, Zarate X, Tomfohr J, Sankey OF, Moore AL, Moore TA, Gust D, Harris G, Lindsay SM (2001) Reproducible measurement of single-molecule conductivity. Science 294 (5542):571–574
19. Leatherman G, Durantini EN, Gust D, Moore TA, Moore AL, Stone S, Zhou Z, Rez P, Liu YZ, Lindsay SM (1998) Carotene as a molecular wire: conducting atomic force microscopy. J Phys Chem B 103(20):4006–4010
20. Chiechi RC, Weiss EA, Dickey MD, Whitesides GM (2008) Eutectic gallium–indium (EGaIn): a moldable liquid metal for electrical characterization of self-assembled monolayers. Angew Chem Int Ed 47(1):142–144
21. Haag R, Rampi MA, Holmlin RE, Whitesides GM (1999) Electrical breakdown of aliphatic and aromatic self-assembled monolayers used as nanometer-thick organic dielectrics. J Am Chem Soc 121(34):7895–7906
22. Wang W, Lee T, Reed MA (2003) Mechanism of electron conduction in self-assembled alkanethiol monolayer devices. Phys Rev B 68(3):035416
23. Reed MA, Zhou C, Muller CJ, Burgin TP, Tour JM (1997) Conductance of a molecular junction. Science 278(5336):252–254
24. Lörtscher E, Weber HB, Riel H (2007) Statistical approach to investigating transport through single molecules. Phys Rev Lett 98(17):176807
25. Lörtscher E, Riel H (2010) Molecular electronics resonant transport through single molecules. CHIMIA Int J Chem 64:376–382
26. Lörtscher E, Ciszek JW, Tour J, Riel H (2006) Reversible and controllable switching of a single-molecule junction. Small 2(8–9):973–977
27. Ballmann S, Hieringer W, Secker D, Zheng Q, Gladysz JA, Görling A, Weber HB (2010) Molecular wires in single-molecule junctions: charge transport and vibrational excitations. ChemPhysChem 11(10):2256–2260
28. Xu B, Tao NJ (2003) Measurement of single-molecule resistance by repeated formation of molecular junctions. Science 301(5637):1221–1223
29. Xu X, Yang X, Zang L, Tao N (2005) Large gate modulation in the current of a room temperature single molecule transistor. J Am Chem Soc 127(8):2386–2387
30. Albrecht T, Guckian A, Ulstrup J, Vos JG (2005) Transistor-like behavior of transition metal complexes. Nano Lett 5(7):1451–1455

31. Haiss W, Nichols RJ, van Zalinge H, Higgins SJ, Bethell D, Schiffrin DJ (2004) Measurement of single molecule conductivity using the spontaneous formation of molecular wires. PCCP 6(17):4330–4337
32. Ward DR, Halas NJ, Ciszek JW, Tour JM, Wu Y, Nordlander P, Natelson D (2008) Simultaneous measurements of electronic conduction and Raman response in molecular junctions. Nano Lett 8(3):919–924
33. Quek SY, Kamenetska M, Steigerwald ML, Choi HJ, Louie SG, Hybertsen MS, Neaton JB, Venkataraman L (2009) Mechanically controlled binary conductance switching of a single-molecule junction. Nat Nano 4(4):230–234
34. Kondo Y, Takayanagi K (2000) Synthesis and characterization of helical multi-shell gold nanowires. Science 289(5479):606–608
35. Hybertsen MS, Venkataraman L, Klare JE, Whalley AC, Steigerwald ML, Nuckolls C (2008) Amine-linked single-molecule circuits: systematic trends across molecular families. J Phys Condens Matter 20(37):374115
36. Chen IWP, Fu M-D, Tseng W-H, Chen C-h, Chou C-M, Luh T-Y (2007) The effect of molecular conformation on single molecule conductance: measurements of π-conjugated oligoaryls by STM break junction. Chem Commun 29:3074–3076
37. Wang K, Rangel NL, Kundu S, Sotelo JC, Tovar RM, Seminario JM, Liang H (2009) Switchable molecular conductivity. J Am Chem Soc 131(30):10447–10451
38. Franco I, George CB, Solomon GC, Schatz GC, Ratner MA (2011) Mechanically activated molecular switch through single-molecule pulling. J Am Chem Soc 133(7):2242–2249
39. Itoh T, McCreery RL (2002) In situ Raman spectroelectrochemistry of electron transfer between glassy carbon and a chemisorbed nitroazobenzene monolayer. J Am Chem Soc 124 (36):10894–10902
40. Gartsman K, Cahen D, Kadyshevitch A, Libman J, Moav T, Naaman R, Shanzer A, Umansky V, Vilan A (1998) Molecular control of a GaAs transistor. Chem Phys Lett 283(5–6):301–306
41. Vilan A, Ussyshkin R, Gartsman K, Cahen D, Naaman R, Shanzer A (1998) Real-time electronic monitoring of adsorption kinetics: evidence for two-site adsorption mechanism of dicarboxylic acids on GaAs(100). J Phys Chem B 102(18):3307–3309
42. Skourtis SS, Beratan DN, Naaman R, Nitzan A, Waldeck DH (2008) Chiral control of electron transmission through molecules. Phys Rev Lett 101(23):238103
43. Yeganeh S, Ratner MA, Medina E, Mujica V (2009) Chiral electron transport: scattering through helical potentials. J Chem Phys 131(1):014707–014709
44. Galperin M, Ratner MA, Nitzan A (2009) Raman scattering from nonequilibrium molecular conduction junctions. Nano Lett 9(2):758–762
45. Jørgensen P, Simmons J (1981) Second quantization-based methods in quantum chemistry. Academic, New York
46. Hohenberg P, Kohn W (1964) Inhomogeneous electron gas. Phys Rev 136(3B):B864
47. Kohn W, Sham LJ (1965) Self-consistent equations including exchange and correlation effects. Phys Rev 140(4A):A1133
48. Parr RG, Yang W (1989) Density-functional theory of atoms and molecules. International Series of Monographs on Chemistry, vol 16. Oxford Science, Oxford
49. Solomon GC, Andrews DQ, Van Duyne RP, Ratner MA (2009) Electron transport through conjugated molecules: when the π system only tells part of the story. ChemPhysChem 10(1):257–264
50. Koentopp M, Chang C, Burke K, Car R (2008) Density functional calculations of nanoscale conductance. J Phys Condens Matter 20(8):083203
51. Anisimov VI, Zaanen J, Andersen OK (1991) Band theory and Mott insulators: Hubbard U instead of Stoner I. Phys Rev B 44(3):943
52. Solovyev IV, Dederichs PH, Anisimov VI (1994) Corrected atomic limit in the local-density approximation and the electronic structure of d impurities in Rb. Phys Rev B 50(23):16861
53. Liechtenstein AI, Anisimov VI, Zaanen J (1995) Density-functional theory and strong interactions: orbital ordering in Mott-Hubbard insulators. Phys Rev B 52(8):R5467

54. Ho Choi S, Kim B, Frisbie CD (2008) Electrical resistance of long conjugated molecular wires. Science 320(5882):1482–1486
55. Xue Y, Datta S, Ratner MA (2002) First-principles based matrix Green's function approach to molecular electronic devices: general formalism. Chem Phys 281(2–3):151–170
56. Ke S-H, Baranger HU, Yang W (2004) Electron transport through molecules: self-consistent and non-self-consistent approaches. Phys Rev B 70(8):085410
57. Cuevas JC, Heurich J, Pauly F, Wenzel W, Schon G (2003) Theoretical description of the electrical conduction in atomic and molecular junctions. Nanotechnology 14(8):R29–R38
58. Damle P, Ghosh AW, Datta S (2002) First-principles analysis of molecular conduction using quantum chemistry software. Chem Phys 281(2–3):171–187
59. Evers F, Weigend F, Koentopp M (2004) Conductance of molecular wires and transport calculations based on density-functional theory. Phys Rev B 69(23):235411
60. Stokbro K, Taylor J, Brandbyge M, Ordejon P (2003) TranSIESTA: a spice for molecular electronics. Ann NY Acad Sci 1006(Mol Electron III):212–226
61. Stokbro K, Taylor J, Brandbyge M, Mozos JL, Ordejon P (2003) Theoretical study of the nonlinear conductance of di-thiol benzene coupled to Au(1-1-1) surfaces via thiol and thiolate bonds. Comput Mater Sci 27(1/2):151–160
62. Palacios JJ, Perez-Jiménez AJ, Louis E, SanFabián E, Vergés JA (2002) First-principles approach to electrical transport in atomic-scale nanostructures. Phys Rev B 66(3):035322
63. Rocha AR, Garcia-Suarez VM, Bailey S, Lambert C, Ferrer J, Sanvito S (2006) Spin and molecular electronics in atomically generated orbital landscapes. Phys Rev B 73(8):085414–085422
64. Frederiksen T, Paulsson M, Brandbyge M, Jauho A-P (2007) Inelastic transport theory from first principles: methodology and application to nanoscale devices. Phys Rev B 75(20):205413–205422
65. Sheng W, Li ZY, Ning ZY, Zhang ZH, Yang ZQ, Guo H (2009) Quantum transport in alkane molecular wires: effects of binding modes and anchoring groups. J Chem Phys 131(24):244712–244719
66. Smeu M, Wolkow RA, Guo H (2009) Conduction pathway of pi-stacked ethylbenzene molecular wires on Si(100). J Am Chem Soc 131(31):11019–11026
67. Flemming I (1978) Frontier orbitals and organic chemical reactions. Wiley, Chichester
68. Caroli C, Combescot R, Nozieres P, Saint-James D (1971) Direct calculation of the tunneling current. J Phys C Solid State Phys 4(8):916
69. Beenakker CWJ, van Houten H (1991) Quantum transport in semiconductor nanostructures. In: Henry E, David T (eds) Solid state physics, vol 44. Academic, San Diego, pp 1–228
70. Nitzan A, Jortner J, Wilkie J, Burin AL, Ratner MA (2000) Tunneling time for electron transfer reactions. J Phys Chem B 104(24):5661–5665
71. Nitzan A (2001) Electron transmission through molecules and molecular interfaces. Annu Rev Phys Chem 52(1):681–750
72. Galperin M, Ratner MA, Nitzan A (2007) Molecular transport junctions: vibrational effects. J Phys Condens Matter 19(10):103201
73. Brandbyge M, Srensen MR, Jacobsen KW (1997) Conductance eigenchannels in nanocontacts. Phys Rev B 56(23):14956
74. Brandbyge M, Kobayashi N, Tsukada M (1999) Conduction channels at finite bias in single-atom gold contacts. Phys Rev B 60(24):17064
75. Heurich J, Cuevas JC, Wenzel W, Schon G (2002) Electrical transport through single-molecule junctions: from molecular orbitals to conduction channels. Phys Rev Lett 88 (25):256803
76. Solomon GC, Gagliardi A, Pecchia A, Frauenheim T, Di Carlo A, Reimers JR, Hush NS (2006) Molecular origins of conduction channels observed in shot-noise measurements. Nano Lett 6(11):2431–2437

77. Wang B, Zhu Y, Ren W, Wang J, Guo H (2007) Spin-dependent transport in Fe-doped carbon nanotubes. Phys Rev B 75(23):235415–235417
78. Paulsson M, Brandbyge M (2007) Transmission eigenchannels from nonequilibrium Green's functions. Phys Rev B 76(11):115117
79. Jacob D, Palacios JJ (2006) Orbital eigenchannel analysis for ab initio quantum transport calculations. Phys Rev B 73(7):075424–075429
80. Bagrets A, Papanikolaou N, Mertig I (2007) Conduction eigenchannels of atomic-sized contacts: ab initio KKR Green's function formalism. Phys Rev B 75(23):235448
81. Solomon GC, Herrmann C, Hansen T, Mujica V, Ratner MA (2010) Exploring local currents in molecular junctions. Nat Chem 2(3):223–228
82. Todorov TN (2002) Tight-binding simulation of current-carrying nanostructures. J Phys Condens Matter 14(11):3049–3084
83. Pecchia A, Di Carlo A (2004) Atomistic theory of transport in organic and inorganic nanostructures. Rep Prog Phys 67(8):1497–1561
84. Stuchebrukhov AA (1996) Tunneling currents in electron transfer reactions in proteins. J Chem Phys 104(21):8424–8432
85. Stuchebrukhov AA (1996) Tunneling currents in electron transfer reaction in proteins. II. Calculation of electronic superexchange matrix element and tunneling currents using nonorthogonal basis sets. J Chem Phys 105(24):10819–10829
86. Stuchebrukhov AA (1997) Tunneling currents in proteins: nonorthogonal atomic basis sets and Mulliken population analysis. J Chem Phys 107(16):6495–6498
87. Sai N, Bushong N, Hatcher R, Di Ventra M (2007) Microscopic current dynamics in nanoscale junctions. Phys Rev B 75(11):115410–115418
88. Ernzerhof M, Bahmann H, Goyer F, Zhuang M, Rocheleau P (2006) Electron transmission through aromatic molecules. J Chem Theory Comput 2(5):1291–1297
89. Emberly EG, Kirczenow G (2001) Models of electron transport through organic molecular monolayers self-assembled on nanoscale metallic contacts. Phys Rev B 64(23):235412
90. Tian J-H, Yang Y, Zhou X-S, Schöllhorn B, Maisonhaute E, Chen Z-B, Yang F-Z, Chen Y, Amatore C, Mao B-W, Tian Z-Q (2010) Electrochemically assisted fabrication of metal atomic wires and molecular junctions by MCBJ and STM-BJ methods. ChemPhysChem 11 (13):2745–2755
91. Quek SY, Venkataraman L, Choi HJ, Louie SG, Hybertsen MS, Neaton JB (2007) Amine-gold linked single-molecule circuits: experiment and theory. Nano Lett 7(11):3477–3482
92. Kiguchi M, Tal O, Wohlthat S, Pauly F, Krieger M, Djukic D, Cuevas JC, van Ruitenbeek JM (2008) Highly conductive molecular junctions based on direct binding of benzene to platinum electrodes. Phys Rev Lett 101(4):046801
93. Yang H, Xie XS (2002) Statistical approaches for probing single-molecule dynamics photon-by-photon. Chem Phys 284(1–2):423–437
94. Gräslund A, Rigler R, Widengren J (eds) (2010) Single molecule spectroscopy in chemistry, physics and biology. Springer Series in Chemical Physics. Springer, Berlin
95. Wang W, Lee T, Kretzschmar I, Reed MA (2004) Inelastic electron tunneling spectroscopy of an alkanedithiol self-assembled monolayer. Nano Lett 4(4):643–646
96. Kushmerick JG, Lazorcik J, Patterson CH, Shashidhar R, Seferos DS, Bazan GC (2004) Vibronic contributions to charge transport across molecular junctions. Nano Lett 4(4):639–642
97. Chen Y-C, Zwolak M, Di Ventra M (2005) Inelastic effects on the transport properties of alkanethiols. Nano Lett 5(4):621–624
98. Troisi A, Ratner MA (2005) Modeling the inelastic electron tunneling spectra of molecular wire junctions. Phys Rev B 72(3):033408
99. Solomon GC, Gagliardi A, Pecchia A, Frauenheim T, Di Carlo A, Reimers JR, Hush NS (2006) Understanding the inelastic electron-tunneling spectra of alkanedithiols on gold. J Chem Phys 124(9):094704–094710

100. Galperin M, Ratner MA, Nitzan A (2004) Inelastic electron tunneling spectroscopy in molecular junctions: peaks and dips. J Chem Phys 121(23):11965–11979
101. Paulsson M, Frederiksen T, Brandbyge M (2006) Inelastic transport through molecules: comparing first-principles calculations to experiments. Nano Lett 6(2):258–262
102. Nakamura H, Yamashita K, Rocha AR, Sanvito S (2008) Efficient ab initio method for inelastic transport in nanoscale devices: analysis of inelastic electron tunneling spectroscopy. Phys Rev B 78(23):235418–235420
103. Troisi A, Ratner MA (2006) Molecular transport junctions: propensity rules for inelastic electron tunneling spectra. Nano Lett 6(8):1784–1788
104. Troisi A, Ratner MA (2006) Propensity rules for inelastic electron tunneling spectroscopy of single-molecule transport junctions. J Chem Phys 125(21):214709–214711
105. Gagliardi A, Solomon GC, Pecchia A, Frauenheim T, Di Carlo A, Hush NS, Reimers JR (2007) A priori method for propensity rules for inelastic electron tunneling spectroscopy of single-molecule conduction. Phys Rev B 75(17):174306–174308
106. Paulsson M, Frederiksen T, Ueba H, Lorente N, Brandbyge M (2008) Unified description of inelastic propensity rules for electron transport through nanoscale junctions. Phys Rev Lett 100(22):226604
107. Troisi A, Ratner MA (2007) Inelastic insights for molecular tunneling pathways: bypassing the terminal groups. PCCP 9(19):2421–2427
108. Troisi A, Beebe JM, Picraux LB, Zee RDv, Stewart DR, Ratner MA, Kushmerick JG (2007) Tracing electronic pathways in molecules by using inelastic tunneling spectroscopy. Proc Natl Acad Sci USA 104(36):14255–14259
109. Kiguchi M (2009) Electrical conductance of single C_{60} and benzene molecules bridging between Pt electrode. Appl Phys Lett 95(7):073301–073303
110. Aviram A, Ratner MA (1974) Molecular rectifiers. Chem Phys Lett 29(2):277–283
111. Carter FL (1983) Molecular level fabrication techniques and molecular electronic devices. J Vac Sci Technol B 1(4):959–968
112. Joachim C, Gimzewski JK, Aviram A (2000) Electronics using hybrid-molecular and mono-molecular devices. Nature 408(6812):541–548
113. Aviram A (1988) Molecules for memory, logic, and amplification. J Am Chem Soc 110 (17):5687–5692
114. Jlidat N, Hliwa M, Joachim C (2008) A semi-classical XOR logic gate integrated in a single molecule. Chem Phys Lett 451(4–6):270–275
115. Renaud N, Ito M, Shangguan W, Saeys M, Hliwa M, Joachim C (2009) A NOR-AND quantum running gate molecule. Chem Phys Lett 472(1–3):74–79
116. Soe W-H, Manzano C, Renaud N, de Mendoza P, De Sarkar A, Ample F, Hliwa M, Echavarren AM, Chandrasekhar N, Joachim C (2011) Manipulating molecular quantum states with classical metal atom inputs: demonstration of a single molecule NOR logic gate. ACS Nano 5(2):1436–1440
117. Metzger RM, Chen B, Höpfner U, Lakshmikantham MV, Vuillaume D, Kawai T, Wu X, Tachibana H, Hughes TV, Sakurai H, Baldwin JW, Hosch C, Cava MP, Brehmer L, Ashwell GJ (1997) Unimolecular electrical rectification in hexadecylquinolinium tricyanoquinodimethanide. J Am Chem Soc 119(43):10455–10466
118. Metzger RM (2003) Unimolecular electrical rectifiers. Chem Rev 103(9):3803–3834
119. Heeger AJ, Sariciftci NS, Namdas EB (2010) Semiconducting and metallic polymers. Oxford University Press, Oxford
120. Sirringhaus H, Tessler N, Friend RH (1998) Integrated optoelectronic devices based on conjugated polymers. Science 280(5370):1741–1744
121. Facchetti A, Yoon MH, Marks TJ (2005) Gate dielectrics for organic field-effect transistors: new opportunities for organic electronics. Adv Mater 17(14):1705–1725
122. Klauk H, Halik M, Zschieschang U, Schmid G, Radlik W, Weber W (2002) High-mobility polymer gate dielectric pentacene thin film transistors. J Appl Phys 92(9):5259–5263

123. Vilan A, Yaffe O, Biller A, Salomon A, Kahn A, Cahen D (2010) Molecules on Si: electronics with chemistry. Adv Mater 22(2):140–159
124. Malicki M, Guan Z, Ha SD, Heimel G, Barlow S, Rumi M, Kahn A, Marder SR (2009) Preparation and characterization of 4′-donor substituted stilbene-4-thiolate monolayers and their influence on the work function of gold. Langmuir 25(14):7967–7975
125. Moore AM, Mantooth BA, Dameron AA, Donhauser ZJ, Lewis PA, Smith RK, Fuchs DJ, Weiss PS (2008) Measurements and mechanisms of single-molecule conductance switching. In: Hasegawa M, Inoue A, Kobayashi N, Sakurai T, Wille L (eds) Frontiers in materials research. Advances in Materials Research, vol 10. Springer, Berlin, pp 29–47
126. Weiss PS (2008) Functional molecules and assemblies in controlled environments: formation and measurements. Acc Chem Res 41(12):1772–1781
127. Landau A, Kronik L, Nitzan A (2008) Cooperative effects in molecular conduction. J Comput Theor Nanosci 5:535–544
128. Reuter MG, Seideman T, Ratner MA (2011) Probing the surface-to-bulk transition: a closed-form constant-scaling algorithm for computing subsurface Green functions. Phys Rev B 83 (8):085412
129. Kushmerick JG, Blum AS, Long DP (2006) Metrology for molecular electronics. Anal Chim Acta 568(1–2):20–27
130. Blum AS, Kushmerick JG, Pollack SK, Yang JC, Moore M, Naciri J, Shashidhar R, Ratna BR (2004) Charge transport and scaling in molecular wires. J Phys Chem B 108 (47):18124–18128
131. Selzer Y, Allara DL (2006) Single-molecule electrical junctions. Annu Rev Phys Chem 57(1):593–623
132. Selzer Y, Cai L, Cabassi MA, Yao Y, Tour JM, Mayer TS, Allara DL (2004) Effect of local environment on molecular conduction: isolated molecule versus self-assembled monolayer. Nano Lett 5(1):61–65
133. Kushmerick JG, Naciri J, Yang JC, Shashidhar R (2003) Conductance scaling of molecular wires in parallel. Nano Lett 3(7):897–900
134. Burtman V, Ndobe AS, Vardeny ZV (2005) Transport studies of isolated molecular wires in self-assembled monolayer devices. J Appl Phys 98(3):034314–034319
135. Muralidharan B, Ghosh AW, Pati SK, Datta S (2007) Theory of high bias coulomb blockade in ultrashort molecules. IEEE Trans Nanotechnol 6(5):536–544
136. Zcaronuticacute I, Fabian J, Das Sarma S (2004) Spintronics: fundamentals and applications. Rev Mod Phys 76(2):323
137. Seneor P, Bernand-Mantel A, Petroff F (2007) Nanospintronics: when spintronics meets single electron physics. J Phys Condens Matter 19(16):165222
138. Wolf SA, Chtchelkanova AY, Treger DM (2006) Spintronics: a retrospective and perspective. IBM J Res Dev 50(1):101–110
139. Emberly EG, Kirczenow G (2002) Molecular spintronics: spin-dependent electron transport in molecular wires. Chem Phys 281(2–3):311–324
140. Bogani L, Wernsdorfer W (2008) Molecular spintronics using single-molecule magnets. Nat Mater 7(3):179–186
141. Rocha AR, Garcia-suarez VM, Bailey SW, Lambert CJ, Ferrer J, Sanvito S (2005) Towards molecular spintronics. Nat Mater 4(4):335–339
142. Katsonis N, Kudernac T, Walko M, van der Molen SJ, van Wees BJ, Feringa BL (2006) Reversible conductance switching of single diarylethenes on a gold surface. Adv Mater 18(11):1397–1400
143. Matsuda K, Yamaguchi H, Sakano T, Ikeda M, Tanifuji N, Irie M (2008) Conductance photoswitching of diarylethene: gold nanoparticle network induced by photochromic reaction. J Phys Chem C 112(43):17005–17010
144. Kronemeijer AJ, Akkerman HB, Kudernac T, van Wees BJ, Feringa BL, Blom PWM, de Boer B (2008) Reversible conductance switching in molecular devices. Adv Mater 20(8):1467–1473

145. van der Molen SJ, Liao J, Kudernac T, Agustsson JS, Bernard L, Calame M, van Wees BJ, Feringa BL, Schönenberger C (2008) Light-controlled conductance switching of ordered metal–molecule–metal devices. Nano Lett 9(1):76–80
146. He J, Chen F, Liddell PA, Andréasson J, Straight SD, Gust D, Moore TA, Moore AL, Li J, Sankey OF, Lindsay SM (2005) Switching of a photochromic molecule on gold electrodes: single-molecule measurements. Nanotechnology 16(6):695
147. Zhang C, He Y, Cheng H-P, Xue Y, Ratner MA, Zhang XG, Krstic P (2006) Current-voltage characteristics through a single light-sensitive molecule. Phys Rev B 73(12):125445
148. Kondo M, Tada T, Yoshizawa K (2005) A theoretical measurement of the quantum transport through an optical molecular switch. Chem Phys Lett 412(1–3):55–59
149. Li J, Speyer G, Sankey OF (2004) Conduction switching of photochromic molecules. Phys Rev Lett 93(24):248302
150. Zhuang M, Ernzerhof M (2005) Mechanism of a molecular electronic photoswitch. Phys Rev B 72(7):073104
151. Zhang C, Du MH, Cheng HP, Zhang XG, Roitberg AE, Krause JL (2004) Coherent electron transport through an azobenzene molecule: a light-driven molecular switch. Phys Rev Lett 92(15):158301
152. Ahn CH, Bhattacharya A, Di Ventra M, Eckstein JN, Frisbie CD, Gershenson ME, Goldman AM, Inoue IH, Mannhart J, Millis AJ, Morpurgo AF, Natelson D, Triscone J-M (2006) Electrostatic modification of novel materials. Rev Mod Phys 78(4):1185
153. Reuter MG, Sukharev M, Seideman T (2008) Laser field alignment of organic molecules on semiconductor surfaces: toward ultrafast molecular switches. Phys Rev Lett 101 (20):208303
154. Ray K, Ananthavel SP, Waldeck DH, Naaman R (1999) Asymmetric scattering of polarized electrons by organized organic films of chiral molecules. Science 283(5403):814–816
155. Ray SG, Daube SS, Leitus G, Vager Z, Naaman R (2006) Chirality-induced spin-selective properties of self-assembled monolayers of DNA on gold. Phys Rev Lett 96(3):036101
156. Liu R, Ke S-H, Baranger HU, Yang W (2005) Intermolecular effect in molecular electronics. J Chem Phys 122(4):044703–044704
157. Lagerqvist J, Chen Y-C, Ventra MD (2004) Shot noise in parallel wires. Nanotechnology 15 (7):S459–S464
158. Yaliraki SN, Ratner MA (1998) Molecule-interface coupling effects on electronic transport in molecular wires. J Chem Phys 109(12):5036–5043
159. Tomfohr J, Sankey OF (2004) Theoretical analysis of electron transport through organic molecules. J Chem Phys 120(3):1542–1554
160. Magoga M, Joachim C (1999) Conductance of molecular wires connected or bonded in parallel. Phys Rev B 59(24):16011
161. Lang ND, Avouris P (2000) Electrical conductance of parallel atomic wires. Phys Rev B 62(11):7325
162. Reuter MG, Ratner MA, Seideman T (2011) unpublished
163. Solomon GC, Andrews DQ, Goldsmith RH, Hansen T, Wasielewski MR, Van Duyne RP, Ratner MA (2008) Quantum interference in acyclic systems: conductance of cross-conjugated molecules. J Am Chem Soc 130(51):17301–17308
164. Cardamone DM, Stafford CA, Mazumdar S (2006) Controlling quantum transport through a single molecule. Nano Lett 6(11):2422–2426
165. Hettler MH, Wenzel W, Wegewijs MR, Schoeller H (2003) Current collapse in tunneling transport through benzene. Phys Rev Lett 90(7):076805
166. Ke S-H, Yang W, Baranger HU (2008) Quantum-interference-controlled molecular electronics. Nano Lett 8(10):3257–3261
167. Patoux C, Coudret C, Launay J-P, Joachim C, Gourdon A (1997) Topological effects on intramolecular electron transfer via quantum interference. Inorg Chem 36(22):5037–5049
168. Sautet P, Joachim C (1988) Electronic interference produced by a benzene embedded in a polyacetylene chain. Chem Phys Lett 153(6):511–516

169. Stadler R, Ami S, Joachim C, Forshaw M (2004) Integrating logic functions inside a single molecule. Nanotechnology 15(4):S115–S121
170. Stafford CA, Cardamone DM, Mazumdar S (2007) The quantum interference effect transistor. Nanotechnology 18(42):424014
171. Yaliraki SN, Ratner MA (2002) Interplay of topology and chemical stability on the electronic transport of molecular junctions. Ann NY Acad Sci 960(Mol Electron II):153–162
172. Segal D, Nitzan A, Davis WB, Wasielewski MR, Ratner MA (2000) Electron transfer rates in bridged molecular systems 2. A steady-state analysis of coherent tunneling and thermal transitions. J Phys Chem B 104(16):3817–3829
173. Weiss EA, Ahrens MJ, Sinks LE, Gusev AV, Ratner MA, Wasielewski MR (2004) Making a molecular wire: charge and spin transport through para-phenylene oligomers. J Am Chem Soc 126(17):5577–5584
174. Luo L, Choi SH, Frisbie CD (2010) Probing hopping conduction in conjugated molecular wires connected to metal electrodes. Chem Mater 23(3):631–645
175. Fleming GR, Scholes GD, Cheng Y-C (2011), Quantum effects in biology (in press)
176. Maassen J, Zahid F, Guo H (2009) Effects of dephasing in molecular transport junctions using atomistic first principles. Phys Rev B 80(12):125423
177. Büttiker M (1988) Coherent and sequential tunneling in series barriers. IBM J Res Dev 32 (1):63–75
178. Heinrich AJ, Lutz CP, Gupta JA, Eigler DM (2002) Molecule cascades. Science 298 (5597):1381–1387
179. Lu Y, Liu M, Lent C (2007) Molecular quantum-dot cellular automata: from molecular structure to circuit dynamics. J Appl Phys 102(3):034311–034317

Top Curr Chem (2012) 313: 39–84
DOI: 10.1007/128_2011_178

Published online: 29 July 2011

Unimolecular Electronic Devices

Robert M. Metzger and Daniell L. Mattern

Abstract The first active electronic components used vacuum tubes with appropriately-shaped electrodes, then junctions of appropriately-doped Ge, Si, or GaAs semiconductors. Electronic components can now be made with appropriately-designed organic molecules. As the commercial drive to make ever-smaller and faster circuits approaches the 3-nm limit, these unimolecular organic devices may become more useful than doped semiconductors. Here we discuss the electrical contacts between metallic electrodes and organic molecular components, and survey representative organic wires composed of conducting groups and organic rectifiers composed of electron-donor and -acceptor groups, and the Aviram-Ratner proposal for unimolecular rectification. Molecular capacitors and amplifiers are discussed briefly. Molecular electronic devices are not only ultimately small (<3 nm in all directions) and fast, but their excited states may be able to decay by photons, avoiding the enormous heat dissipation endured by Si-based components that decay by phonons. An all-organic computer is an ultimate, but more distant, goal.

Keywords Aviram-Ratner theory · Cold gold evaporation · Electron-acceptor groups · Electron-donor groups · Langmuir-Blodgett film · Langmuir-Blodgett monolayer · Orbital-mediated tunneling · Rectifier · Scanning tunneling microscopy · Schottky barrier · Schottky-Mott theory · Self-assembled film · Self-assembled monolayer · Unimolecular amplifier · Unimolecular electronic devices

R.M. Metzger
Laboratory for Molecular Electronics, Department of Chemistry, The University of Alabama, Tuscaloosa, AL 35487-0336, USA
e-mail: rmetzger@ua.edu

D.L. Mattern (✉)
Department of Chemistry and Biochemistry, The University of Mississippi, University, MS 38677, USA
e-mail: mattern@olemiss.edu

Contents

1 Introduction

Molecules, acting either singly or in parallel as a monolayer array, can serve as electronic components. This ability has spawned the field of "Unimolecular Electronics" (UE) [1, 2], which, with some luck, could eventually lead to ultra-tiny and ultra-fast electronic circuitry with dimensions of 1–3 nm – a true nanotechnology. UE was inspired by organic crystalline metals [3, 4] and superconductors [4, 5] and organic conducting polymers [6–10], and it exploits the valence molecular orbitals of molecules, that is, molecular Highest Occupied Molecular Orbitals (HOMOs) and Lowest Unoccupied Molecular Orbitals (LUMOs) where electrons can be most easily influenced. UE differs from the robust area of molecule-based electronics, involving electrically-conducting crystals and polymers [1], in that UE focuses on nonpolymeric organic molecules, as will the present chapter [11, 12].

The push to make ever smaller and denser arrays of electronic devices may be inspired by technological feasibility, but it is driven by commercial considerations and by Moore's empirical "law" [13]: ever since the 1960s, the minimum distance or "design rule" (DR) between components in integrated circuits has halved every 2 years, and therefore the speed of the digital circuits has doubled [13]. At present, 3-GHz computers use 65-nm DRs, 35-nm DRs are found in research, and the challenges of DRs < 22 nm are under discussion [14]. However, Moore's "second law" suggests that the cost of higher integration increases exponentially. Going down to DR $= 3$ nm using inorganic materials and inorganic metal electrodes will be difficult and expensive [14].

This opens the door for UE, with its organic molecules as the circuit building blocks, with typical sizes of 2–3 nm. But there are challenges beyond the synthesis of appropriate organic candidates. Even when carefully designed, organic molecules may be air-sensitive, and may lose their integrity above about 150 °C, while inorganic systems (or pure carbon nanotubes) may continue to perform as hot as 600 °C. Furthermore, the "molecule | metal" interface must be well understood in order to connect the molecular circuit to the macroscopic world.

UE provides an important potential advantage beyond small size. The excited states in Si-based electronics decay by phonons, and thus a huge heat dissipation problem faces nanoscale inorganic electronics at DR = 3 nm. In contrast, UE devices may be able to decay from their excited states by photon emission [15]. If the photon decay channel can be maximized, UE devices will have a great heat advantage over inorganic ones.

2 Contacts to Metal Electrodes

To extract useful results from a molecular electronic device, or just to measure its electronic characteristics, connections must be made to macroscopic probes. That is, metallic electrodes must interface to different ends of the molecule of interest. An experiment may interrogate a single molecule, or may measure a one-molecule-thick layer, i.e., a *monolayer*, of the molecules of interest, provided all the molecules are oriented in the same direction. In either case, several questions arise. What is the nature of the contact between metal and molecule(s)? What metal should be chosen, and what should be the form or shape of this electrode?

There are two ways to attach a single molecule or a monolayer of molecules to a metal electrode: *physisorption* (where the connection is maintained by noncovalent forces) and *chemisorption* (where the connection is maintained by coordinating or covalent bonds). Monolayer devices often begin with the physisorption of a monolayer onto one (the "bottom") electrode. One could imagine deposition of vapor-phase molecules onto a metal electrode substrate, but such depositions usually yield poor, disordered packing and random orientations. Much better is the physisorption of an ordered *Langmuir* [*or Pockels-Langmuir* (*PL*)] *monolayer* formed by amphiphilic molecules floating at a water–air interface. The monolayer is laterally compressed until the molecules are close-packed. The *Langmuir-Blodgett* (*LB*) or vertical transfer method [16, 17] then lets the monolayer adhere to an electrode (or any solid surface) being gradually removed from (or inserted into) the water, while the *Langmuir-Schaefer* (*LS*) or quasi-horizontal transfer method [18] lays the electrode flat onto the interface so that the monolayer sticks to it.

Ordering of a PL film presents some restrictions on the chemical nature of the molecules. The molecules will typically need to be made amphiphilic by attaching either pendant nonpolar (e.g., alkyl) groups to yield a hydrophobic end, or pendant polar (e.g., carboxylic acid) or even ionic (e.g., carboxylate anion) groups to make a hydrophilic end, complicating the synthetic task. Further, the pendant groups may have a cross section that is not well matched with the size of the electroactive portions of the molecule. As a consequence, it may not be possible for the monolayer to be tightly-packed at all depths. If the second ("top") electrode is introduced by vapor deposition, stalactite- or stalagmite-like shards or filaments or whiskers of metal may partially penetrate the monolayer, yet dominate the electrical characteristics [19, 20]. Also, such a monolayer may not be laterally uniform, leading to poorly-reproducible measurements. Finally, physisorbed molecules can move easily after deposition, either as they seek a thermodynamic steady state on the surface, or in response to an applied electric field (a 2-V bias across a monolayer 2 nm thick makes a 1 GV/m field). That means that LB and LS films can reorient over days, weeks, or months, changing their properties.

A variety of chemisorption techniques can covalently bind molecules to specific metal surfaces [21]. Most commonly, thiols (H–S–R) or thioesters (like thioacetates, CH_3CO–S–R) coordinate to gold or mercury, serving as "alligator clips" (by analogy to the clips that connect macroscopic wires and terminals). In addition, carboxylates can bind to oxide-covered aluminum or silver, chlorosilanes to silicon oxide, amines to platinum, etc. The resulting films are called "*self-assembled monolayers*" (*SAMs*) because they form spontaneously when the molecules are allowed access to the surface [22, 23]. The binding of SAMs to metals often involves a homolytic bond scission in the ligand, and the formation of a covalent or partially ionic bond to the substrate. For example, the chemisorption of thiols to Au involves the bonding of R-S• to Au; the S–Au bond is fairly polar [24]. The thiol's H could escape the substrate as H_2 or H_2O [24]. Because thiols easily oxidize to disulfides under ambient conditions, they are often supplied as protected thioacetates. In situ base-assisted cleavage of the thioacetate produces thiolate anion, which then coordinates to Au. The Au–S bond is normal to the Au plane, but the binding of alkanethiols to Au as thiolates involves a tilt angle (typically 30°) between the Au–S bond and the rest of the molecule. For aromatic thiols, the tilt angle is typically closer to the normal [24]. For thiolate–Au(111) SAMs in high vacuum, there is evidence that the Au surface reorganizes to control the SAM structural phases, and that a single Au atom coming out of the Au(111) surface plane may bond to the thiolate [25–27].

The advantages of SAMs are that they are sturdily anchored at a fixed distance from the metal substrate, may be more robust than Langmuir films, and can be convenient to prepare. A disadvantage is that uniform monolayer coverage, so easily achieved kinetically for LB films, is more difficult to obtain in SAMs. This is because SAMs are created by random attack on the electrode surface, in contrast to Langmuir films, which are transferred when they are close-packed.

3 Electrode Metal and Schottky Barriers

Most metals are covered by an oxide (impervious and insulating, or more often, as with Al, defect-ridden). In contrast, *gold* has no oxide, and has the advantage of making SAMs with thiols. However, Au atoms migrate somewhat after deposition to minimize total energy, and migrate even more under an electric field ("electromigration"), so the surface may not be stable during measurement [28]. Further, an Au film freshly deposited on a substrate from Au vapor is hydrophilic, but after about 20 min of exposure to room air, becomes hydrophobic because of adventitious adsorbates from air. These impurities can be displaced later during SAM formation. *Palladium* and *platinum* are less susceptible to spontaneous migration than Au, and stable nanogaps of 0.5–2.1 nm can be formed [29]. *Magnesium* has a low work function, comparable to the electron affinities of some good organic one-electron acceptors; however, Mg oxidizes rapidly in O_2 to MgO, and dissolves as $Mg(OH)_2$ in the presence of H_2O; so if Mg is used, it needs a protective cover layer, e.g., of Ag. *Aluminum* also has a relatively low work function, and forms a protective but very disordered oxide Al_2O_3 layer in air.

The interface between a metal electrode and an organic molecule creates a set of interfacial dipoles as charges reorganize across the interface. These dipoles constitute a "Schottky barrier" to electron transport. The Schottky barrier [30] is formed because the chemical potential, or partial molar Gibbs free energy, or Fermi energy E_F, must become the same if one accosts two materials of different free energy, e.g., a metal and a semiconductor; it is conceptually related to the Schottky–Nordheim energy barrier to electron emission from a metal surface (where the local metal work function is reduced). Charge will flow toward the interface, and semiconductor valence and conduction bands will "bend" until the equilibrium in free energies is established. The charge clustering becomes a set of interfacial dipoles, and the Schottky barrier becomes a rectifying junction.

The Schottky–Mott theory predicts a current $I = (4\ \pi\ e\ m\ k_B{}^2/h^3)\ T^2 \exp(-e\ \Delta/k_B\ T)[\exp(e\ n\ V/k_B\ T)-1]$, where e is the electronic charge, m is the effective mass of the carrier, k_B is Boltzmann's constant, T is the absolute temperature, n is a "filling factor," Δ is the Schottky barrier height (see Fig. 1), and V is the applied voltage [31]. In Schottky–Mott theory, Δ should be the difference between the Fermi level of the metal and the conduction band minimum (for an n-type semiconductor-to-metal interface) or the valence band maximum (for a p-type semiconductor-metal interface) [32, 33]. Certain experimentally observed variations of Δ were for decades ascribed to "pinning" of states, but can now be attributed to local inhomogeneities of the interface, so the Schottky–Mott theory is secure. The opposite of a Schottky barrier is an "ohmic contact," where there is only an added electrical resistance at the junction, typically between two metals.

Inorganic Schottky barrier rectifiers are commercially available. Of course, for a monolayer, one cannot speak of band bending as in a bulk semiconductor, but the ideas of a barrier, and of dipoles across it, are valid.

To avoid having different-sized Schottky barriers at the two interfaces, the same metal (or metals with almost the same work functions) should be used for both electrodes. For example, the different work functions of Pt and Mg made studies of "glass | Pt | molecule | Mg | Ag" sandwiches hard to interpret [34]. In that case, Mg probably reacted with the end of the molecule containing the strong acceptor TCNQ to form a TCNQ-salt Schottky barrier that dominated the electrical asymmetry [34]. With a different molecule lacking TCNQ, the dominating Schottky barrier effect was eliminated [35, 36].

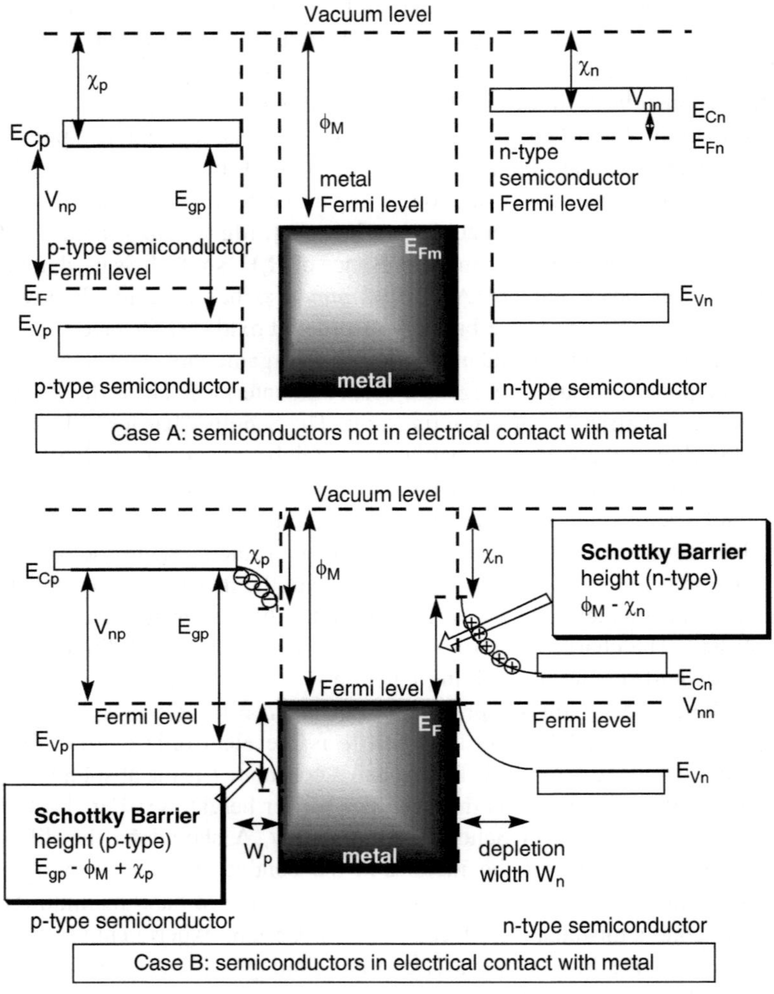

Fig. 1 Idealized energy band diagram and Schottky barrier between metal (*center*) an n-type semiconductor (*right*), and a p-type semiconductor (*left*), not in electrical contact (Case A) and after contact between metal and semiconductors (Case B) (adapted from [32]). Here ϕ_M is the work function of the metal, E_F is the common Fermi level, E_V is the top of the valence band, E_C is the bottom of the conduction band, χ is the electron affinity of the semiconductor, and E_{gp} is the energy gap inside the bulk semiconductor. The band bending is shown, as well as the Schottky barrier heights, and the charges in the "space charge region" responsible for the band bending. A set of corresponding image charges of opposite charge will exist within the metal: hence the Schottky dipole layer

4 Bottom Electrode Surface

A uniform monolayer surface depends upon it having been deposited on a flat electrode substrate. Impressive flatness can be achieved with the semi-metal graphite, a lamellar structure with weak van der Waals forces between the layers. A clean and flat surface can be exposed by the "Scotch® tape technique". A graphite sample is pressed between sticky tapes; when the tapes are separated, the graphite separates along the layers. The fresh surface that is exposed is atomically flat over large areas (e.g., 20 × 20 nm), except for steps and corrugations. Iteration of this process can eventually create a one-molecule-thick graphene layer [37].

For metals and crystals, cleavage can attempt similar feats, but the results are not as good. Metal surfaces formed by cleavage are usually not atomically flat. When an Au wire is flame-annealed in a hydrogen-air flame, the Au(111) face is formed preferentially, since it has a lower surface energy than the Au(100) or Au(110) faces, but these Au(111) faces resemble New Mexico "mesas": the atomically flat region may be only 50 × 50 nm, and is surrounded by one- or two-atom steps leading down to the "plain," and then on to the next mesa.

The Si wafer industry has achieved very flat Si surfaces by electropolishing: a root-mean-square (rms) roughness of 0.3 nm is customary, even for highly-doped degenerate Si. For a 100-nm Au layer deposited atop a 10-nm Ti adhesion layer atop an electropolished Si wafer, an rms roughness of 0.4 nm was measured by AFM [38].

5 Top Electrode

The second (or "top") electrode used to interrogate a UE monolayer that has been deposited on the underlying ("bottom") electrode can be, most simply, the *probe tip* of a scanning probe microscope, i.e., a Scanning Tunneling Microscope (STM) or a Conducting-Probe Atomic Force Microscope (CP-AFM), brought close to the monolayer surface. Voltage can be controlled easily and the resulting currents measured. Disadvantages include that the monolayer is sampled over only a very small portion of its expanse at a time, so that the measurement may not be representative.

The second metal electrode can also be a *macroscopic pad*, typically 0.25–1 mm^2 in area, that can give measurements representative of a population of molecules. Disadvantages include the danger that the larger area sampled may include defects in the monolayer with electrical short circuits that perturb measurements. Also, construction of a second electrode pad is not trivial. Laying a preformed pad on a monolayer can disrupt the integrity of the monolayer. Letting a pad accumulate by sputtering is not feasible because the metal vapor is too hot and damages the monolayer. Deposition from evaporated low-melting metals (Mg, Ca, Al, or Pb) is possible, especially if the monolayer assembly is externally cryo-cooled to circumvent thermal damage [39, 40]. In our experience, half of the pads made in this way avoided electrical shorting [38, 41]. In contrast, higher-melting metals, such as Au, are not cooled quickly enough by the cryocooled substrate holder, so that Au atoms reach the monolayer hot, and every "Au | monolayer | Au" sandwich is shorted. Adding a low pressure of room-temperature Ar to the

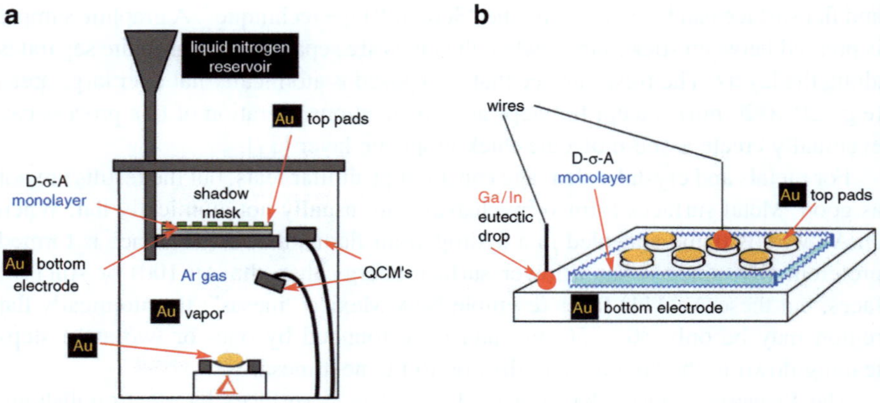

Fig. 2 (**a**) Edwards E308 evaporator. One quartz-crystal thickness monitor is pointed towards the Au source to monitor Au vapor deposition on chamber walls; the other monitors Au deposited through the shadow mask atop the organic layer. In the "cold Au" deposition, a small amount of Ar gas is added to the chamber to cool the Au atoms to room temperature before they physisorb atop the cryocooled organic monolayer. (**b**) Geometry of an "Au | monolayer | Au pad" sandwich, with electrical connections made using a Ga/In eutectic

Fig. 3 Geometry of the crossed-wire technique

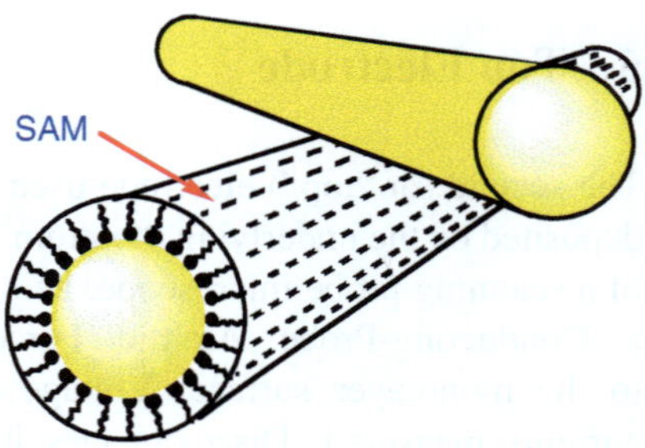

evaporation chamber allows the hot Au atoms to be cooled by several collisions with Ar atoms, so that they arrive at the sample surface at room temperature, in what is called the "cold-gold" technique (Fig. 2a). The rms roughness of a cold-gold electrode was measured as 1 nm, twice rougher than the bottom Au electrode [38].

Preformed gold *nanoparticles* have been used as the top electrode for interrogating SAMs of dithiols diluted within an alkylthiol matrix. The alkyl tails of the alkylthiols do not coordinate to the nanoparticles, but a dithiol's second SH group, sticking above the alkyl level, can bond to the nanoparticle, which is then contacted with an Au-coated AFM probe [42].

Another electrode geometry is the *crossed-wire* technique (Fig. 3). Here, the bottom electrode is a wire, coated cylindrically with a SAM of UE molecules. The second electrode is a second wire, oriented at right angles, and slowly brought into soft mechanical contact with the monolayer. Wires of 10 μm diameter would interrogate a patch of about 10^3 molecules [43–45].

6 Break Junctions

An alternative method to position two electrodes at nanometer distances apart is the *mechanically-controlled break junction* (MCBJ) technique. An ultra-thin, notched Au wire on a flexible substrate can be broken reliably by pushing on the Au with a piezoelectric piston, cracking the Au (Fig. 4). This produces a gap between the Au shards whose size can be finely varied to ± 1 Å by a piston or control rod [46, 47]. When UE molecules with thiol groups on both ends are present in a surrounding solution, the gap can be adjusted until the molecules can span it. A dilute solution means the number of spanning molecules will be small, and the least-common-multiple of current flow among many junctions indicates those spanned by a single molecule [47].

A break junction can also be created by passing a current (0.5–1.0 V) through an Au nanowire (<20 nm diameter) defined by electron-beam lithography and shadow evaporation. Such *electromigrated break junctions* (EMBJs) have yielded reproducible 1–3 nm gaps between electrodes [48–50].

Another type of break junction, the "in situ" or *STM break junction* (STM-BJ), involves pushing an STM tip through a SAM to make contact with the underlying electrode (Fig. 5). The probe tip is then quickly withdrawn, stretching a bridge of metal atoms until the bridge breaks. As the gap widens, SAM molecules can fill it, giving distinctive current flows, until the gap becomes too wide for molecules to

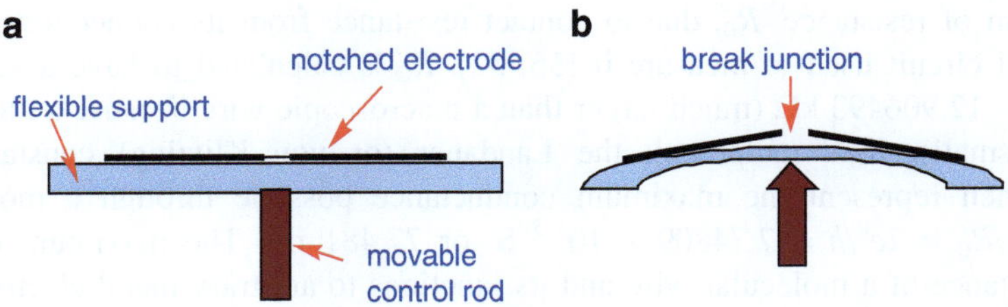

Fig. 4 Design of the Mechanically-Controlled Break Junction (MCBJ) technique

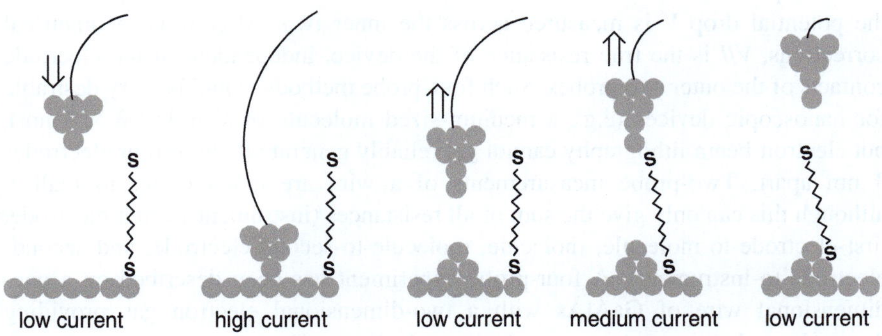

Fig. 5 Steps in the STM Break Junction (STM-BJ) technique

bridge, and the current ceases [51, 52]. Similar measurements can be made by varying the gap distance without actually making tip-to-electrode contact first [the $I(s)$ method] [53], while in the $I(t)$ method, the gap is held constant and current changes are monitored as bridging molecules spontaneously attach and detach [54]. The $I(t)$ method favors the formation of single-molecule bridges.

7 Molecular Resistors

Among the two-terminal devices that can be imagined for UE [capacitors, inductors, rectifiers, negative differential resistance (NDR) devices], the simplest is a molecular wire, that is, a molecule capable of conducting electricity: a nanoconductor or, equivalently, a nanoresistor. Even the most conductive of molecular wires has a minimum resistance.

The electrical resistance R of a device (in ohms, Ω) is given by Ohm's Law [55]: $R = V/I$, where V is the applied potential (in volts, V), and I is the resulting current (in amperes, A). Its reciprocal, the conductance, is given by $G \equiv 1/R = I/V$ (in siemens, $S = \Omega^{-1}$). This law is valid in macroscopic wires, where electron scattering off impurities and lattice defects dominates the conductivity. For nanoscopic objects like a molecular wire, the current I is determined by Landauer's formula [56], which incorporates the concept of a minimum resistance. That is, resistance is quantized, and even a perfectly conductive molecule retains a quantum of resistance, R_0, due to contact resistance from its connections to the external circuit used to measure it [56, 57]. R_0 is calculated to have a value of $h/2e^2 = 12.906493$ kΩ (much larger than a macroscopic wire because a molecule is so small). R_0's reciprocal, the Landauer (or von Klitzing) constant G_0, must then represent the maximum conductance possible through a molecule; $G_0 \equiv 1/R_0 = 2e^2/h = 7.74809 \times 10^{-5}$ S, or 77,481 nS. The maximum overall conductance of a molecular wire and its junctions to arbitrary metal electrodes is (1/2) G_0, assuming two carriers of opposite spin.

The resistance R of a macroscopic device is best measured by four probes: an electrical potential drives a fixed current I between the outermost two probes, and the potential drop V is measured across the inner two. After some geometrical corrections, V/I is the true resistance of the device, independent of the electrode contacts of the outer two probes. Such four-probe methods would be very desirable for nanoscopic devices (e.g., a medium-sized molecule of size $3 \times 3 \times 3$ nm), but electron-beam lithography cannot yet reliably generate three or four electrodes 3 nm apart. Two-probe measurements of a wire are much easier to realize, although this can only give the sum of all resistances (instrument-to-first-electrode, first-electrode-to-molecule, molecule, molecule-to-second-electrode, and second-electrode-to-instrument). A four-probe experiment has been described on a one-dimensional wire of GaAlAs with a two-dimensional electron gas providing the probes, showing a resistance-free conductance when the outer contacts are bypassed [57, 58].

The *IV* measurements on molecules and monolayers have been carried out almost exclusively using direct current (DC); frequency-dependent alternating current (AC) impedance measurements have rarely been performed, even though a rich spectroscopy may reveal itself, if the IV measurements were followed as a function of frequency ν.

8 Molecular Wires

An early measurement of current through a molecule was the report in 1995 of the resistance of a single C_{60} molecule, **1** (Fig. 6), deposited on an Au substrate and located and measured by an STM probe [59]. The conductivity was a respectable 18 nS. Most molecules studied as wires are more linear, with a coordinating atom at one or both ends.

For example, 1,4-benzenedithiol (**2**) was suspended between the two shards of an MCBJ to create a molecular wire consisting of a single benzene ring [47]. The conductance of this aromatic molecule was 45 nS [47]. When measured by the STM-BJ technique, a larger value of 833 nS was obtained [60]. When 1,4-xylenedithiol (**3**) was also measured by the STM-BJ technique, a smaller value, compared to **2**, of 47 nS was obtained [60]. Finding a smaller conductance when insulating methylene (CH_2) groups separate the aromatic ring from the metal electrodes is consistent with the intuitively sensible idea that the more polarizable and delocalized electrons of π systems should be better conductors than σ-bond frameworks.

This idea is borne out by additional experiments. Alkanethiols should be poor conductors (although better than no wire at all). An octanethiol (**4**) SAM, interrogated by CP-AFM, gave a miniscule conductivity of ca. 0.1 pS [42]. The advantage of having a chemisorption contact, rather than a physisorption one, is shown by the results with octanedithiol **5**, distinguished in a matrix of **4** by its contact to an Au nanoparticle. Here, the conductance jumped to 1.1 nS – a remarkable increase, although still well below that of conjugated systems [42]. Measurement with the *I*(*t*) STM single-molecule technique gave a comparable value [54]. When measured by the STM-BJ technique, **5** gave a larger conductance value of 20 nS [51]. In that study, **5** was compared with dithiols of difference length: hexanedithiol was more conductive (95 nS) while decanedithiol was less so (1.6 nS), consistent with the idea that saturated chains are relatively insulating.

The inconvenient finding that different techniques can give quite different conductances for the same molecule is compounded by the inconvenience that measurements of the same molecule with the same technique may give results, clustering around two or three different values, representing up to three different "conductance groups" or conductance maxima. For example, **5** also gives conductances around 3.8 nS [61]. These results can be interpreted as reflecting the strength of the S–Au attachment. The low-conductance value occurs when

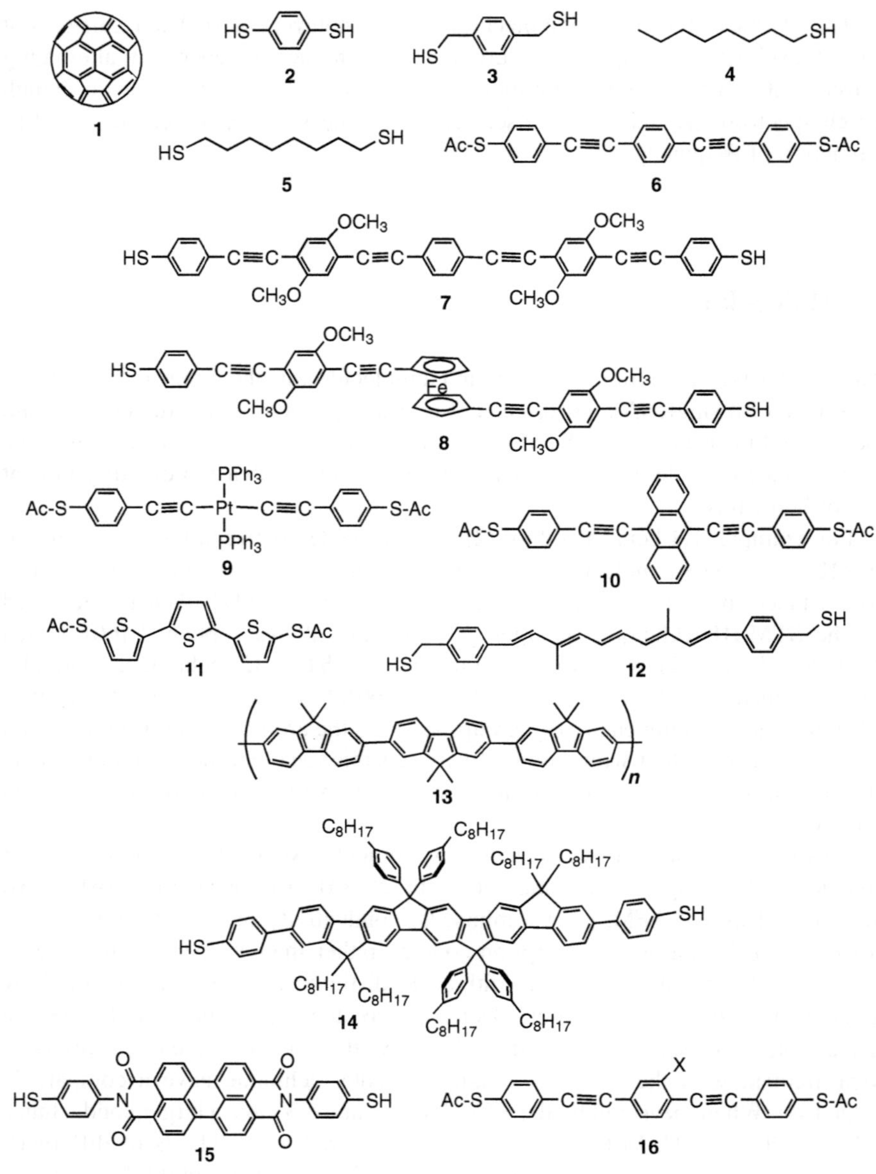

Fig. 6 Representative molecular wires

each end of the molecule connects to a single gold atom. When one end is more highly coordinated to Au – e.g., at a defect on a rough surface, or attached to the side, rather than the end of an STM tip – the medium-conductance value is obtained. Multiple Au coordinations at both ends give the high-conductance value [61, 62].

Longer unsaturated "oligophenylethynyl" (OPE) wires have been made by alternating benzene rings with ethynyl (C≡C triple bond) groups. For example, **6** has a conductance of 13 nS, measured by STM-BJ [63]. In an EMBJ geometry, the longer OPE **7** gives two orders of magnitude lower conductance than the analog with a ferrocene core, **8** [64]. In this comparison, the swiveling flexibility of the ferrocene may allow it to make better contact with the two break-junction electrodes, improving the conductance. The insulating platinum atom at the center of **9** lowers its MCBJ conductance by three orders of magnitude, compared to **10** [65, 66].

Thiophene chains also make serviceable wires. The terthiophene **11**, measured by MCBJ, gives two conductance maxima, at 10 nS and 80 nS [67]. Even simple polyene chains can be wires: the carotenoid derivative **12** has a conductance of 2 nS, measured by STM-BJ [68].

Dimethylfluorene polymers **13** were made by polymerization of a dibromotrimer on the Au(111) surface, leaving a debrominated radical at each end. A single end can be picked up by an STM tip, and, as the tip is raised, the conductance falls, as each fluorene unit lifts off the surface [69]. The conductance begins at about 10 nS, and drops roughly two orders of magnitude each time a unit is added to the wire. The monomer units are not in the same plane, and hence are not fully conjugated, contributing to the poor conductance. This experiment cannot be duplicated in solution, because the polymer would not be soluble, but the similar, decorated fluorenoid **14** has been studied with the STM-BJ method and has a conductance of 1 nS [70].

The dimer, trimer, and tetramer of **14** were also studied, and the conductance was found to be independent of temperature, but inversely dependent on oligomer length for **14** and the dimer [70]. Conversely, the conductance of the trimer and tetramer were thermally activated, but without a wire-length dependence. Thus, the longer trimer could be made more conductive than the dimer by raising the temperature. The rationale is that the shorter wire passes electrons by tunneling, while the longer oligomers are too long for tunneling, and must transport electrons by incoherent charge-hopping. The transition between transport mechanisms appears to take place at a length of around 6 nm. Similar studies have been reported on highly-substituted oligothiophenes [71] and amino-terminated OPEs [72].

The conductance of the perylenebisimide (PBI) **15** was measured by the STM-BJ technique as 1 nS [73]. Note that the thiophenol handles are not conjugated to the central core, contributing to the small value. Electron transport was temperature-independent, indicating a tunneling mechanism. However, when a gate electrode reduced the core to its radical anion, the conductance became thermally activated, indicating that electron transport then follows a hopping mechanism into and out of the core.

The conductance of the OPE nitro-**16** ($X = NO_2$) was monitored by the STM-BJ method, as the nitro group was reduced to NO and NH_2, and then protonated to NH_3. The resulting data gave an inverse linear Hammett plot with the meta-σ substituent parameter, indicating that substituent electron release into the aromatic core increases conductance [63].

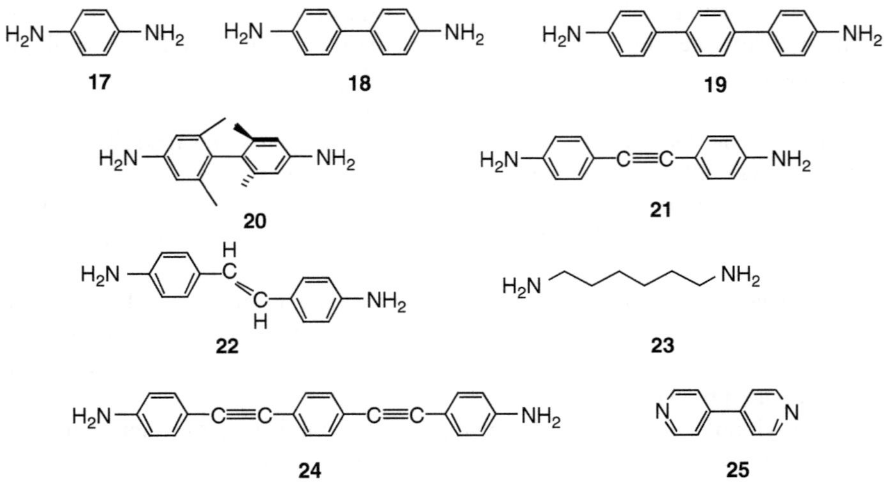

Fig. 7 Representative molecular wires with nitrogen connectors

Several molecules have also been studied with nitrogen rather than sulfur "alligator clips" to gold (Fig. 7). Using the MCBJ technique, **17** showed a healthy 496 nS of conductance; the effect of lengthening the wire can be seen in the biphenyl analog **18** (90 nS) and the terphenyl analog **19** (14 nS). The hindered biphenyl **20** has nearly orthogonal benzene rings; loss of conjugation lowers its conductance to 6 nS [74].

The prototypical OPE **21** has a conductance of 62 nS by the STM-BJ technique; the analog with an ethene bridge, **22**, has a similar value (77 nS). By way of comparison, the saturated **23** has a lower conductance of 9 nS [75]. Another STM-BJ study of six OPEs included **21** (43 nS) and **24** (10 nS), an analog of **6**. As the OPE wire length increased, the conductance diminished [72]. The STM-BJ technique was also used to examine bipyridine, **25**. Here the alligator clip nitrogens are part of the aromatic bridge, and the conductance is large, at 770 nS [51].

9 Schottky and Asymmetric Rectifiers

A rectifier, or diode, passes electrical current in one direction (the forward bias direction), but blocks it in the other direction (reverse bias). For a molecule between two electrodes in a "metal | molecule | metal" sandwich, there are three distinct processes that can give rise to such an asymmetrical conduction.

The first process is due to Schottky barriers [30], which are electrical dipole moments that form at the "metal | molecule" interfaces, as discussed above [34, 40]. The second process arises if the electrically-active portion of the molecule is placed asymmetrically within the "metal | molecule | metal" sandwich. This geometry is common, because a long alkyl "tail" is often needed to make the molecule amphiphilic so that it will form well-ordered Langmuir-Blodgett monolayers [76–78].

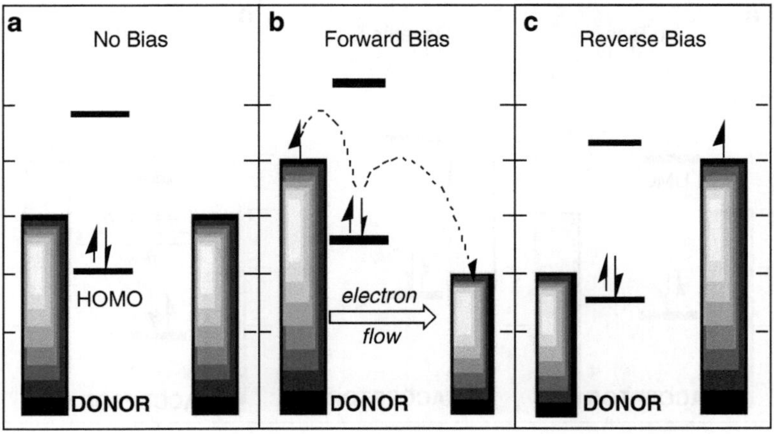

Fig. 8 Rectification via HOMO

To understand how an asymmetric geometry leads to rectification, consider the electroactive group(s) in the molecule that serve as "way-stations" for the flowing electrons. A donor group, with high-energy electrons in the molecule's HOMO, may relinquish an electron to a low-potential electrode. Conversely, an acceptor group, with a low-energy vacancy in the molecule's LUMO, may be willing to take on an electron from a high-potential electrode.

Figure 8 shows asymmetric rectification via HOMO. In Fig. 8a, the electrodes have no bias, and the HOMO orbital of a donor group is fairly high in energy. Forward bias, shown in Fig. 8b, consists of the proximal electrode being raised in potential and the distal electrode lowered. The energy of the molecule's orbitals reflects the altered potential; since the HOMO is nearer to the high-potential side, its energy is raised. When the HOMO energy is intermediate between the two electrodes, electrons can flow from left to right with a decrease in energy. But in Fig. 8c, when the same amount of bias is applied in the reverse direction, the HOMO is lowered by being closer to the low-potential electrode, so that no downhill pathway is available for reverse electron flow.

Similarly, Fig. 9 shows asymmetric rectification via LUMO. Here, an acceptor supplies a low-lying but vacant orbital, as in Fig. 9a. When the proximal electrode has high potential, the LUMO energy rises high enough to avoid providing a pathway for current flow, as in Fig. 9b. But when the near electrode has low potential, as in Fig. 9c, the LUMO energy lowers enough to be intermediate in energy between the two electrodes, and electrons can flow downhill from right to left.

10 The Aviram-Ratner Proposal for Unimolecular Rectification

UE began with Aviram and Ratner's 1973 proposal for a molecular rectifier; this is the third rectification process [79]. An Aviram-Ratner rectifier has donor and acceptor groups in one molecule, separated by an insulating bridge of sigma

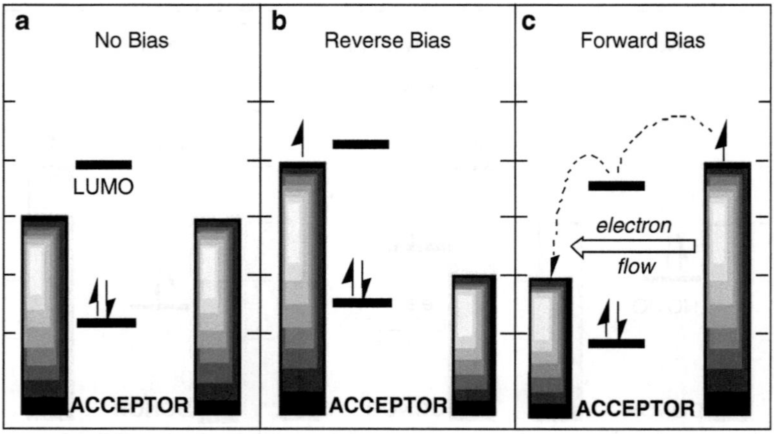

Fig. 9 Rectification via LUMO

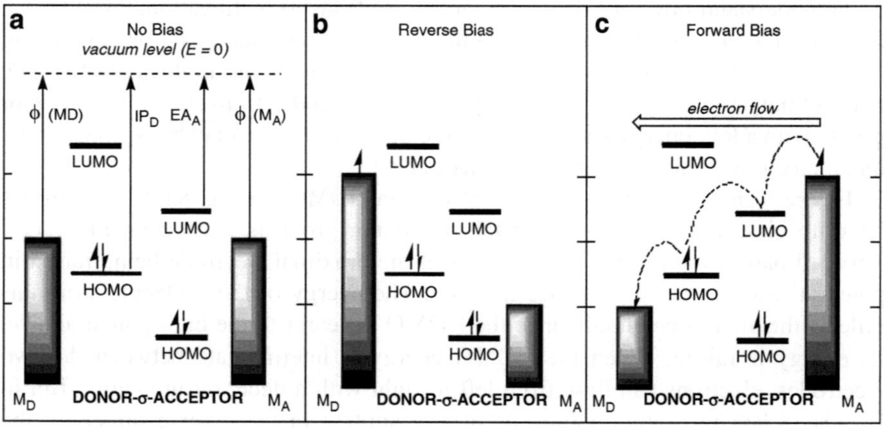

Fig. 10 Aviram-Ratner rectification via HOMO and LUMO. (**a**) A D-σ-A molecule is sandwiched between two metal electrodes. M_D is the electrode proximal to the donor, M_A is the electrode proximal to the acceptor, ϕ is the electrode metal work function, IP_D is the ionization potential of the donor, EA_A is the electron affinity of the acceptor. (**b**) No pathway for current exists when a voltage is applied in the reverse bias direction. (**c**) Under a comparable voltage to (**b**) but in the forward bias direction, rectification results from electrons flowing from M_A to LUMO to HOMO to M_D

bonds. Figure 10 shows how the rectification arises. Under no bias, as in Fig. 10a, the electrode potentials are the same. Under reverse bias, as in Fig. 10b, the M_D potential is raised and the M_A potential is lowered: no direct pathway exists for electron flow from M_D to M_A. Under forward bias, as in Fig. 10c, the M_D potential is lowered and the M_A potential is raised, so there is now a direct downhill pathway for electrons to flow from M_A to the acceptor's LUMO to the donor's HOMO to M_D. This may be considered true unimolecular rectification, and has been the impetus for much work, including the focus of our labs.

If M_D takes a donor HOMO electron while M_A gives an electron to the acceptor LUMO, the molecule's ground state D^0-σ-A^0 will transform into the electronic excited state D^+-σ-A^-. This highly-polar zwitterionic state will decay back to the less-polar ground state D^0-σ-A^0 by inelastic tunneling through the molecule, completing the passage of an electron from M_A to M_D [79]. This decay can be enhanced by intramolecular charge transfer (ICT) or intervalence transfer (IVT) mixing of the donor and acceptor states, which would be indicated by the existence of an extra ICT or IVT absorption band. If the D and A moieties are too far apart to communicate, then D^+-σ-A^- will not decay, and no rectification will occur. On the other hand, if D and A are too close, they will form a single, mixed ground state, so that electron flow is the same in both directions, and no rectification will occur. The right length for σ is probably between two and six atoms.

The Aviram-Ratner D-σ-A molecule is analogous to a "pn" junction rectifier: the electron-rich donor region D would be similar to the electron-rich semiconducting "n" region, while the electron-poor A region would be similar to a semiconductor's "p" region [79]. However, note that under forward bias the preferred direction of Aviram-Ratner electron flow is from A to D, while in a pn junction rectifier the preferred direction is from n to p.

It is worth noting that a metal | D-σ-A | metal assembly may yield a combination of Schottky, asymmetric, and Aviram-Ratner effects. Pure unimolecular rectifiers are rare [78]. As an example of the possible complexities involved, consider how far from the center of the D-σ-A molecule the D and A groups reside. We can use the factor ρ to represent the fractional distance across the electrode–electrode gap for each group. In Fig. 11, we assume that the center of the donor is 40% of the distance from M_D to M_A (i.e., $\rho_D = 0.4$) while the center of the acceptor is 60% of the distance ($\rho_A = 0.6$). We show the bias in Fig. 11 by assuming M_D is at ground and the potential of M_A varies as follows: Fig. 11a, −2 V; Fig. 11b, −1 V; Fig. 11c, no bias; Fig. 11d, +1 V; Fig. 11e, +2 V. Under no bias (Fig. 11c), we arbitrarily choose a donor HOMO level of −0.5 eV and an acceptor LUMO level of +0.5 eV. But the MO energy levels change under the influence of an electric field: the closer to the biased electrode, the stronger the influence. We model this by changing each MO by an amount = bias × ρ.

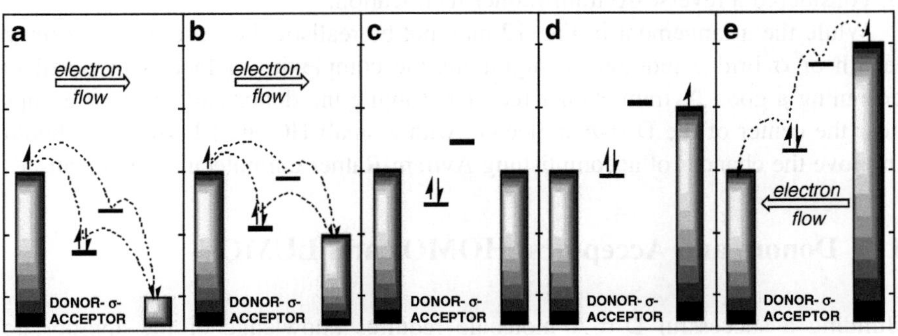

Fig. 11 Aviram-Ratner rectification with centered donor and acceptor groups

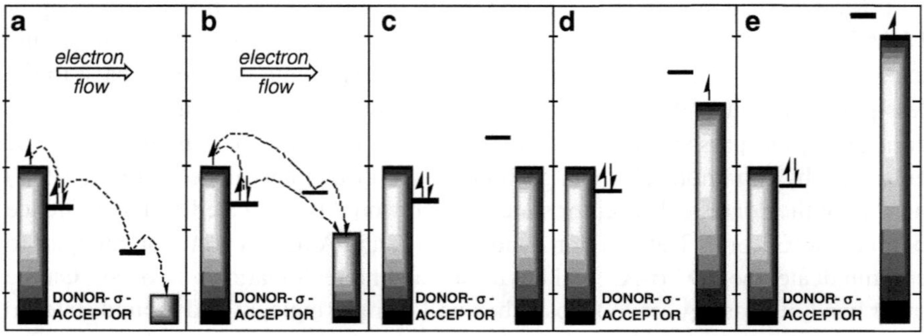

Fig. 12 Aviram-Ratner rectification with edge-positioned donor and acceptor groups

Under +1 V of forward bias (Fig. 11d), there is no pathway for current flow. At +2 V, however, the orbitals have adjusted to give a downhill path from M_A to acceptor to donor to M_D, and Aviram-Ratner current flows. On the reverse bias side, however, two pathways exist for current flow at -1 V (Fig. 11b) as well as -2 V (Fig. 11a). These pathways (Fig. 11a, b) are asymmetric rectification via HOMO and via LUMO, and they are in the anti-Aviram-Ratner direction, i.e., from donor to acceptor. This could allow for anti-Aviram-Ratner rectification under moderate biases. Note, however, that the electrons in Fig. 11a, b must tunnel over longer distances than those in Fig. 11e, because there is only one "way-station," instead of two. The Aviram-Ratner current flow under the higher bias of Fig. 11e could therefore be much more intense than the reverse flow of Fig. 11b or 11a.

Now consider a case where the donor and acceptor are at the far ends of the D-σ-A molecule: let $\rho_D = 0.1$ and $\rho_A = 0.9$ (Fig. 12). The energy levels under no bias (Fig. 12c) are the same as in Fig. 11(c). Now, under "forward" bias (Fig. 12d) of even +2 V (Fig. 12e) there is no current flow, because the raising LUMO stays higher in energy than M_A, and the HOMO stays lower than M_D. At -1 V of bias (Fig. 12b) there are two anti-Aviram-Ratner asymmetric pathways that should, however, be slow because of long tunneling distances. At -2 V of bias (Fig. 12a), the acceptor LUMO has assumed a lower energy than the donor HOMO. Now there is a direct electron pathway from donor to acceptor via two way stations: this might be considered a reverse-Aviram-Ratner rectification.

While the arrangement in Fig. 12 may not be realistic, because of the extreme length of σ bridge required, it highlights the complexity of factors involved in designing a good Aviram-Ratner rectifier. Having the donor and acceptor groups near the center of the D-σ-A molecule, with a small HOMO-LUMO gap, should improve the chances of accomplishing Aviram-Ratner unimolecular rectification.

11 Donors and Acceptors; HOMOs and LUMOs

Building a successful D-σ-A molecule requires knowledge of the appropriate HOMO and LUMO energies and the work function energies of the electrodes. A donor's ionization potential IP_D corresponds to its HOMO energy, and an acceptor's electron affinity EA_A corresponds to its LUMO energy. Usually, good electron donors (with low IP_D) are poor electron acceptors, while good acceptors (with large EA_A) are poor donors. An electrode's zero-bias energy is represented by its metal's work-function φ. The work function is related to the metal's Fermi level E_F, and is moderately dependent (0.1–0.3 eV) on the Miller indices of the exposed crystallographic face of the metal.

These relationships are represented graphically in Fig. 13, which also shows some representative donor and acceptor structures. In practice, the range of IP_D and

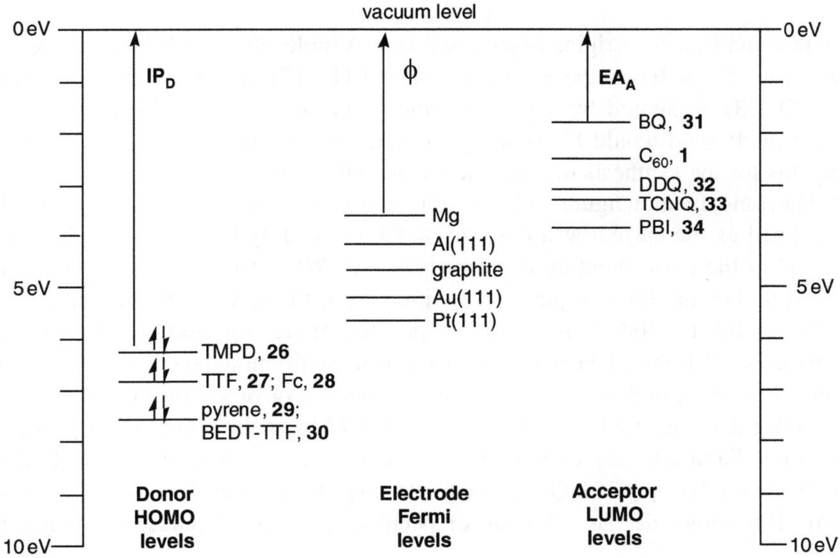

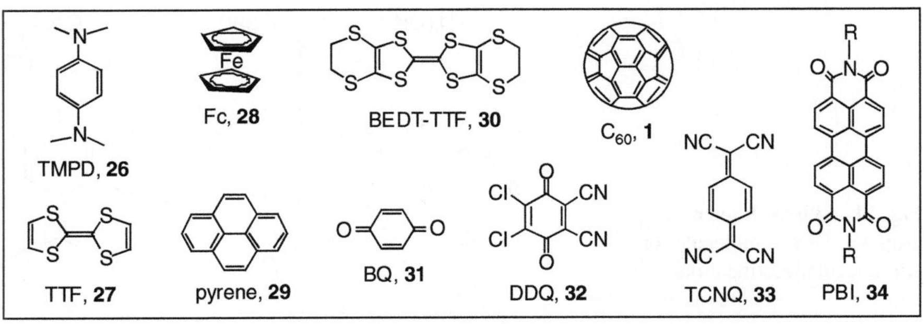

Fig. 13 Representative donor, electrode, and acceptor energy levels. First gas-phase ionization potentials IP_D of five donors: *N,N,N',N'*-tetramethylphenylenediamine (TMPD, **26**) = 6.25 eV [80]; tetrathiafulvalene (TTF, **27**) = 6.83 eV [81]; ferrocene (Fc, **28**) = 6.72 [82]; pyrene (**29**) = 7.55 [83]; bisethylendithio-tetrathiafulvalene (BEDT-TTF, **30**) = 7.6 eV [84]. Bulk work functions ϕ of four metals and one semimetal: Mg = 3.66 eV [85]; Al(111) = 4.24 eV [86]; graphite = 4.3 eV [87]; Au(111) = 5.31 eV [88]; and Pt(111) = 5.7 eV [89]. Gas-phase electron affinities EA_A of five acceptors: para-benzoquinone (BQ, **31**) = 1.9 eV [90]; buckminsterfullerene (C_{60}, **1**) = 2.6 to 2.8 eV [91]; 2,3-chloro-5,6,-dicyano-para-benzoquinone (DDQ, **32**) = 3.13 eV [92]; 7,7,8,8-tetracyanoquinodimethan (TCNQ, **33**) = 2.8 eV [93] corrected to 3.33 eV [94]; perylenetetracarboxylic bisimide (PBI, **34**) = 3.9 eV [95]

EA_A is limited because the molecules, as well as their cations or anions, must be stable in ambient air or solvent: donors that are too powerful as reducers, and acceptors that are too powerful as oxidizers, will not persist until their rectification is measured. Further, the donor's HOMO must not be so high, or its LUMO so low, that spontaneous electron transfer occurs to convert the molecule into a zwitterion.

12 Molecular Rectifiers

Aviram and Ratner's originally proposed D-σ-A molecule (or "Gedankenmolekül"), **35** (Fig. 14), included the excellent donor TTF (**27**) and the excellent acceptor TCNQ (**33**), separated by a rigid bicyclooctane bridge [79]. This molecule was never made (and would likely not have made a good monolayer), but it was the impetus for the synthesis of many new candidate rectifiers.

The fundamental figure of merit for rectification, the *rectification ratio, RR*, is defined as the current at a positive bias V divided by the absolute value of the current at the corresponding negative bias $-V$: $RR \equiv I(V)/|\, I(-V)\,|$. Commercial doped Si, Ge, or GaAs pn junction rectifiers have RR between 10 and 100.

Beginning in 1986, Sambles developed techniques for studying the electrical properties of LB multilayers and (harder to work with) monolayers, by sandwiching them between a high-work-function noble metal (Ag or Pt) on one side, and, to minimize damage, the low-work-function metal Mg on the other [34, 40]. In 1990, Ashwell, Sambles, and co-workers reported rectification from the zwitterionic molecule $C_{16}H_{33}$-γQ-3CNQ, **36a** (Figs. 15 and 16) between Pt and Mg electrodes [35]. This molecule has a couple of peculiarities. First, because the donor and

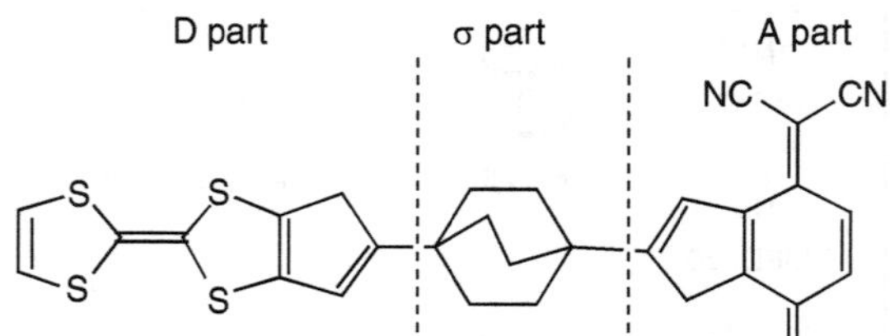

35

Fig. 14 Aviram-Ratner proposed D-σ-A molecule for unimolecular rectification

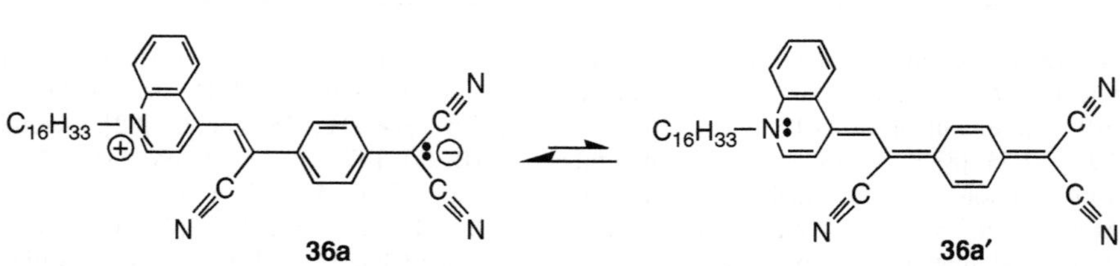

Fig. 15 First confirmed unimolecular rectifier

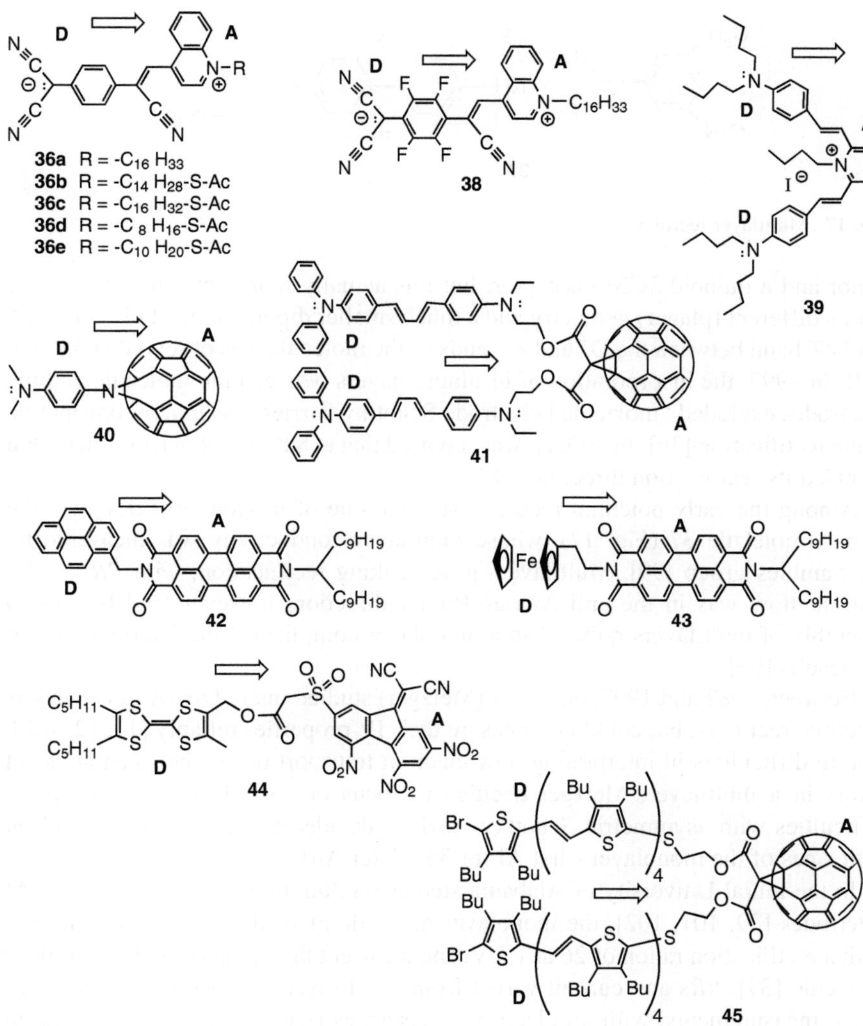

Fig. 16 Eleven unimolecular rectifiers studied at the University of Alabama as monolayers between Au electrodes (except for **36d** and **36e**: see text). The *arrows* indicate the experimentally observed direction favored electron transfer; however the direction is reversed for **36d** and **36e** (see text)

acceptor are connected by a C=C bridge, it is formally a zwitterionic D^+-π-A^- molecule. However, due to an angle twist between D and A, the C=C bridge is not conjugating, so the bridge functions as a σ bridge. Second, the molecule is ionic, with a dipole moment of 43 D [39]. The 3CNQ portion is clearly related to the strong acceptor TCNQ (**33**), but it has grabbed an electron and become the anionic, donor part of the molecule. The acceptor is a quinolinium cation. It is possible to write an apparent resonance form, **36a'** (Fig. 15), in which there is an anilinoid

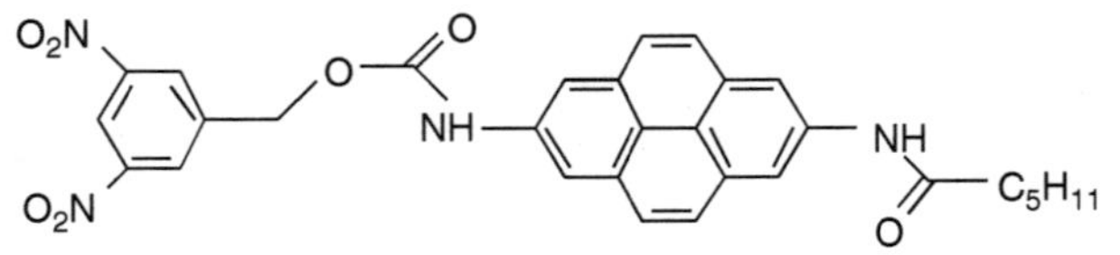

37

Fig. 17 Multilayer rectifier

donor and a quinoid 3CNQ acceptor, but this actually represents an *excited state* with a different (planar) geometry and a much smaller dipole moment of ~6 D [96]. An IVT band between the D^+ and A^- ends of the molecule was observed at 535 nm [97]. In 1993, the incorporation of insulating layers between the molecule and the electrodes excluded "molecule | electrode Schottky barriers" as major contributors to the rectification [36]. In 2002, it was reported that at 8 K, a monolayer film of **36a** reversed its rectification direction [98].

Among the early potential rectifiers studied, one of us (Mattern) designed the D-σ-A molecule **37** (Fig. 17), whose multilayer conductivity was measured by the Sambles group [99]. Multilayers gave striking rectification, with $RR = 130$. Current flow was in the anti-Aviram-Ratner direction, however, and having an assembly of multilayers rather than a monolayer complicated the interpretation of the results [99].

Between 1982 and 1997, one of us (Metzger) studied many D-σ-A molecules as potential rectifiers, but could not measure their IV properties reliably [11, 12, 100]. Due to difficulties in interpreting how electron transport occurs between adjacent layers in a multilayer, Metzger decided to focus on monolayers, and to avoid difficulties with asymmetric Schottky barriers, decided to use the same metal on both sides of the monolayer (first Al for **36a**, later Au).

In the initial University of Alabama studies on **36a**, utilizing oxide-covered Al electrodes [39, 101, 102], the monolayer had a dramatically asymmetric current, with a rectification ration of 26 at 1.5 V, and a current flow of about 0.3 electrons/s/molecule [39]. *RR*s and current varied from pad to pad, because these were two-probe measurements, with all electrical resistances (wire to paste to electrode to monolayer) in series. Further, as high potentials were repeatedly scanned, the *IV* curves became less asymmetric, with *RR*s gradually decreasing: the monolayers appeared to be losing their asymmetric Langmuir film orientation. A minority of monolayers of **36a** rectified in the reverse direction [102]. With the change to oxide-free Au electrodes, the asymmetry persisted, with *RR*s as high as 27, and the current per molecule increased dramatically to as high as 9×10^4 electrons/s (Fig. 18a) [38, 41]. Figure 19 shows the current *I* and log *I* vs *V* for **36a** [38]. As before, however, the *RR* decreased upon repeated scanning. In some cells, the current increased until breakdown occurred at 5.0 V, i.e., the cells suffered dielectric breakdown only at a field close to 2 GV/m [38].

Since 1997, 11 of the molecules studied at the University of Alabama, including **36a**, have exhibited rectification as monolayers between Au electrodes [19, 38, 41, 97, 103–109]. The structures are shown in Fig. 16. With the exceptions of **40** and

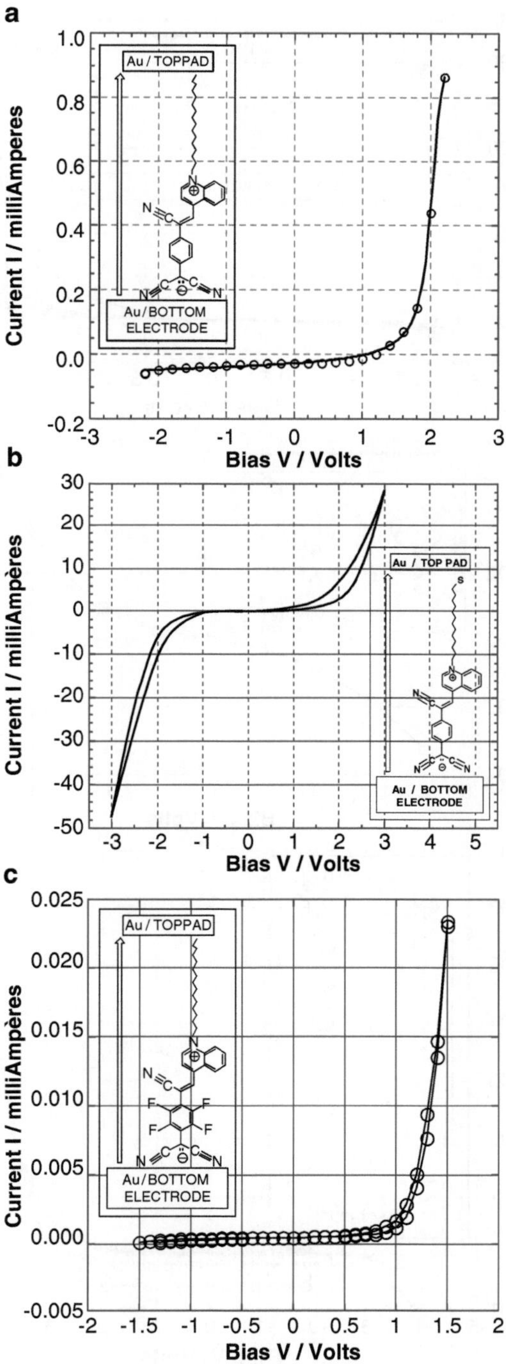

Fig. 18 (continued)

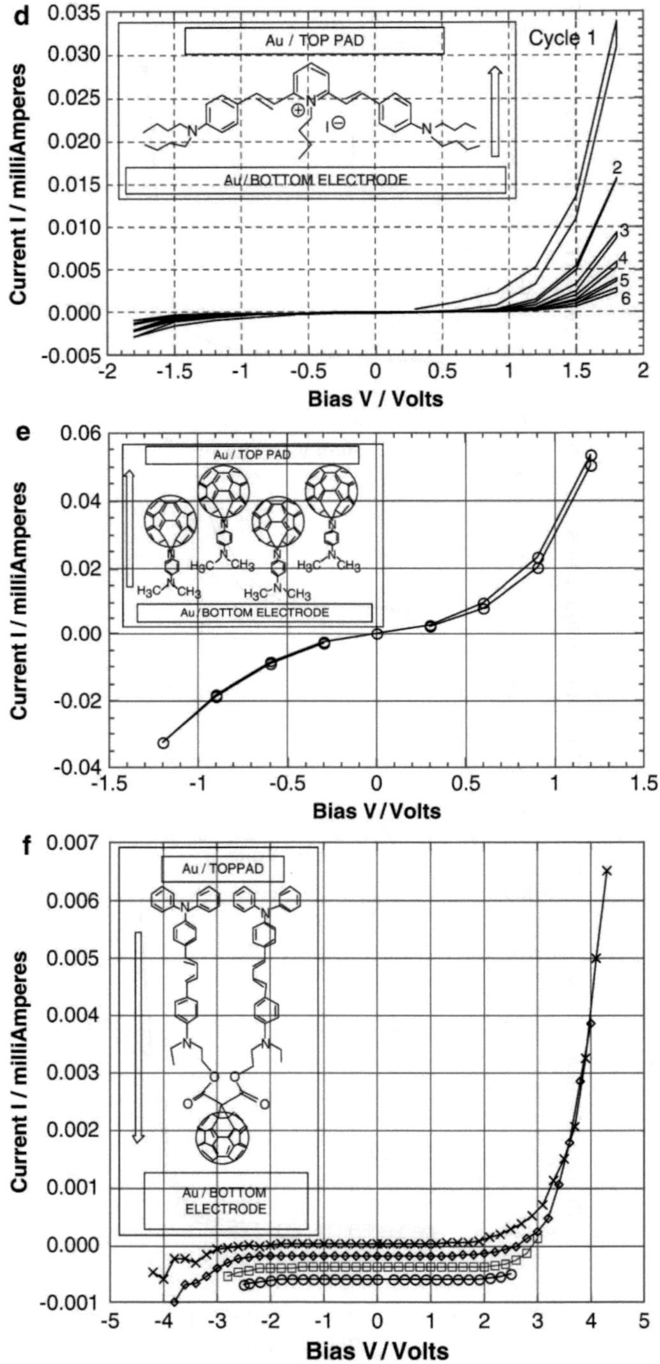

Fig. 18 (continued)

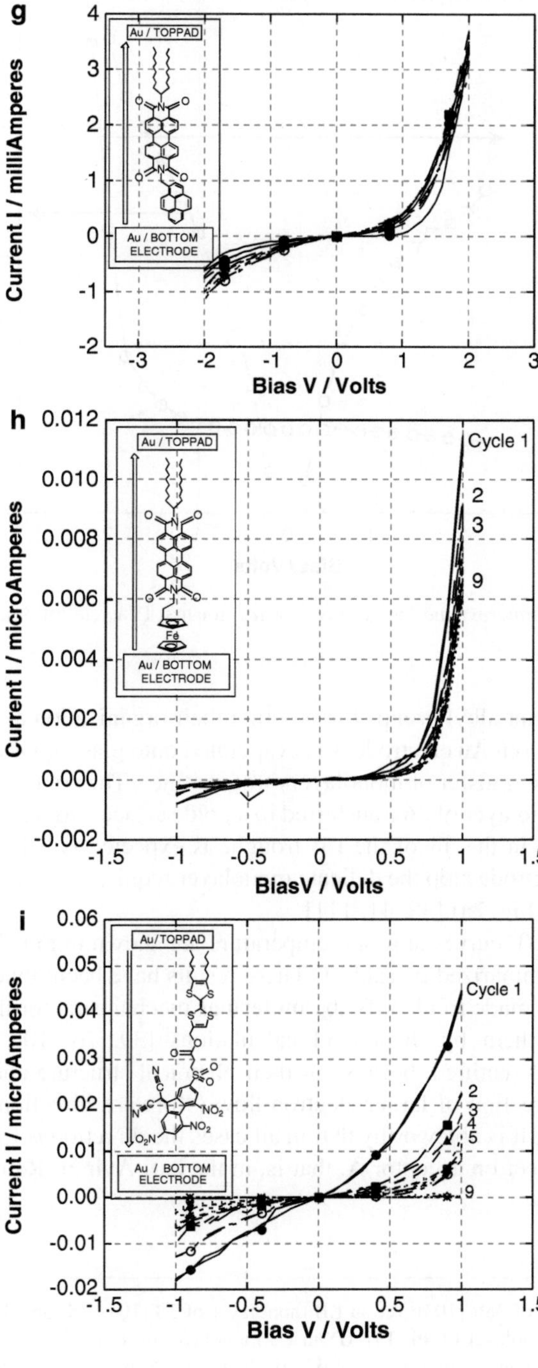

Fig. 18 The rectification of MOM (Metal | Organic | Metal) sandwiches consisting of three elements: (1) a macroscopic bottom Au or Al electrode, (2) an 0.3 mm^2 top Al or "cold Au" electrode pad, and, between them: (3) (**a**) an LB monolayer of **36a** [38]; (**b**) an LB monolayer

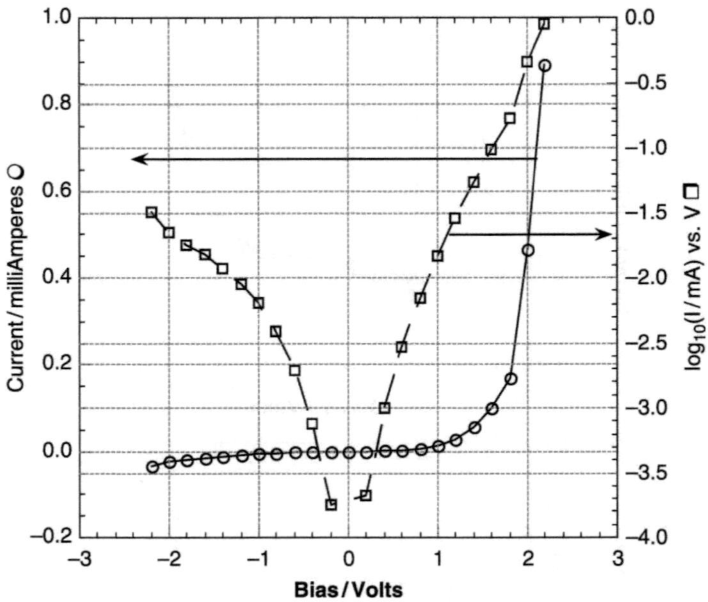

Fig. 19 I/V curve (*squares*) and $\log_{10}$I vs V (*circles*) for first IV cycle for **36a** [38]

41, all have pendant alkyl groups for assisting their organization as monolayers. In each case, the bottom Au electrode was evaporated onto glass, quartz, or very flat Si substrates, and the transfer of monolayers was routine. (The contact angle of a drop of water on a monolayer of **36** transferred to a gold surface was 92°, consistent with its C_{16} tail being in the air of the LB trough, as expected [110].) Depositing the second metal electrode atop the delicate monolayer required institution of the "cold gold" technique (Fig. 2a) [38, 41, 111].

Characteristic *IV* curves at room temperature are shown in Fig. 18, and some of the results are summarized in Table 1. These results have been reviewed often [11, 12]. Efforts were made to identify the molecular mechanisms for the rectification, and to buttress them by theoretical calculations [39, 76, 106, 112]. Not all compounds tested rectified, because of their chemical structure and/or monolayer structure. The direction of larger electron flow ("forward direction") is shown by arrows in Fig. 16; it is noteworthy that in all cases the direction is from the electron donor D to the electron acceptor A, that is, in the anti-Aviram-Ratner direction.

Fig. 18 (continued) of **36b** [103]; (**c**) an LB monolayer of **38** [104]; (**d**) an LB monolayer of **39** [104]; (**e**) an LB monolayer of **40** [19]; (**f**) an LS monolayer of **41** [105] (the repeated scans are offset for clarity); (**g**) an LB monolayer of **42** [106]; (**h**) an LS monolayer of **43** [106]; (**i**) an LB monolayer of **44** [107]

Table 1 Summary data for 11 unimolecular rectifiers studied at the University of Alabama

#	Type	IVT (nm)	LB or LS	RR	Survives cycling	AR or anti-AR	Ref.
36a	D^+-π-A^-	530	LB	2–27	No	anti-AR	[38]
36b	D^+-π-A^-	530	LS	5	Yes	anti-AR	[103]
36c	D^+-π-A^-	530	LS	5	Yes	anti-AR	[103]
38	D^+-π-A^-	504	LB	3–64	No	anti-AR	[97]
39	D^+ I^-	480	LB	8–60	No	anti-AR	[104]
40	D-σ-A	720	LB	2	No	anti-AR	[19]
41	D-σ-A	–	LS	10	Yes	anti-AR	[105]
42	D-σ-A	–	LB	2–5	No	anti-AR	[106]
43	D-σ-A	595	LB	30	Yes	anti-AR	[106]
44	D-σ-A	1,220	LB	13	No	anti-AR	[107]
45	D-σ-A	–	LS	6–60	Yes	anti-AR at low V, AR at high V	[108]

All compounds were measured at room temperature in air between Au electrodes inside a Faraday cage. IVT = maximum of the optical InterValence Transfer or intramolecular charge transfer band. *RR* Rectification Ratio. "Survives cycling" means RR does not decrease with cycling of IV measurements. "AR or anti-AR" indicates whether electron flow is from A to D (AR) or from D to A (anti-AR)

For the rectifiers listed in Table 1, the current was found to decay with successive measurements of the same junction for **36a**, **38**, **39**, **40**, **42**, and **44** between Au electrodes (where the monolayers were not sufficiently rigid, there probably was room in the "Au | monolayer | Au" sandwich for molecular reorientation under applied bias) (see Fig. 18d, g, and i for examples). In contrast, the current did not decay at all in subsequent cycles for **36b** (Fig. 18b) or **36c**, where the molecules were chemisorbed onto the Au electrode with an S anchor, or for **41** (Fig. 18f), **43**, or **45** (where the monolayer was sufficiently rigid and closely packed to resist reorientation).

The unwelcome gradual decreases in the electrical conductivity and in the RR of an LB monolayer of **36a**, from an initial value of 27 [38, 39] to close to 1 upon repeated cycling, was seen in about half the rectifiers studied in Table 1. This prompted us to try combining the LB and SAM techniques, by using variants of **36a** with thioacetyl anchors at the end of their C_{14} (**36b**) or C_{16} (**36c**) alkyl tails. LB films should put these tails in the air, where the deposited top Au pads should strongly coordinate to the sulfur groups. LS films would directly coordinate to the bottom electrode laid on top of it. In either case, the Au–S bond should stabilize the monolayer and prevent reorientation [103, 109]. The thioacetate analog with a C_{11} tail did not make a satisfactorily-ordered LB film [109], but **36b** and **36c** (Fig. 20) made LB films, rectified, and survived cycling without loss of RR [103].

For most of the molecules, the electron flow is from D to A, which is supported by the "anti-Aviram-Ratner mechanism" of Fig. 11a. However, there may be yet another possibility, shown in Fig. 21: implicit in the analyses of Figs. 9–11 has been the Aviram and Ratner assumption [79] that "auto-ionization" is a less efficient competing process. If the electric field induces intramolecular ionization first by sufficiently altering the orbital energies, then the direction of electron flow may occur in the anti-AR direction (Fig. 21)!

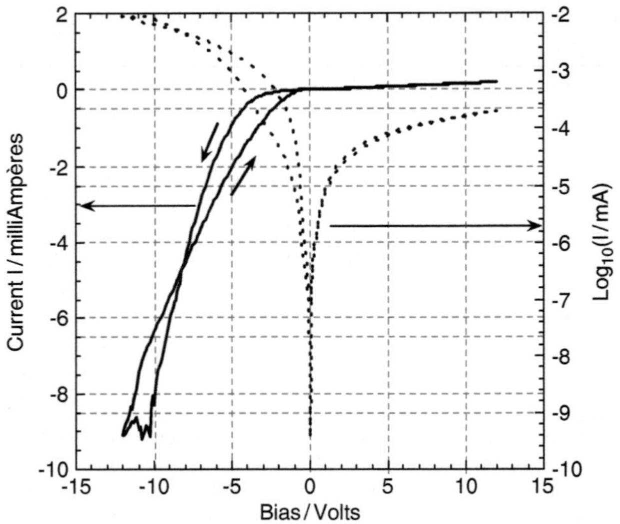

Fig. 20 I/V curve and Log_{10}I vs V (first cycle) for an LB upstroke film of **36c** [103]

The issue of whether the ground state of **36a** is zwitterionic (**36a**) or quinoid (**36a'**) has received renewed attention, and was recently reviewed [113]. The length of the alkyl tail R, the solvation, and the packing mode within a monolayer are important. For the very short tail R = CH_3, theory at the semiempirical AM1 and PM3 level, with solvation corrections, predicts that the molecule is largely ionic, with dipole moments increasing upon "solvation" in acetonitrile [112]. For the long tail R = $C_{16}H_{33}$, in experiments, in a monolayer, in a multilayer, and also in dichloromethane solution, the zwitterionic form **36a** is the ground state [39, 110]. It is important to note that in monolayers and multilayers the chromophore in **36a** is tilted about 45° from the normal to the film, thus stabilizing the zwitterionic state **36a** (rather than **36a'**) through favorable intermolecular electrostatic interactions. If the tail is intermediate in length, as in **36d** (R = C_8H_{16}) and **36e** (R = $C_{10}H_{20}$), the chromophore is no longer tilted, but aligned, the quinonoid form (similar to **36a'**) is now the ground state, and the rectification direction is *opposite* to the arrow direction given in Fig. 16 [114].

The four fluorines present in **38** should have made its $3CNQF_4$ group a much poorer donor than the 3CNQ in **36a**; however, **38** also rectified, with electron flow in the same D to A direction as in **36a** (Fig. 18c) [97].

The butylpyridinium iodide **39** transfers to hydrophilic substrates as a Z-type LB multilayer [104]. The initial *RR* (for a monolayer) is as high as 60; rectification shows a decrease upon successive cycles (Fig. 18d). The favored direction of electron flow is from iodide to pyridinium, i.e., in the direction of "back-charge-transfer." The rectification may be attributed to an interionic electron transfer, or to an intramolecular electron transfer [104].

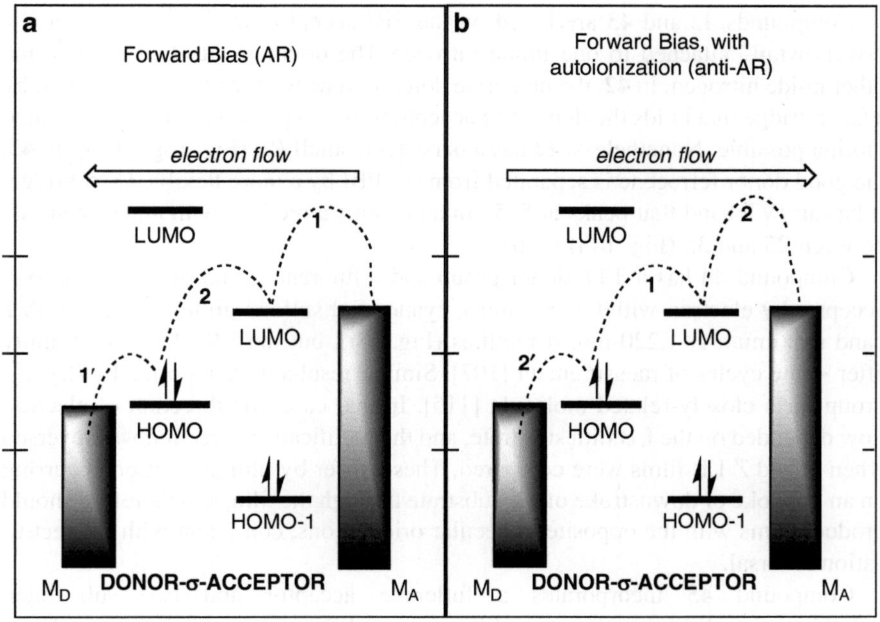

Fig. 21 (**a**) Rectification in the AR sense: electron transfers occur first between electrode M_D and the HOMO of the D-σ-A molecule (**1'**), and between the LUMO and M_A (**1**), thus establishing the zwitterionic excited state D^+-σ-A^-, which then (**2**) relaxes back into the neutral state. (**b**) Rectification in the anti-AR sense if "auto-ionization" occurs first (**1**), forming the excited state D^+-σ-A^- by an interaction of the intense electric field and the molecule, followed by transfers to and from the electrodes (**2**) and (**2'**), resulting in the electron passing from M_D to M_A. The molecular energy levels, which in reality must perforce shift dramatically during the electron transfer process, are drawn here for simplicity as if the molecule were still isolated in the "gas phase"

Dimethylaminophenylazafullerene, **40**, consists of the weak electron donor dimethylaniline bonded to the moderate electron acceptor N-capped-C_{60}, with an IVT peak at 720 nm [19]. The Langmuir film is very rigid, i.e., the slope of the isotherm is relatively large. However, the molecular area is 50 Å^2 at the LB film transfer pressure, whereas the true molecular area of C_{60} is close to 100 Å^2. Therefore it is thought that **40** transfers onto Au in a staggered arrangement, as shown in the insert of Fig. 18e. Angle-resolved N1s XPS spectra confirm that the two N atoms are closer to the bottom Au electrode than is the C_{60} cage [19]. Molecule **40** is a moderate rectifier with $RR \approx 2$. It can also exhibit a tremendous, *spurious* apparent rectification ratio as high as 20,000. This is probably an example of partial penetration ("electromigration") of Au, forming stalagmites that create an electrical short [19].

The bis(triphenylamine)-substituted fullerene **41** makes a dense and stiff monolayer that transfers onto an Au substrate by LS but not by LB [105]. It rectifies with $RR = 10$ (Fig. 18f) and the RR does not decrease at all upon successive cycling [105].

Compounds **42** and **43** are based on the PBI acceptor, made soluble by a C_{19} "swallowtail" attached to one imide nitrogen The donor is attached through the other imide nitrogen. In **42**, the moderate donor pyrene is separated from the PBI by a C_1 σ bridge that holds the donor and acceptor close to perpendicular, with limited motion possible. Nonetheless, **42** has a persistent, small *RR* (Fig. 18g) [106]. In **43**, the good donor ferrocene is separated from the PBI by a more flexible C_2 σ bridge. It has an IVT band that peaks at 595 nm and forms a rectifier with a persistent *RR* between 25 and 35 (Fig. 18h) [106].

Compound **44** has a TTF donor group and a fluorene group made into a good acceptor by electron-withdrawing nitro, cyano, and sulfone groups. It has an IVT band maximum at 1,220 nm. It rectifies (Fig. 18i), but the *RR*s decrease to unity after ~nine cycles of measurement [107]. Similar results were reported by Bryce's group for a closely-related molecule [115]. In that case, the direction of electron flow depended on the LB film substrate, and the rectification direction was reversed when X and Z LB films were compared. These differ by film deposition occurring on an upstroke or downstroke of the substrate through the film, and therefore should produce films with the opposite molecular orientations, consistent with the rectification reversal.

Compound **45** incorporates a fullerene acceptor and two substituted polythiophene-ethenyl tails; it rectifies as an LS monolayer. Interestingly, it rectifies in one direction for smaller biases between 0 and ± 2.0 V, but in the reverse direction at higher biases, perhaps indicating a change in conduction mechanism, as discussed above. It also rectifies between Al and Pb electrodes at 4.2 K (Fig. 22a) [108].

Compound **45** was analyzed by *inelastic electron tunneling spectroscopy* (IETS) using a home-built instrument [108]. In IETS, a low-frequency (~100 Hz) AC source and a low-bias DC voltage are applied together to a thin film at low temperatures (4.2–100 K). The second derivative of the current as a function of the voltage, measured using second harmonic detection, reveals peaks that correspond to intramolecular vibrations [116]. This verifies that the current is traveling through the molecule. In addition, an elastic component measured at slightly higher bias can reveal "orbital-mediated tunneling" [117, 118], an enhanced current due to resonance between the Fermi level of an electrode and an accessible molecular orbital of the molecule [108]. The IETS spectrum of **45** at 4.2 K shows the expected intramolecular vibrations (Fig. 22b), and a wide-scan IETS spectrum shows the signature of rectification: an asymmetric signal (higher intensity at positive bias) occurs at the same potential as the enhanced current due to rectification (Fig. 22c): this is orbital-mediated tunneling, and is a *definitive proof that the rectification is due to through-bond tunneling, not through-space tunneling* [108].

Bryce, Petty, and co-workers studied **46** (Fig. 23), containing a TTF donor and a TCNQ acceptor like the original Aviram-Ratner Gedankenmolekül **35**. Although **46** gave strong Langmuir films, the TCNQ group lies flat on the water surface rather than end-on, and rectification could not be observed [119].

Weber, Mayor, and co-workers created the D-π-A compound **47**, where four fluorine atoms serve to make an acceptor end. The π link is a combination of ethyne

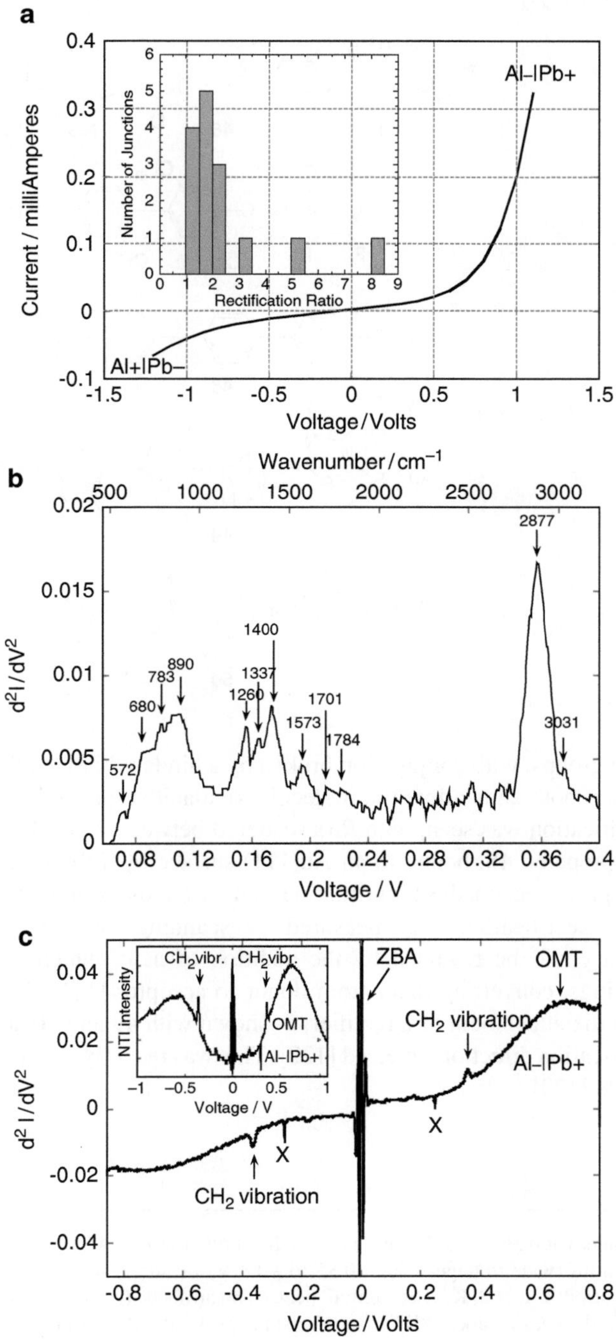

Fig. 22 (**a**) Rectification at 4.2 K due to an LS monolayer of **45** between Al and Pb electrodes: the onset is at 0.65 V. (**b**) IETS of **45** at 4.2 K: the most prominent peaks are the CH_2 vibrations

Fig. 23 More donor-acceptor molecules

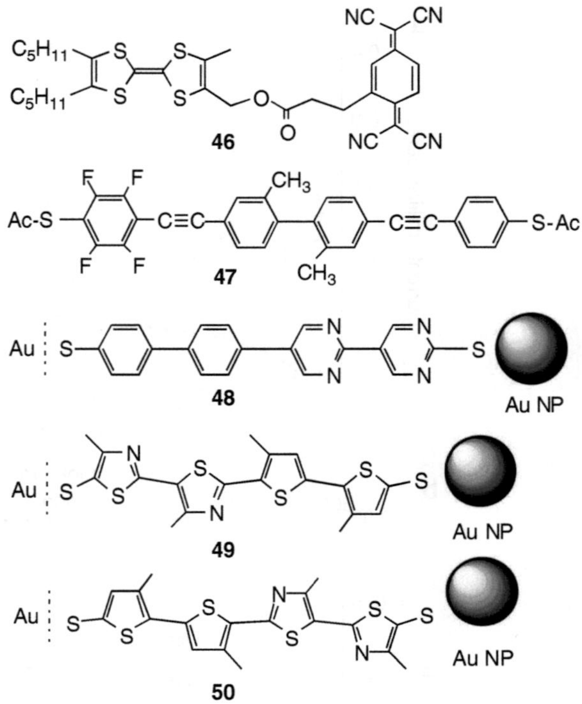

and aromatic groups, with conjugation broken by a hindered biphenyl at the center. Thioacetates at both ends allow the molecule to attach suspended between an Au MCBJ. Rectification was seen, with *RR*s scattered between 2 and 10 [120].

Yu et al. prepared **48**, bound to an Au(111) surface by a thiol on the biphenyl end, then deprotected on the bipyrimidine end to expose a thiol to bind an Au nanoparticle. Rectification was measured by scanning tunneling spectroscopy (STS), and it could be reversed by the addition of acid, which protonated the pyrimidine rings, converting them from donor to acceptor [121]. Yu's group has also inverted the attachment of a rectifier, as shown with **49** and **50**, and confirmed that the rectification direction reversed [122]. This was recently verified by STM-BJ measurements [123].

Fig. 22 (continued) at 0.36 V (2,877 cm^{-1}) due to the butyl groups of **45**. (**c**) Inelastic and elastic tunneling spectrum (wide voltage scan, −0.85 to +0.8 V) of an LS monolayer of **45** between Al and Pb electrodes at 4.2 K. The central peaks are artifacts (zero-bias anomalies due to metal phonons); the peaks marked "X" are artifacts; the peaks at ±0.36 V are the prominent CH_2 vibrations [see (**b**)]; the broad peak centered at +0.65 V, marked OMT, is orbital-mediated tunneling, an elastic current at 0.65 V [same as the onset in (**a**)] in resonance with the LUMO [108]

Ashwell and co-workers studied several zwitterionic systems (Fig. 24) by STS, starting with **36a** [124]; for **36a**, and for several other zwitterionic systems [21, 125, 126], the addition of acid stopped the rectification, while addition of base restored it. Ashwell and co-workers saw a reversal of the direction of rectification

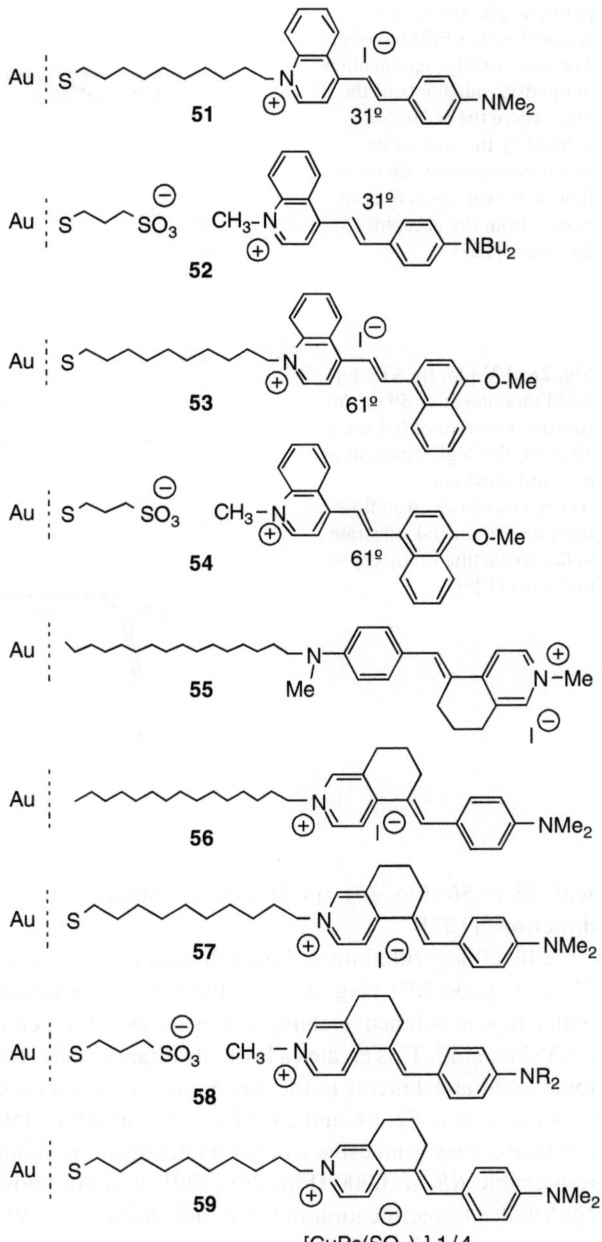

Fig. 24 Some Ashwell rectifiers

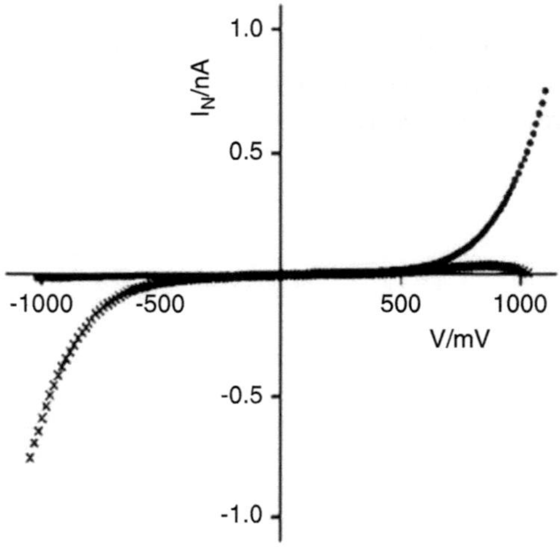

Fig. 25 STS curves (normalized I/V plots of LB monolayers of isomers **55** (*crosses*) and **56** (*dots*), deposited on an Au film over HOPG (highly oriented pyrolytic graphite), and scanned with a Pt/Ir nanotip. The films exhibit rectification in opposite quadrants of the plot, where the polarity is defined by the sign of the substrate electrode. Electron flow at forward bias in each case is from the acceptor to the donor [127]

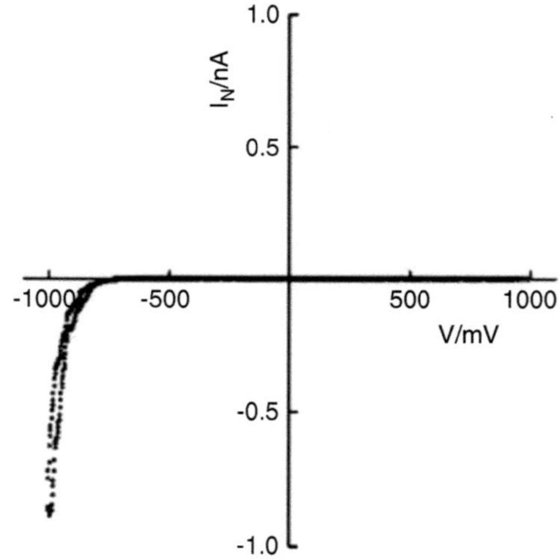

Fig. 26 I/V plot by STS for SAM monolayer of **59** on Au (set-point current of 0.8 nA at 40 mV). The higher current in the third quadrant corresponds to electron flow from the Au-coated substrate to the contacting tip (acceptor to donor) [129]

with **55** vs **56** (Fig. 25): the D-π-A chromophore is oriented in mutually opposite directions [127].

Ashwell's pyridinium and quinolinium D-π-A rectifiers, like **51**, **53**, **55**, **56**, and **57**, have good RRs (Fig. 25), but the RR can be raised by an order of magnitude with a new attachment technique. First, a short thiol ending in a sulfate anion forms a SAM on gold. This creates a lawn of sulfate anions. Then a cationic rectifier forms ionic-bond attachments to the anionic lawn, making a second SAM of the rectifier molecules, as in **52**, **54**, and **58**, which gave an *RR* of 450 [128]. A surprising further advance comes from using a bulky phthalocyanine counteranion **59** which yielded a remarkable *RR* of 3,000 (Fig. 26) [129]. It is not known how much ionic motion contributes to rectification in these molecules (or in **39**).

Gayathri and Patnaik found rectification in a C_{60} dialkoxybenzene dyad by STM, with the RR as high as 158 [130].

13 Capacitors

In macroscopic circuits, a capacitor consists of two conductor plates, close to each other, but not in contact, that respond to a voltage by accumulating, rather than passing, charge. In the nanoscopic world, then, we might think of a nanocapacitor as a way of collecting, and later providing, charge to a circuit. But while a macroscopic plate can tolerate the accumulation of billions of electrons (or holes), a molecular-sized device will be limited to a handful of charges confined in its nano-sized volume.

A quantum limit called the "Coulomb blockade" [131] involves the capacitance (C, typically 1 fF) of a two-dimensional confined region, or quantum dot. If an electron is confined to such a dot, then adding another electron will cost a "charging energy" of e^2/C. If $(e^2/2C) < k_B T$, a Coulomb blockade occurs, and no more charges can be added until a threshold voltage $V_{CB} = k_B T/e$ is reached. This causes a flat region of no current increase in the I/V curve from $V = 0$ to $V = V_{CB}$ (at 300 K, $V_{CB} = 0.026$ volts). When $V > V_{CB}$, the original charge can move off the quantum dot, and a finite current can be observed; the maximum capacitance of the dot has been exceeded. Coulomb blockade behavior has been observed in single molecules such as **60**, where the complexed cobalt ion plays the role of the quantum dot (Fig. 27) [49].

A single electron avoids the problem of forcing multiple charges into close proximity: Shkrob and Schlueter have calculated that the cavity of one conformation of calix[4]cyclohexanol, a cyclic tetramer of 2-methylcyclohexanol, represents a deep electron trap. Essentially, the electron would be "solvated" by four hydroxy groups in a tetrahedral arrangement. Such a molecule could serve as a nanocapacitor for a unit charge [132].

Bocian, Lindsey and co-workers studied sandwich complex nanocapacitors comprised of porphyrin and phthalocyanine ligands separated by lanthanide metals [133]. A triple-decker sandwich of phthalocyanine-Eu-phthalocyanine-Eu-porphyrin, with two phenylethynyl linker wires from the porphyrin, potentially has up to nine accessible oxidation states (−4 to +4). SAMs of monomers, dimers, trimers, and oligomers of this sandwich, anchored at one or both ends by thioacetyl groups, gave charge densities up to 10^{-10} mol cm^{-2}, electron-transfer rates up to 10^5 electrons s^{-1}, and charge-dissipation half-lives in the 10–50 s range.

On a larger scale, it is possible to store charge in gold nanoparticles. Chen described 1.8 nm diameter nanoparticles coated with a layer of alkylthiols,

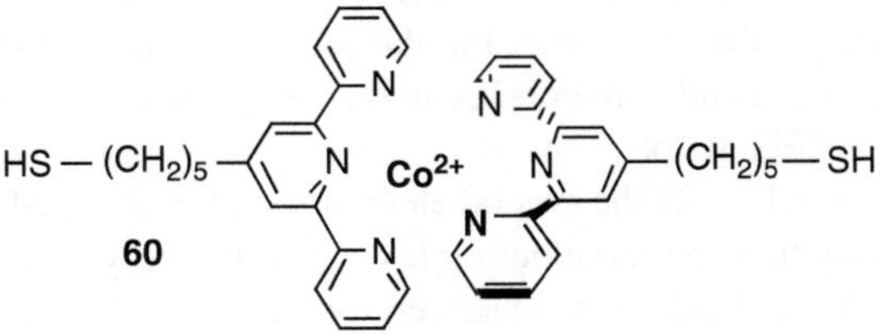

Fig. 27 Quantum-dot molecule **60** involved in Coulomb blockade between electromigrated Au electrodes [49]

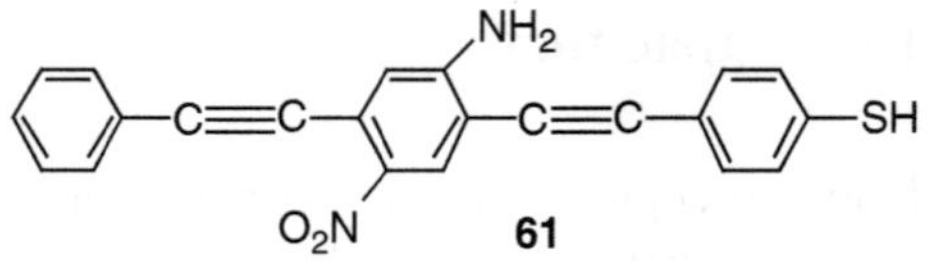

Fig. 28 About 5,000 molecules of **61**, covalently attached to Au, exhibit NDR at 60 K, but not at 300 K [136]

including occasional dithiols that allow the nanoparticles to form a SAM on Au. The dielectric nature of the alkyl coating allows the nanoparticles to exhibit a molecular capacitance of the order of attofarads [134].

Finally, Eklund's group has described double-walled nanotubes (DWNT), with ca. 0.7 nm inner diameter and 1.4 nm outer diameter, that can form a three-layer cylindrical capacitor with a shell of bromine anions forming the outer electrode. Holes segregate on the outer nanotube, and saturation bromine doping creates 1 hole per 26 carbon atoms in the DWNT [135].

14 NDR Devices

A monolayer of about 5,000 molecules of **61** (Fig. 28) in a "nanopore," attached to Au on one side and topped by a Ti electrode on the other, provided rectification (due to the dissimilar metal electrodes) but also NDR within a definite range of applied bias, i.e., increasing voltage was associated with a decrease in current, so (dI/dV) is negative. NDR was dramatic at 60 K, but vanished at room temperature [136]. NDR is also observed in inorganic two-terminal Esaki tunnel diodes [137], which do have a limited commercial application: signal amplification is possible within such two-terminal devices, because in the output circuit a finite negative resistance R_T from the tunnel diode may be matched in series by an equal but opposite positive load resistance R_L. In the output circuit the equation for power amplification contains the circuit resistance in the denominator: if this resistance $R_T + R_L = 0$, then infinite gain is seen. However, efforts to commercialize the organic NDR effect failed, because the devices proved to be unreliable.

15 Field-Effect Transistors/Gates

Field-effect transistors (FETs) have dominated the semiconductor industry, largely displacing the earlier bipolar junction transistor (BJT) because of its negligible gate current and convenience in the design of integrated circuits. Figure 29 sketches how an FET works.

In Fig. 29 the crucial elements are the p-doped Si layer, which can be replaced by any thin semiconducting layer, and the very thin insulator below the gate electrode. FET behavior was observed for polymeric LB films three decades ago [138], and by

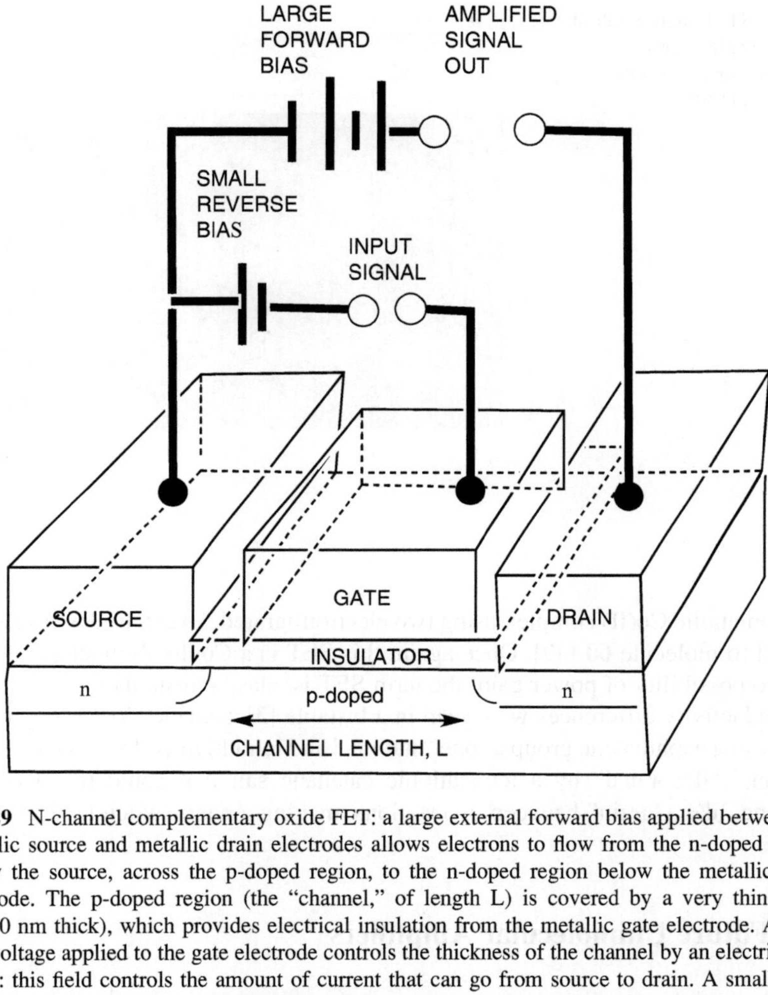

Fig. 29 N-channel complementary oxide FET: a large external forward bias applied between the metallic source and metallic drain electrodes allows electrons to flow from the n-doped region below the source, across the p-doped region, to the n-doped region below the metallic drain electrode. The p-doped region (the "channel," of length L) is covered by a very thin oxide (10–50 nm thick), which provides electrical insulation from the metallic gate electrode. A back bias voltage applied to the gate electrode controls the thickness of the channel by an electric field effect: this field controls the amount of current that can go from source to drain. A small input signal applied to the gate-to-source leg is then amplified into a large output signal between source and drain, drawing power from the external DC bias

low-temperature STM for a single-walled carbon nanotube curled over parallel Au lines, with the STM acting as a gate electrode (Fig. 30) [139].

Lee and Reed have modified the MCBJ measurements for 1,4-benzenedithiol (**2**), and have used the Si electrode substrate as a gate electrode to find that the conductivity through the benzenedithiolate bonded to two Au shards showed evidence of FET behavior [140]. However, these results are controversial (van der Zant HSJ, 2010, private communication).

Using an EMBJ between two Au electrodes, the organometallic equivalent of a no-gain single-electron transistor (SET) was realized at 0.1 K with an

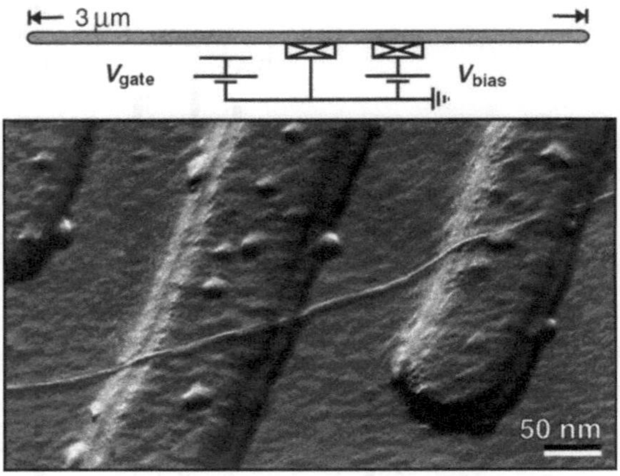

Fig. 30 STM micrograph of a single-walled carbon nanotube curled atop Au electrodes [139]

organometallic Co(II) complex using two electromigrated Au electrodes covalently bonded to molecule **60** [49]. Once again, this SET is a Coulomb blockade device with no possibility of power gain: the term SET is, alas, a misnomer.

Conductivity differences were seen in a bistable [3]catenane closed-loop molecule, with a naphthalene group as one "station," and tetrathiafulvalene as the second "station," "decorated" by a tetracationic catenane salt that could travel on the catenane, like a "train" between two stations, making a nano-switch [141].

16 Future Unimolecular Amplifiers

A three-sided molecule, designed to control the current pathway within it by judicious choice of three moieties with different electron affinities and/or ionization potentials, when covalently bonded to three metal electrodes 3 nm apart, could be the unimolecular equivalent of a BJT.

Many suitable molecules can be designed, with end-groups chosen for SAM formation with dissimilar metal electrodes. For instance, two Au and one Al electrode could be used. The molecule D_1D_2A would have two –SH terminations to bond to Au, and one –COOH end group (on D_2) to bond to Al. The electron path between the two Au electrodes would traverse a donor moiety D_1 with low IP_D and an acceptor moiety A, while the path from the Al electrode to the second Au electrode would traverse a weaker donor moiety D_2 (with larger IP_D) and the common acceptor moiety A. The larger electron current would flow between the two Au electrodes, because the intra-molecular electron mobility would be larger $D_1 \rightarrow A$, while the smaller electron current would flow $D_2 \rightarrow A$. The smaller

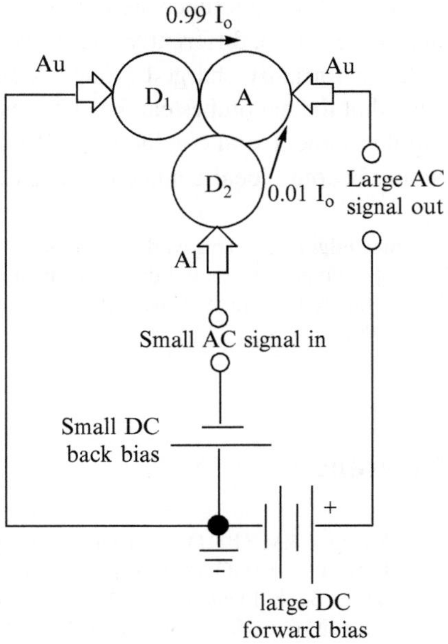

Fig. 31 Proposed unimolecular amplifier D_1D_2A, in a circuit analogous to a grounded-emitter junction transistor, grounded-source FET, or grounded-cathode triode circuit. The *arrows* show the direction of preferred electron flow. The two Au and one Al electrode tips must be about 3 nm apart

current path would be either under zero bias or slightly back-biased, and carry the input signal; the larger current, under forward bias, would carry the amplified output signal (Fig. 31). Whether the Al electrode would function as a gate electrode, affecting conductivity by an electric field, or as a drain electrode, affecting conductivity by an electric potential, is not yet clear. Metal-molecule interface issues would be important, just as in two-electrode resistors or rectifiers. The details of how the electron current from D_1 to A would be significantly larger than the current from D_2 to A must still be worked out.

At present making the first unimolecular amplifier is mainly hampered by the experimental difficulty in placing three electrodes 3 nm apart from each other. As discussed above, MCBJs and EMBJs have yielded reproducible 1–3 nm gaps between two electrodes; the third "gate" electrode can be an STM tip or a CP-AFM tip or an oxide-covered Al electrode.

17 Future Organic Interconnects

Once a sufficient set of resistors, capacitors, rectifiers, and amplifiers have been demonstrated with conventional metal electrodes, one can initiate a new project, of assembling all-organic polymeric electrodes to replace the inorganic metals. This would lead to the all-organic computer! The controlled electrochemical growth of conducting oligomer filaments has already been demonstrated [142]. Alas, single-walled carbon nanotubes are not yet fully chemically processable. For that we may need defect-free, differently end-derivatized SWCNT, e.g., A_n-SWCNT-B_m, with n polar or formally charged groups A and m polar or oppositely charged groups B, such that the nanotubes can be chemically separated by chromatography by charge, dipole moment, and conductivity: if this can be achieved, then the A_n-SWCNT-B_m would become ideal connectors in UE.

Acknowledgments This work was achieved by the diligence and insight of so many colleagues, students, and post-doctoral fellows, to whom we owe an immense debt of gratitude, and facilitated by several grants from the United States National Science Foundation (the most recent being NSF-08-48206).

References

1. Metzger RM (1991) Prospects for truly unimolecular devices. In: Metzger RM, Day P, Papavassiliou GC (eds) Lower-dimensional systems and molecular electronics. NATO ASI Series B248. Plenum press, New York, pp 659–666
2. Tour JM, Kozaki M, Seminario JM (1998) Molecular-scale electronics: a synthetic/computational approach to digital computing. J Am Chem Soc 120:8486–8493
3. Ferraris J, Cowan DO, Walatka V Jr, Perlstein JH (1973) Electron transfer in a new highly conducting donor acceptor complex. J Am Chem Soc 95:948–949
4. Cowan DO, Fortkort JA, Metzger RM (1991) Design constraints for organic metals and superconductors. In: Metzger RM, Day P, Papavassiliou GC (eds) Lower-dimensional systems and molecular electronics. NATO ASI Ser, vol B248. Plenum Press, New York, pp 1–22
5. Jérôme D, Mazaud A, Ribault M, Bechgaard K (1980) Superconductivity in a synthetic organic conductor, $(TMTSF)_2PF_6$. J Phys Lett 41:L95–L97
6. Shirakawa H, Louis EJ, MacDiarmid AG, Chiang CK, Heeger AJ (1977) Synthesis of electrically conducting organic polymers: halogen derivatives of polyacetylene, $(CH)_x$. J Chem Soc Chem Commun 578–580
7. Chiang CK, Fincher CR Jr, Park YW, Heeger AJ, Shirakawa H, Louis EJ (1977) Electrical conductivity in doped polyacetylene. Phys Rev Lett 39:1098–1101, erratum (1978) Phys Rev Lett 40:1472
8. Heeger AJ (2001) Semiconducting and metallic polymers: the fourth generation of polymeric materials (Nobel lecture). Angew Chem Int Ed 40:2591–2611
9. MacDiarmid AG (2001) Synthetic metals: a novel role for organic polymers (Nobel lecture). Angew Chem Int Ed 40:2581–2590
10. Shirakawa H (2001) The discovery of polyacetylene film: the dawning of an era of conducting polymers (Nobel lecture). Angew Chem Int Ed 40:2574–2580
11. Metzger RM (2008) Unimolecular electronics. J Mater Chem 18:4364–4396
12. Metzger RM (2003) Unimolecular electrical rectifiers. Chem Rev 103:3803–3834
13. Moore GE (1965) Cramming more components onto integrated circuits. Electronics 38 (8):114
14. ITRS-2007, "International Technology Roadmap for Semiconductors" (2007) version (http://www.itrs.net/reports) section on emerging research materials (sub-22 nm)
15. Hoffmann G, Libioulle L, Berndt R (2002) Tunneling-induced luminescence from adsorbed organic molecules with submolecular lateral resolution. Phys Rev B 65:212107
16. Blodgett KB (1935) Films built by depositing successive monomolecular layers on a solid surface. J Am Chem Soc 57:1007–2022

17. Blodgett KB, Langmuir I (1937) Built-up films of barium stearate and optical properties. Phys Rev 51:964–982
18. Langmuir I, Schaefer VJ (1938) Activities of urease and pepsin monolayers. J Am Chem Soc 60:1351–1360
19. Metzger RM, Baldwin JW, Shumate WJ, Peterson IR, Mani P, Mankey GJ, Morris T, Szulczewski G, Bosi S, Prato M, Comito A, Rubin Y (2003) Large current asymmetries and potential device properties of a Langmuir-Blodgett monolayer of dimethyanilinoazafullerene sandwiched between gold electrodes. J Phys Chem B107:1021–1027
20. Stewart DR, Ohlberg DAA, Beck PA, Chen Y, Williams RS, Jeppesen JO, Nielsen KA, Stoddart JF (2004) Molecule-independent electrical switching in Pt/organic monolayer/Ti devices. Nano Lett 4:133–136
21. Ashwell GJ, Tyrrell WD, Whittam AJ (2005) Molecular rectification: self-assembled monolayers in which donor-(π-bridge)-acceptor moieties are centrally located and symmetrically coupled to both gold electrodes. J Am Chem Soc 126:7102–7110
22. Sagiv J (1980) Organized monolayers by adsorption. 1. Formation and structure of oleophobic mixed monolayers on solid surfaces. J Am Chem Soc 102:92–98
23. Bain CD, Troughton EB, Tao YT, Evail J, Whitesides GM, Nuzzo RG (1989) Formation of monolayer films by the spontaneous assembly of organic thiols from solution onto gold. J Am Chem Soc 111:321–335
24. Love JC, Estroff LA, Kriebel JK, Nuzzo RG, Whitesides GM (2005) Self-assembled monolayers of thiolates on metals as a form of nanotechnology. Chem Rev 105:1103–1169
25. Yu M, Bovet N, Satterly CJ, Bengió S, Lovelock KRJ, Milligan PK, Jones RG, Woodruff DP, Dhanak V (2006) True nature of an archetypical self-assembly system: mobile Au-thiolate species on Au(111). Phys Rev Lett 97:166102
26. Woodruff DP (2008) The interface structure of N-alkylthiolate self-assembled monolayers on coinage metal surfaces. Phys Chem Chem Phys 10:7211–7221
27. Maksymovych P, Voznyy O, Dougherty DB, Sorescu Dan DC, Yates JT (2010) Gold adatom as a key structural component in self-assembled monolayers of organosulfur molecules on Au(111). Prog Surf Sci 85:206–240
28. Black JR (1969) Electromigration–a brief survey and some recent results. IEEE Trans El Dev ED 16:338–347
29. Prins F, Hayashi T, de Vos van Steenwijk BJA, Gao B, Osorio EA, Muraki K, van der Zant HSJ (2009) Room-temperature stability of Pt nanogaps formed by self-breaking. Appl Phys Lett 94:123108
30. Schottky W (1938) Halbleitertheorie der Sperrschicht. Naturwissenschaften 26:843
31. Mott NF (1938) Note on the contact between a metal and an insulator or semiconductor. Proc Camb Philol Soc 34:568–572
32. Cowley AM, Sze SM (1965) Surface states and barrier height of metal-semiconductor systems. J Appl Phys 36:3212–3220
33. Sze SM (1981) Physics of semiconductor devices, 2nd edn. Wiley, New York, Chap. 5
34. Geddes NJ, Sambles JR, Jarvis DJ, Parker WG, Sandman DJ (1992) The electrical properties of metal-sandwiched Langmuir-Blodgett multilayers and monolayers of a redox-active organic molecular compound. J Appl Phys 71:756–768
35. Ashwell GJ, Sambles JR, Martin AS, Parker WG, Szablewski M (1990) Rectifying characteristics of Mg | ($C_{16}H_{33}$-Q3CNQ LB film) | Pt structures. J Chem Soc Chem Commun 1374–1376
36. Martin AS, Sambles JR, Ashwell GJ (1993) Molecular rectifier. Phys Rev Lett 70:218–221
37. Novoselov AS, Geim AK, Morozov SV, Jiang D, Zhang Y, Dubonos SV, Grigorieva IV, Firsov AA (2004) Electric field effect in atomically thin carbon films. Science 306:666–669
38. Metzger RM, Xu T, Peterson IR (2001) Electrical rectification by a monolayer of hexadecylquinolinium tricyanoquinodimethanide measured between macroscopic gold electrodes. J Phys Chem B105:7280–7290

39. Metzger RM, Chen B, Höpfner U, Lakshmikantham MV, Vuillaume D, Kawai T, Wu X, Tachibana H, Hughes TV, Sakurai H, Baldwin JW, Hosch C, Cava MP, Brehmer L, Ashwell GJ (1997) Unimolecular electrical rectification in hexadecylquinolinium tricyanoquinodimethanide. J Am Chem Soc 119:10455–10466
40. Geddes NJ, Sambles JR, Jarvis DJ, Parker WG, Sandman DJ (1990) Fabrication and investigation of asymmetric current-voltage characteristics of a metal/Langmuir-Blodgett monolayer/metal structure. Appl Phys Lett 56:1916–1918
41. Xu T, Peterson IR, Lakshmikantham MV, Metzger RM (2001) Rectification by a monolayer of hexadecylquinolinium tricyanoquinodimethanide between gold electrodes. Angew Chem Int Ed 40:1749–1752
42. Cui XD, Primak A, Zarate X, Tomfohr J, Sankey OF, Moore AL, Moore TA, Gust D, Harris G, Lindsay SM (2001) Reproducible measurement of single-molecule conductivity. Science 294:571–574
43. Kushmerick JG, Holt DB, Pollack SK, Ratner MA, Yang JC, Schull TL, Naciri J, Moore MH, Shashidhar R (2002) Effect of bond-length alternation in molecular wires. J Am Chem Soc 124:10654–10655
44. Kushmerick JG, Holt DB, Yang JC, Naciri J, Moore MH, Shashidhar R (2002) Metal-molecule contacts and charge transport across monomolecular layers: measurement and theory. Phys Rev Lett 89:86802
45. Beebe JM, Kim B-S, Gadzuk JW, Frisbie CD, Kushmerick JG (2006) Transition from direct tunneling to field emission in metal-molecule-metal junctions. Phys Rev Lett 97:026801
46. Muller CJ, van Ruitenbeek JM, de Jongh LJ (1992) Experimental observation of the transition from weak link to tunnel junction. Physica C191:485–504
47. Reed MA, Zhou C, Muller CJ, Burgin TP, Tour JM (1997) Conductance of a molecular junction. Science 278:252–253
48. Park H, Li AKL, Alivisatos AP, Park J, McEuen PL (1999) Fabrication of metallic electrodes with nanometer separation by electromigration. Appl Phys Lett 75:301
49. Park J, Pasupathy AN, Goldsmith JI, Chang C, Yaish Y, Petta JR, Rinkoski M, Sethna JP, Abruña HD, McEuen PL, Ralph DC (2002) Coulomb blockade and the Kondo effect in single-atom transistors. Nature 417:722–725
50. Strachan DR, Smith DE, Johnston DE, Park T-H, Therien MJ, Bonnell DA, Johnson AT (2005) Controlled fabrication of nanogaps in ambient environment for molecular electronics. Appl Phys Lett 86:043109
51. Xu B, Tao NJ (2003) Measurement of single-molecule resistance by repeated formation of molecular junctions. Science 301:1221–1223
52. Zhou X-S, Wei Y-M, Liu L, Chen Z-B, Tang J, Mao B-W (2008) Extending the capability of STM break junction for conductance measurement of atomic-size nanowires: an electrochemical strategy. J Am Chem Soc 130:13228–13230
53. Haiss W, van Zalinge H, Higgins SJ, Bethell D, Höbenreich H, Schiffrin DJ, Nichols RJ (2003) Redox state dependence of single-molecule conductivity. J Am Chem Soc 125:15294–15295
54. Haiss W, Nichols RJ, van Zalinge H, Higgins SJ, Bethell D, Schiffrin DJ (2004) Measurement of single molecule conductivity using the spontaneous formation of molecular wires. Phys Chem Chem Phys 6:4330–4337
55. Ohm GS (1827) Die Galvanische Kette, Mathematisch Bearbeitet. Riemann, Berlin
56. Landauer R (1957) Spatial variation of currents and fields due to localized scatterers in metallic conduction, IBM. J Res Dev 1:223–231
57. Chang AM (2001) Resistance of a perfect wire. Nature 411:39–40
58. de Picciotto R, Störmer HL, Pfeiffer LN, Baldwin KW, West KW (2001) Four-terminal resistance of a ballistic quantum wire. Nature 411:51–54
59. Joachim C, Gimzewski JK, Schlittler RR, Chavy C (1995) Electronic transparence of a single C_{60} molecule. Phys Rev Lett 74:2102–2105

60. Xiao X, Xu B, Tao NJ (2004) Measurement of single molecule conductance: benzenedithiol and benzenedimethanethiol. Nano Lett 4:267–271
61. Haiss W, Martin S, Leary E, van Zalinge H, Higgins SJ, Bouffier L, Nichols RJ (2009) Impact of junction formation method and surface roughness on single-molecule conductance. J Phys Chem C113:5823–5833
62. Nichols RJ, Haiss W, Higgins SJ, Leary E, Martin S, Bethell D (2010) The experimental determination of the conductance of single molecules. Phys Chem Chem Phys 12:2801–2815
63. Xiao X, Nagahara LA, Rawlett AM, Tao N (2005) Electrochemical gate-controlled conductance of single oligo(phenylene ethynylene)s. J Am Chem Soc 127:9235–9240
64. Getty SA, Engtrakul C, Wang L, Liu R, Ke S-H, Baranger HU, Yang W, Fuhrer MS, Sita LR (2005) Near-perfect conduction through a ferrocene-based molecular wire. Phys Rev B 71:241401
65. Mayor M, von Hänisch C, Weber HB, Reichert J, Beckmann B (2002) A *trans*-platinum(II) complex as a single-molecule insulator. Angew Chem Int Ed 41:1183–1186
66. Reichert J, Ochs R, Beckmann D, Weber HB, Mayor M, von Löhneysen H (2002) Driving current through single organic molecules. Phys Rev Lett 88:176804
67. Kergueris C, Bourgoin J-P, Palacin S, Esteve D, Urbina C, Magoga M, Joachim C (1999) Electron transport through a metal-molecule-metal junction. Phys Rev B59:12505–12513
68. He J, Chen F, Li J, Sankey OF, Terazono Y, Herrero C, Gust D, Moore TA, Moore AL, Lindsay SM (2005) Electronic decay constant of carotenoid polyenes from single-molecule measurements. J Am Chem Soc 127:1384–1385
69. Lafferentz L, Ample F, Yu H, Hecht S, Joachim C, Grill L (2009) Conductance of a single conjugated polymer as a continuous function of its length. Science 323:1193–1197
70. Hines T, Diez-Perez I, Hihath J, Liu HM, Wang ZS, Zhao JW, Zhou G, Muellen K, Tao NJ (2010) Transition from tunneling to hopping in single molecular junctions by measuring length and temperature dependence. J Am Chem Soc 132:11658–11664
71. Yamada R, Kumazawa H, Tanaka S, Tada H (2009) Electrical resistance of long oligothiophene molecules. Appl Phys Exp 2:025002
72. Lu Q, Liu K, Zhang H, Du Z, Wang X, Wang F (2009) From tunneling to hopping: a comprehensive investigation of charge transport mechanism in molecular junctions based on oligo(p-phenylene ethynylene)s. ACS Nano 3:3861–3868
73. Li X, Hihath J, Chen F, Masuda T, Zang L, Tao N (2007) Thermally activated electron transport in single redox molecules. J Am Chem Soc 129:11535–11542
74. Venkataraman L, Klare JE, Nuckolls C, Hybertsen MS, Steigerwald ML (2006) Dependence of single-molecule junction conductance on molecular conformation. Nature 442:904–908
75. Widawsky JR, Kamenetska M, Klare J, Nuckolls C, Steigerwald ML, Hybertsen MS, Venkataraman L (2009) Measurement of voltage-dependent electronic transport across amine-linked single-molecular-wire junctions. Nanotechnology 20:434009
76. Krzeminski C, Delerue C, Allan G, Vuillaume D, Metzger RM (2001) Theory of rectification in a molecular monolayer. Phys Rev B 64:085405
77. Chabinyc ML, Chen X, Holmlin RE, Jacobs H, Skulason H, Frisbie CD, Mujica V, Ratner MA, Rampi MA, Whitesides GM (2002) Molecular rectification in a metal-insulator-metal junction based on self-assembled monolayers. J Am Chem Soc 124:11731–11736
78. Mujica V, Ratner MA, Nitzan A (2002) Molecular rectification: why is it so rare? Chem Phys 281:147–150
79. Aviram A, Ratner MA (1974) Molecular rectifiers. Chem Phys Lett 29:277–283
80. Batley M, Lyons LE (1968) Photoelectric emission from donor-acceptor solids and donor molecules. Mol Cryst 3:357–374
81. Dvorák V, Nemek I, Zyka J (1967) Electrochemical oxidation of some aromatic amines in acetonitrile medium II. benzidine, N,N,N',N'-tetramethylbenzidine, and 1,4-phenylenediamine derivatives. Microchem J 12:324–349
82. Evans S, Green MLH, Jewitt B, Orchard AF, Pygall CF (1972) Electronic spectra of metal complexes containing π-cyclopentadienyl and related ligands: part I. – He(I)

photoelectron spectra of some closed-shell metallocenes. J Chem Soc Faraday Trans II 68: 1847–1865
83. Lianos P, Georghiou S (1979) Complex formation between pyrene and the nucleotides GMP, CMP, TMP and AMP. Photochem Photobiol 29:13–21
84. Lichtenberger DL, Johnston RL, Hinkelmann K, Suzuki T, Wudl F (1990) Relative electron donor strengths of tetrathiafulvene derivatives: effects of chemical substitutions and the molecular environment from a combined photoelectron and electrochemical study. J Am Chem Soc 112:3302–3307
85. Garron R (1964) C R Hebd Seances Acad Sci 258:1458
86. Grepstad JK, Garland PO, Slagsvold BJ (1976) Anisotropic work function of clean and smooth low-index faces of aluminium. Surf Sci 57:348–362
87. Takahashi T, Tokailin H, Sagawa T (1985) Angle-resolved ultraviolet photoelectron spectroscopy of the unoccupied band structure of graphite. Phys Rev B32:8317–8324
88. Potter HC, Blakeley JM (1975) LEED, Auger spectroscopy, and contact potential studies of copper–gold alloy single crystal surfaces. J Vac Sci Technol 12:635–642
89. Demuth JE (1977) Chemisorption of C_2H_2 on Pd(111) and Pt(111): formation of a thermally activated olefinic surface complex. Chem Phys Lett 45:12–17
90. Kebarle P, Chowdhury S (1987) Electron affinities and electron-transfer reactions. Chem Rev 87:513–534
91. Yang SH, Pettiette CL, Conceicão CJO, Smalley RE (1987) UPS of buckminsterfullerene and other large clusters of carbon. Chem Phys Lett 139:233–238
92. Chen ECM, Wentworth WE (1975) A comparison of experimental determinations of electron affinities of π-charge-transfer-complex acceptors. J Chem Phys 63:3183–3191
93. Compton RN, Cooper CD (1977) Negative ion properties of tetracyanoquinodimethan: electron affinity and compound states. J Chem Phys 66:4325–4329
94. Jin C, Haufler RE, Hettich RL, Bashick CM, Compton RN, Puretzky AA, Dem'yanenko AV, Tuinman AA (1994) Synthesis and characterization of molybdenum carbide clusters Mo_nC_{4n} ($n = 1$ to 4). Science 263:68–71
95. Mikroyannidis JA, Stylianakis MM, Sharma GD, Balraju P, Roy MS (2009) A novel alternating phenylenevinylene copolymer with perylene bisimide units: synthesis, photophysical, electrochemical, and photovoltaic properties. J Phys Chem C113:7904–7912
96. Baldwin JW, Chen B, Street SC, Konovalov VV, Sakurai H, Hughes TV, Simpson CS, Lakshmikantham MV, Cava MP, Kispert LD, Metzger RM (1999) Spectroscopic studies of hexadecylquinolinium tricyanoquinodimethanide. J Phys Chem B103:4269–4277
97. Honciuc A, Otsuka A, Wang Y-H, McElwee SK, Woski SA, Saito G, Metzger RM (2006) Polarization of charge-transfer bands and rectification in hexadecylquinolinium 7,7,8-tricyanoquinodimethanide and its tetrafluoro analog. J Phys Chem B110:15085–15093
98. Okazaki N, Sambles JR, Jory MJ, Ashwell GJ (2002) Molecular rectification at 8 K in an Au | $C_{16}H_{33}$Q-3CNQ LB film | Au structure. Appl Phys Lett 81:2300–2302
99. Brady AC, Hodder B, Martin AS, Christopher JR, Ewels P, Jones R, Briddon PR, Musa AM, Panetta CA, Mattern DL (1999) Molecular rectification with M|(D-s-A LB film)|M junctions. J Mater Chem 9:2271–2275
100. Metzger RM (1999) The prospects for unimolecular rectification. In: Sasabe H (ed) Hyperstructured molecules I: chemistry, physics, and applications. Gordon & Breach Science Publishers, Amsterdam, pp 19–39
101. Chen B, Metzger RM (1999) Rectification between 370 K and 105 K in hexadecylquinolinium tricyanoquinodimethanide. J Phys Chem B103:4447–4451
102. Vuillaume D, Chen B, Metzger RM (1999) Electron transfer through a monolayer of hexadecylquinolinium tricyanoquinodimethanide. Langmuir 15:4011–4017
103. Jaiswal A, Rajagopal D, Lakshmikantham MV, Cava MP, Metzger RM (2007) Unimolecular rectification and other properties of $CH_3C(O)S\text{-}C_{14}H_{28}Q^+\text{-}3CNQ^-$ and $CH_3C(O)S\text{-}C_{16}H_{32}Q^+\text{-}3CNQ^-$ organized by self-assembly, Langmuir-Blodgett, and Langmuir-Schaefer techniques. Phys Chem Chem Phys 9:4007–4017

104. Baldwin JW, Amaresh RR, Peterson IR, Shumate WJ, Cava MP, Amiri MA, Hamilton R, Ashwell GJ, Metzger RM (2002) Rectification and nonlinear optical properties of a Langmuir-Blodgett monolayer of a pyridinium dye. J Phys Chem B106:12158–12164
105. Honciuc A, Jaiswal A, Gong A, Ashworth K, Spangler CW, Peterson IR, Dalton LR, Metzger RM (2005) Current rectification in a Langmuir-Schaefer monolayer of fullerene-bis-[4-diphenylamino-4"-(N-ethyl-N-2"'-ethyl)amino-1,4-diphenyl-1,3-butadiene] malonate between Au electrodes. J Phys Chem B109:857–871
106. Shumate WJ, Mattern DL, Jaiswal A, Burgess J, Dixon DA, White TR, Honciuc A, Metzger RM (2006) Spectroscopic and rectification studies of three donor-sigma-acceptor compounds, consisting of a one-electron donor (pyrene or ferrocene), a one-electron acceptor (perylenebisimide), and a C_{19} swallowtail. J Phys Chem B110:11146–11159
107. Shumate WJ (2005) Ph.D. dissertation, University of Alabama
108. Honciuc A, Metzger RM, Gong A, Spangler CW (2007) Elastic and inelastic electron tunneling spectroscopy of a new rectifying monolayer. J Am Chem Soc 129:8310–8319
109. Jaiswal A, Amaresh RR, Lakshmikantham MV, Honciuc A, Cava MP, Metzger RM (2003) Electrical rectification in a monolayer of zwitterions assembled by either physisorption or chemisorption. Langmuir 19:9043–9050
110. Xu T, Morris TA, Szulczewski GJ, Amaresh RR, Gao Y, Street SC, Kispert LD, Metzger RM, Terenziani F (2002) A spectroscopic study of hexadecylquinolinium tricyanoquinodimethanide as a monolayer and in bulk. J Phys Chem B106:10374–10381
111. Okazaki N, Sambles JR (2000) Extended abstracts of the international symposium on organic molecular electronics. Nagoya, Japan, p 66
112. Kwon O, McKee ML, Metzger RM (1999) Theoretical calculations of methylquinolinium tricyanoquinodimethanide (CH_3Q-3CNQ) using a solvation model. Chem Phys Lett 313:321–331
113. Metzger RM (2011) The many faces of quinolinium tricyanoquinodimethanide. Gale PA, Steele JW (eds) Supramolecular chemistry: from molecules to nanomaterials. Anzenbacher P (ed) Section 7: Supramolecular devices. Wiley, London (in press)
114. Girlando A, Sissa C, Terenziani F, Painelli A, Chwialkowska A, Ashwell GJ (2007) In situ spectroscopic characterization of rectifying molecular monolayers self-assembled on gold. Chem Phys Chem 8:2195–2201
115. Ho G, Heath JR, Kontratenko M, Perepichka DF, Arseneault K, Pézolet M, Bryce MR (2005) The first studies of a tetrathiafulvalene-σ-acceptor molecular rectifier. Chem Eur J 11:2914–2922
116. Wang W, Lee T, Reed MA (2004) Elastic and inelastic electron tunneling in alkane self-assembled monolayers. J Phys Chem B108:18398–18407
117. Mazur U, Hipps KW (1995) Resonant tunneling bands and electrochemical reduction potentials. J Phys Chem 99:6684–6688
118. Mazur U, Hipps KW (1999) Orbital-mediated tunneling, inelastic electron tunneling, and electrochemical potentials for metal phthalocyanine thin films. J Phys Chem B103:9721–9727
119. Perepichka DF, Bryce MR, Pearson C, Petty MC, McInnes EJL, Zhao JP (2003) A covalent tetrathiafulvalene-tetracyanoquinodimethane diad: extremely low HOMO-LUMO gap, thermoexcited electron transfer, and high-quality Langmuir-Blodgett films. Angew Chem Int Ed 42:4636–4639
120. Elbing M, Ochs R, Keontopp M, Fischer M, von Hänisch C, Weigend F, Evers F, Weber HB, Mayor M (2005) A single-molecule diode. Proc Natl Acad Sci USA 102:8815–8820
121. Morales GM, Jiang P, Yuan S, Lee Y, Sanchez A, You W, Yu L (2005) Inversion of the rectifying effect in diblock molecular diodes by protonation. J Am Chem Soc 127:10456–10457
122. Jiang P, Morales GM, You W, Yu LP (2004) Synthesis of diode molecules and their sequential assembly to control electron transport. Angew Chem Int Ed 43:4471–4475
123. Díez-Pérez I, Hihath J, Lee Y, Yu L, Adamska L, Kozhushner MA, Oleynik II, Tao N (2009) Rectification and stability of a single molecular diode with controlled orientation. Nat Chem 1:635–641

124. Ashwell GJ, Tyrrell WD, Whittam AJ (2003) Molecular rectification: self-assembled monolayers of a donor–(π-bridge)–acceptor chromophore connected via a truncated Au–S–$(CH_2)_3$ bridge. J Mater Chem 13:2855–2857
125. Ashwell GJ, Chwialkowska A, Herrmann High LR (2004) Rectifying Au-S-C_nH_{2n}-P3CNQ derivatives. J Mater Chem 14:2848–2851
126. Ashwell GJ, Berry M (2005) Hybrid SAM/LB device structures: manipulation of the molecular orientation for nanoscale electronic applications. J Mater Chem 15:108–110
127. Ashwell GJ, Robinson BJ, Amiri MA, Locatelli D, Quici S, Roberto D (2005) Dipole reversal in Langmuir–Blodgett films of an optically nonlinear dye and its effect on the polarity for molecular rectification. J Mater Chem 15:4203–4205
128. Ashwell GJ, Chwialkowska A (2006) Controlled alignment of molecular diodes via ionic assembly of cationic donor-(pi-bridge)-acceptor molecules on anionic surfaces. Chem Commun 1404–1406
129. Ashwell GJ, Urasinska B, Tyrrell WD (2006) Molecules that mimic Schottky diodes. Phys Chem Chem Phys 8:3314–3319
130. Gayathri SS, Patnaik A (2006) Electrical rectification from a fullerene[60]-dyad based metal-organic-metal junction. Chem Commun 1977–1979
131. Averin DV, Likharev KK (1986) Coulomb blockade of tunneling, and coherent oscillations in small tunnel junctions. J Low Temp Phys 62:345–372
132. Shkrob IA, Schlueter JA (2006) Can a single molecule trap the electron? Chem Phys Lett 431:364–369
133. Schweikart KH, Malinovskii VL, Yasseri AA, Li J, Lysenko AB, Bocian DF, Lindsey JS (2003) Synthesis and characterization of bis(S-acetylthio)-derivatized europium triple-decker monomers and oligomers. Inorg Chem 42:7431–7446
134. Chen S (2004) Chemical manipulations of nanoscale electron transfers. J Electroanal Chem 574:153–165
135. Chen G, Bandow S, Margine ER, Nisoli C, Kolmogorov AN, Crespi VH, Gupta R, Sumanasekera GU, Iijima S, Eklund PC (2003) Chemically doped double-walled carbon nanotubes: cylindrical molecular capacitors. Phys Rev Lett 27:257403
136. Chen J, Reed MA, Rawlett AM, Tour JM (1999) Large on-off ratios and negative differential resistance in a molecular electronic device. Science 286:1550–1552
137. Esaki L (1958) New phenomenon in narrow germanium p-n junctions. Phys Rev 109:603–604
138. Paloheimo J, Kuivalainen P, Stubb H, Vuorimaa E, Yli-Lahti P (1990) Molecular field-effect transistors using conducting polymer Langmuir-Blodgett films. Phys Lett 56:1157–1159
139. Tans SJ, Devoret MH, Dai H, Thess A, Smalley RE, Geerligs LJ, Dekker C (1997) Individual single-wall carbon nanotubes as quantum wire. Nature 386:474–477
140. Song H, Kim Y, Jang YH, Reed MA, Lee T (2009) Observation of molecular orbital gating. Nature 462:1039–1043
141. Collier CP, Mattersteig G, Wong EW, Beverly K, Sampaio J, Raymo FM, Stoddart JF, Heath JR (2000) A [2]catenane-based solid-state electronically reconfigurable switch. Science 289:1172–1175
142. He H, Zhu J, Tao NJ, Nagahara LA, Amlani I, Tsui R (2001) A conducting polymer nanojunction switch. J Am Chem Soc 123:7730–7731

Top Curr Chem (2012) 313: 85–120
DOI: 10.1007/128_2011_221

Published online: 28 September 2011

Active and Non-Active Large-Area Metal–Molecules–Metal Junctions

Barbara Branchi, Felice C. Simeone, and Maria A. Rampi

Abstract The study of charge transport processes through organic molecules by using molecular junctions has generated great attention in the last few years, partially triggered by the possibility of developing molecular electronic devices to be implemented somehow into current silicon-based technology. As experimental tools, a large variety of conceptually and geometrically different metal–molecule(s)–metal junctions has been proposed. While the intrinsic conductivity of a molecule is still elusive, parameters crucial for molecular electronics have been extracted by using a variety of junctions. Significantly, the results extracted from molecular junctions and those obtained by the kinetic approach in supramolecular D–B–A systems are complementary. For the sake of a practical discussion, a distinction is made between "active junctions" and "non-active junctions". Active junctions are those aimed at switching the electrical response by an external stimulus acting "in situ" to modify the electronic structure of the molecular system. Non-active junctions are those aimed at studying different conduction regimes by incorporating molecules of different electronic structures. Depending on their geometry, the junctions can incorporate different numbers of molecules. Large area molecular junctions present two main advantages: (1) a simpler assembly, by requiring less sophisticated fabrication and (2) a higher versatility, relative to single molecule junctions, towards potential applications in organic electronics. The present chapter focuses on the fabrication of a variety of large-area molecular junctions and summarizes and compares the experimental results.

Keywords Charge transport · Electrochemical molecular junctions · Electron tunnelling · Electron hopping · Molecular electronics · Photoactive molecular junctions

---

B. Branchi, F.C. Simeone and M.A. Rampi (✉)
Dipartimento di Chimica, Università di Ferrara, Ferrara 44100, Italy
e-mail: rmp@unife.it

Contents

Abbreviations

AZO	Azobenzene
BJs	Break junctions
c-AFM	Conductive atomic force microscopy
CT	Charge transfer
CWJs	Cross wires based junctions
DAE	Diarylethene
D–B–A	Donor–molecular bridge–acceptor supramolecular system
E-GaIn	Metal eutectic Ga–In (75% Ga 25% In)
ET	Electron transfer
HBC	Hexabenzocoronene
HS-C_{10}-Ru	$HS(CH_2)_{10}CONHCH_2pyRu(NH_3)_5(PF_6)_2$
IJs	Interlayer based junctions
LAJs	Large area junctions
LB	Langmuir–Blodgett
M–B–M	Metal–molecular bridge–metal (junction)
MCMWs	Metal centre molecular wires
OPh	Oligophenylene
PEDOT:PSS	Poly(3,4-ethylenedioxythiophene): poly(4-styrenesulphonic acid)
PPV	Poly[(*m*-phenylenevinylene)-*co*-(2,5-dioctoxy-*p*-phenylenevinylene)]
STMJs	Scanning tunnelling microscope junctions

1 Introduction

"Molecular electronics" was born in the 1990s as "a branch of nanotechnology dealing with the study and application of molecular building blocks for the fabrication of electronic components" [1–3]. Today, after more than 15 years of intense experimental and theoretical studies, the first observations emerging from reviewing the results are that (1) the "intrinsic" conductance of molecules is still elusive and (2) so far, there is no indication that electronic components based on organic molecules can compete with both stability and operational speed of silicon-based electronic devices. Possible applications of molecular electronics may come to light in the future; in the meanwhile, it is certainly worthwhile to continue searching for technological applications and to improve fundamental scientific understanding.

The process of charge transfer (CT) through organic molecules is central not only in molecular electronics – it is also a fundamental feature in a large number of biological phenomena (e.g. photosynthesis [4]). Electron transfer (ET) processes have been extensively studied over the last 60 years, both experimentally and theoretically, by using different approaches. Since the early 1970s, many experimental studies have been dedicated to measuring ET processes through organic molecules in supramolecular D–B–A systems, where D and A are, respectively, electron donor and acceptor molecular units, covalently linked by a molecular bridge, B [5]. These studies have underlined the importance of the electronic structure of B in facilitating ET between D and A. Since then, as a colloquial way to emphasize this role, it has been customary to refer to the bridge as a "molecular wire", and to discuss its ability to "conduct" electrons. It should be clear that these expressions imply an analogy that does not exist between pathways for electron transport through organic molecules and through metallic conductors. It would be more correct to say that the molecular wires facilitate electron transport relative to vacuum.

In 1971, the experimental work of Kuhn [6, 7] contrasted the ET studies carried out on D–B–A systems in solution by measuring current flowing through fatty acids organized in Langmuir–Blodgett (LB) films when sandwiched between two metal electrodes. Even before these mailstone papers, few experimental and theoretical studies had been carried out on the process of tunnelling through molecules incorporated between metal films [8]. Surprisingly, after Kuhn's pioneering work only a few authors focused on this approach [9, 10]. In 1974, the theoretical paper of Aviram and Ratner envisaged the use of organic molecules sandwiched between two electrodes to obtain the function of a conventional electronic rectifier [11]. In 1990 Ashwell and Sambles [12, 13] and later in 1997 Metzger et al. [14] reported experimental results obtained from junctions incorporating LB films based on hexadecylquinolinium tricyanoquinodimethanide, proving the rectification effect predicted theoretically. In the late 1990s, (1) the combination of nanofabrication [15], (2) a more familiar use of scanning probe microscopies [16, 17] and (3) the ability to form stable chemical bonds between molecules and metal surfaces triggered the fabrication of metal–molecules–metal, M–B–M junctions, and opened the door to experimental "molecular electronics". In the last decade the number of works dedicated to characterize the electrical properties of organic molecules in M–B–M junctions has increased exponentially [18–25].

As experimental tools, a large variety of conceptually and geometrically different M–B–M junctions has been proposed to the scientific community. To name only a selection of the most popular ones, we mention (1) break junctions (BJs) [26, 27], (2) nanopores-based junctions NPJs [28], (3) STM (STMJs) [29–31] and conducting AFM based junctions (c-AFMJs) [32], (4) crossed wires junctions (CWJ) [33] and (5) large-area junctions (LAJs) [34]. As depicted in Fig. 1, each type of the mentioned junctions measures the electrical characteristics of molecules in different number and different environments: (1) as single unit in BJs and STMJs, (2) as set of molecules in smaller or larger number respectively in c-AFMJs and CWJs and (3) in very large organized aggregates as self-assembled monolayer (SAM) in LAJs.

Among the mentioned junctions, certainly BJs and STMJs are those which provide the most informative data for fundamental understanding with a molecular-level resolution. At the opposite end, junctions able to characterize the electrical behaviour of large ensembles of organic molecules contribute to the field with results which complement those extracted from single-molecule experiments and with information useful for application of organic electronic devices. Since each one of these different junctions shows both advantages and limitations, the design of conceptually new molecular junctions able to give reproducible electrical measurements and convenient for application is still underway [35].

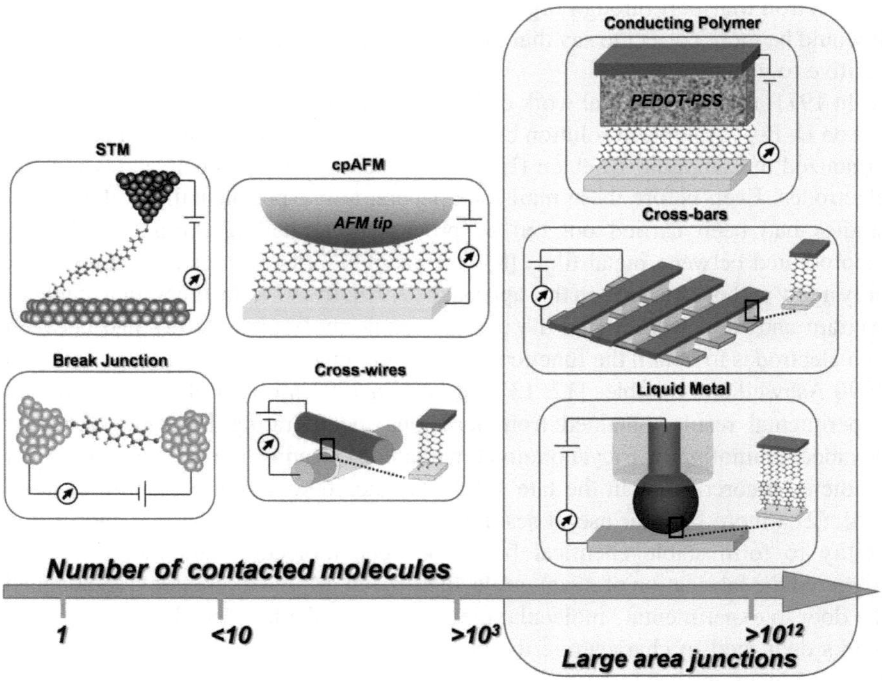

Fig. 1 Schematic representation of the most commonly used molecular electronic junctions ordered as a function of the number of contacted molecules

After a brief discussion of fundamentals of charge transport mechanisms, this chapter summarizes and discusses the most significant results obtained by using different junctions and in particular LAJs. In order to facilitate a systematic discussion, we make a functional distinction between "non-active" and "active" junctions: we will refer to "active junctions" as those aimed at changing the electrical response by means of an external stimulus acting in situ to modify the molecular electronic structure; "non-active junctions" are those used to measure and compare the electrical properties inherent to the different electronic structure of incorporated molecules, without any modification induced by an external signal.

2 Charge-Transfer Mechanisms

Experimental works [36, 37], supported by theory [38, 39], have shown that the charge-transfer processes through organic molecules in D–B–A systems and charge transport in M–B–M junctions are strongly related (Fig. 2a). The processes occurring between electron donor (D) and acceptor units (A) – A and D being either molecules or electrodes (M) – involve an electronic mediation of the molecular energy levels. The nature of the mechanism dominating these processes depends in first approximation on the distance in energy between the D/A units and the closest molecular orbitals, as outlined in the following.

When the current in molecular junctions is dominated by electron tunnelling (Fig. 2b), it can be described as a first approximation by Simmons's theory [40, 41]. In this model, the current depends on (1) the height Φ of the potential barrier, which is determined by the interactions of the electron with the medium and (2) the thickness d of the barrier (Fig. 3).

The presence of molecules inside the tunnel gap decreases the tunnel barrier Φ with respect to the vacuum. However, the Simmons model does not take into account the discrete spacing between the energy levels of the molecular bridge, and the hypothesis of a rectangular barrier, as sketched in Fig. 3, certainly neglects interfacial effects. On the other hand, experimental results clearly demonstrate a correlation of the electron transport rate with the molecular HOMO–LUMO gap. In order to account for these findings, a so-called "two band" model has been used [42]. A more refined model, that takes into account the active role played by the molecule, introduces the coupling of the tunnelling electron with the off-resonance electronic states of the molecular bridge. This model, known as "superexchange" coupling and originally proposed by McConnell for D-B-A systems [43], is also useful for the description of the electrical behaviour of molecular junctions, where the charge transport is actually assisted by the molecular bridge (through-bond). According to this model, the rate of the charge transfer process in D–B–A systems depends on the ΔE offset between the energy levels of D/A units and those of the bridge, while in a molecular junction the transport process depends on the difference between the electrode Fermi level and the molecular levels. Therefore molecules with lower LUMOs provide states which are energetically more

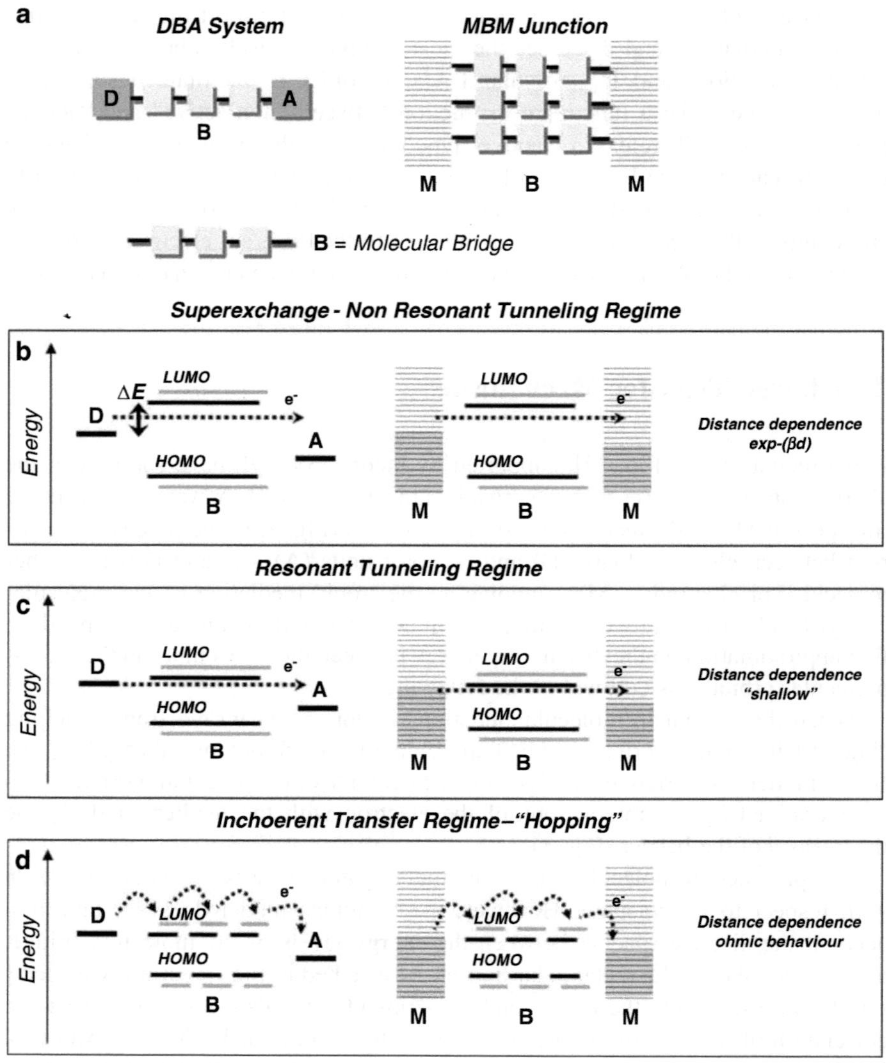

Fig. 2 (**a**) Schematic representation of the energy levels diagrams for a DBA system and a MBM junction in which the electron transfer process is dominated (**b**) by superexchange or non-resonant tunnelling, (**c**) by resonant tunnelling or (**d**) by "hopping"

accessible for a superexchange coupling [5]. When the HOMO is closer than the LUMO to the Fermi level of the electrodes, the transport will occur via holes [5].

All these theories provide the basis for using, as first approximation, the simple phenomenological equations to describe the ET rate constant k in D–B–A systems as $k = k_o\, e^{-\beta d}$ and the current flow I as $I = I_o e^{-\beta d}$ in molecular junctions, where d is the length of the molecule, and β is a decay factor. Although the decay

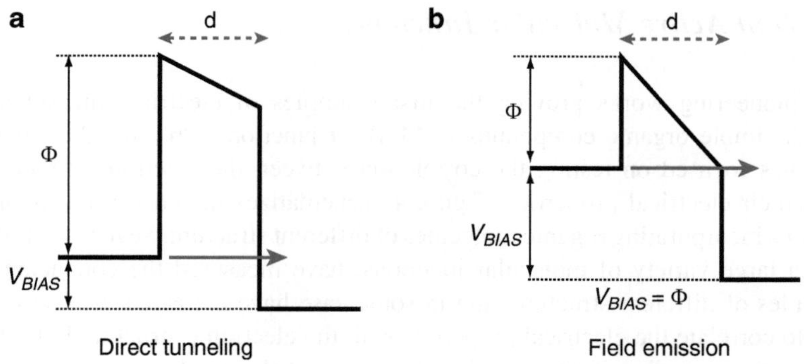

Fig. 3 Schematic representation of the Simmons model for (**a**) electron tunnelling and (**b**) field-emission regime

parameter β can be defined only for a tunnelling process, its value is often used as a first approximation to distinguish between different transport mechanisms [44–46]. It is worth underlining that in the Simmons model $\beta \propto \Phi^{1/2}$, while in a superexchange treatment $\beta \propto \ln(\Delta E)$, where $\Delta E = E_{\text{LUMO}} - E_{\text{F}}$. We stress here that the tunnelling barrier Φ of the Simmons model and ΔE of the superexchange treatment have different physical origins, and therefore different values, but certainly for a bridge assisted transport, it is roughly true that $\Phi \propto \Delta E$.

Recently, Frisbie and co-workers, on the basis of a Simmons-like approach, extracted the tunnel barrier height from *i–V* curves obtained from c-AFM based junctions incorporating a variety of molecules [47, 48]. The observed discontinuity in the *i–V* curves at well-defined values of applied bias, has been interpreted as a transition from tunnelling to field-emission regime. As sketched in Fig. 3b, the process of field-emission occurs when, upon application of increasing bias, the barrier becomes triangular, and hence partially classically accessible. The voltage at which the transition occurs is suggested to provide a measure of the barrier height ($V_{\text{bias}} = \Phi$). However, a deeper analysis of such kinds of results suggests that the change in the electrical behaviour of the molecular junction must be rather attributed to a resonant tunnelling through molecular levels [49]. This discussion introduces some concern about the range of validity of the Simmons model, which is more adequate for description of direct through-space tunnelling, where the barrier Φ is determined by electrostatic interaction of the tunnelling electron with the molecular bridge [50, 51].

Figures 2c and 2d schematize the energy diagram for molecular orbitals of B in close proximity with the energy levels of the donor/acceptor units. In this case the ET process can occur via a coherent resonant tunnelling mechanism or via hopping between localized states of the molecular bridge [52]. Experimentally, these processes will exhibit a shallow dependence on d and a quasi-ohmic behaviour, allowing for charge transport at distances not reachable by tunnelling. These two mechanisms can be more rigorously disentangled by the temperature dependence: while hopping is a thermally activated process, tunnelling is not [52].

2.1 Non-Active Molecular Junctions

After pioneering works proving the first examples of electrical measurements through simple organic compounds in M–B–M junctions [26, 28, 53], a number of studies focused on testing the correlation between the electronic structure of B and their electrical properties. Figure 4a schematizes the energy levels of two junctions incorporating organic molecules of different structure. Several authors, by using a large variety of molecular junctions, have measured the conductance of molecules of different structure, and in some case have extracted the factor β, in order to correlate the electrical properties with the electronic structure [54–56].

From these studies, it emerges that the values of the conductance for the same organic molecule, obtained with different junctions, are widely scattered [19, 54]: for alkane chains they span over eight orders of magnitude [55]. As discussed below, this observation, rather than surprising or problematic, is actually very informative [56]. In general, this disagreement has been attributed to the molecules/metal electrode interactions, that can varies when using different anchoring groups on different electrodes [57–61]. Indeed, the metal/molecules contact has an intrinsic resistance, which is related to the electronic structure of the anchoring group and to the charge distribution at the molecule/metal interface. Long ago it was demonstrated that LB films are less conductive than SAMs anchored to the electrodes via covalent bonds; more recently a number of authors have indicated how anchoring groups of different nature such as –SH, –SeH, –SCN and $–NH_2$ can deeply affect the molecular conductance value [62–65].

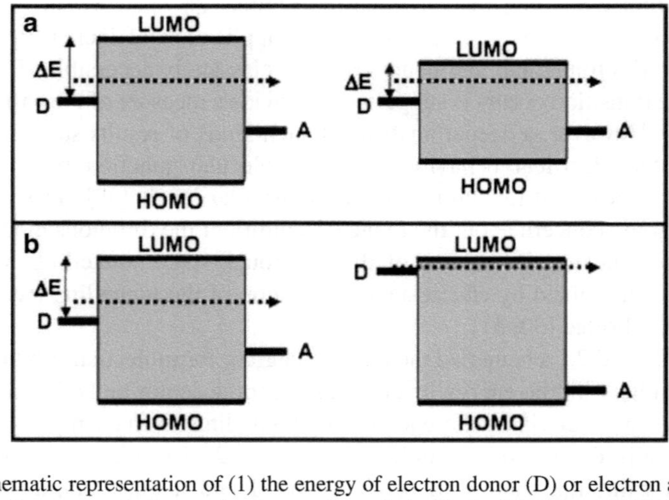

Fig. 4 Schematic representation of (1) the energy of electron donor (D) or electron acceptor (A) units (regardless as to whether molecules or electrodes), (2) the HOMO and LUMO molecular orbitals, and (3) the energy gap ΔE between D/A and the molecular orbitals. (**a**) ΔE is changed by changing the electronic structure of the molecular bridge. (**b**) ΔE is changed by changing the energy levels of the donor or acceptor units

In addition, it is important to remind that in many type of junctions the molecules are not even in physical contact with one of the electrodes, and therefore an additional, unknown, resistive contribution is measured. For instance (1) in STM-based junctions the STM tip and the molecules are separated by a gap, (2) in Hg based LAJs the top electrode is covered by alkane monolayers (Fig. 5a, b), (3) in E-GaIn-based junction the electrode carries always a film of oxide (Fig. 5c) and (4) in the interlayer-based junctions (IJs) there is a film between the molecules and the top electrode (Fig. 1).

Furthermore, a different molecular conductance for the same molecule can be expected, depending on whether the leads are contacting (1) an isolated molecule, either by "fishing for it" with the STM tip [29–31] or in a break junction, (2) a single molecule embedded in a SAM [66, 67] or (3) a number of molecules contacted through metal nanoparticles [53, 68, 69]. In fact, the conductance of the same molecule measured when embedded in a SAM or as an isolated unit in BJs shows values differing by several orders of magnitude [70]. This difference can be understood by considering that the SAM is a 2D supramolecular system, where lateral intermolecular interactions can perturb the molecular electronic structure, and therefore affect the electrical properties. On this basis, individual and collective charge transport phenomena can be very different, and the size of the deviations depends on the type of molecule [71].

In contrast with the large differences observed for molecular conductance, the values of the decay factor β calculated from measurements performed by different types of molecular junctions, for molecules of similar electronic structures, are in good agreement, as shown in Table 1. As expected, β decreases by increasing the molecular conjugation. In addition and significantly, Table 1 shows that the β parameters extracted from D–B–A systems and M–B–M junctions for the same type of B, are similar. In fact the β value reflects the intrinsic conductivity of the core of B, independently of other parameters, such as the molecular environment and/or interaction of the anchoring group with the electrodes.

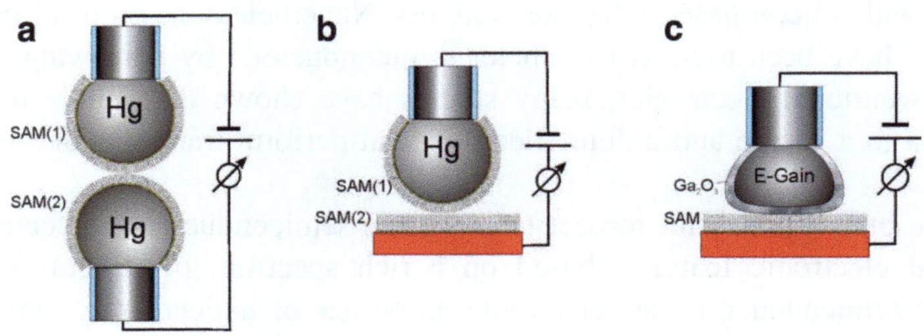

Fig. 5 Schematic representation of LAJs based on liquid metal electrodes. (**a**) The two Hg drops junction. The drops are extruded from two microsyringes and covered singularly by similar or different SAMs before being brought in contact. (**b**) An Hg-drop electrode covered by SAM(1) (usually formed by hexadecane thiol) is brought in electrical contact with a SAM(2) formed on a solid metal surface. (**c**) A drop of In/Ga eutectic alloy (E-GaIn) contacts a SAM formed on a solid electrode surface

Table 1 Values of the decay factor, β, for aliphatic and oligophenylene chains obtained with different experimental systems

Experimental system	Aliphatics $\beta/Å^{-1}$	Oligophenylenes $\beta/Å^{-1}$
D–B–A Molecular systems	0.8–1 [72, 73]	0.4–0.6 [74, 75]
LAJ	1.4 [6]	
	0.89 [76]	0.66 [77]
	0.87 [77]	
	0.86 [78]	0.61 [79]
	0.9 [79]	
	0.57–0.66[a] [80]	
	0.43 [81]	
STM	1.2 [82]	
cp-AFM	0.94 [83]	0.42 [83]

[a]Bias dependent

Recently, it has been reported that by incorporation into non-active molecular junctions of Bs with low-lying energy sites, the conduction mechanism can be dominated by electron hopping [32, 47, 84, 85].

In addition and importantly, even in non-active junctions, when the electrode Fermi level matches the molecular bridge energy levels, resonance phenomena can generate electrical behaviours similar to those of conventional electronic devices, such as rectification [86–89] and negative differential resistance (NDR) [90, 91].

2.2 Active Molecular Junctions

The early studies on non-active molecular junctions showed that the electrical properties of organic molecules are generally dominated by a non-resonant tunnelling mechanism, that is, a mechanism occurring in insulating materials. Therefore, electronic devices based on molecular systems seemed unlikely to compete with silicon-based integrated circuits. Nonetheless, as proof of principle, molecules have been used as nanometer semiconductors: by borrowing concepts from conventional electronics, many studies have shown that single molecules connected to a source and a drain electrode can perform transistor-like functions [92–94].

On the other hand, with respect to inorganic semiconductors, molecules offer additional electronic features, based on a rich spectrum of structures. Unlike inorganic semiconductors, the electronic properties of molecular systems can be tailored by a proper design and/or by functionalization with a large variety of chemical groups, and most importantly, their electronic structure can be changed by external magnetic, electromagnetic or electrochemical signals. Figures 4a and b suggest the possibility to switch current flowing through molecular junctions by changing ΔE via an external trigger [95, 96], as described below.

2.2.1 Photoactive Molecular Junctions

In principle, a variation of ΔE (see Fig. 4) can be achieved by inducing a change in the electronic structure of B by light absorption. Therefore, in molecular junctions incorporating photoactive compounds, light can be used as external trigger to tune the current flow. For a molecular junction to behave as a photoswitch, three conditions of different nature are required: (1) the geometry of the junction should allow for irradiation "in situ" of the molecular systems when sandwiched between the electrodes, (2) irradiation of the chromophore must produce isomers of intrinsic different conductance and (3) the excited state of the chromophore should not be quenched by the metal electrodes [97]. In some cases, the quenching of chromophore excited states by the electrodes' metal surfaces via energy or electron transfer process can be avoided by careful control of the electronic coupling of the photoactive molecular cores with the metal leads (see below) [98, 99].

To date, two types of chromophores have been successfully incorporated in molecular junctions: diarylethenes (DAE) [100] and azobenzene (AZO) derivatives [101].

DAE-based compounds have attracted great attention because of their unique characteristic, consisting of a reversible photo-induced isomerization under irradiation at two different wavelengths (see Fig.13 and a detailed discussion in Sect. 4.2.1).

The open form (O) of the DAE moiety, under irradiation in the UV region, undergoes a photoisomerization to the closed DAE form (C); a reversible photoisomerization to O isomer takes place under irradiation of C in the visible region. The O and C forms exhibit a different degree of conjugation, and therefore different HOMO–LUMO gaps. DAE-based compounds have been incorporated into a number of different types of junctions (BJs [102, 103], STMJs [104], PBJs [99]): the results, often controversial, have finally demonstrated that, under alternating irradiation, a partially reversible switch in current takes place. However, basic information are still missing about the effects of the electrode metal surface on the photophysical properties of the DAE, as the quantum yields of the photoisomerizations, the efficiency of the photoconversion to the photostationary state and the reversibility of the photoisomerization processes.

AZO-based compounds have been extensively studied. An isomerization from the thermodynamically more stable *trans* to the *cis* conformation can be induced by irradiation with UV light, and reversed upon heating or irradiation with visible light [101, 105] (see Fig.14a). Rigid and fully conjugated AZO compounds, in spite of forming densely packed SAMs on Au surfaces, have shown a high yield of photoconversion [34, 106]. These AZO compounds have been incorporated either in LAJ or in STM junctions [107, 108], showing that the current density increases and decreases reversibly over one order of magnitude upon alternating irradiation at $\lambda = 370$ nm and $\lambda = 450$ nm respectively (details are discussed in Sect. 4.2.1).

2.2.2 Electrochemical Molecular Junctions

A second type of active junctions is that based on electrochemical systems, where the current flowing through junctions incorporating electroactive molecules depends on the molecular redox state. While in two-electrode junctions the relative positions of the Fermi levels of the electrodes with respect to the molecular energy levels is undefined, in an electrochemical junction the Fermi level can be changed relatively to the redox potential of the electroactive species by an applied potential measured against an external reference electrode. Therefore, the Fermi level of the electrodes can be tuned and become isoenergetic with the molecular orbitals, as schematized in Fig. 4b. In this case the reference electrode acts as a gate of a conventional transistor, while the two electrodes sandwiching the molecules are the source and the drain [109, 110].

After the pioneering work of Tao [111], a number of studies reported how to control the current flow via an external reference electrode in electrochemical STMJs and LAJs [93, 112–143].

It has been reported that the electrical properties of single molecules incorporating redox groups (e.g. viologens [114, 119, 120, 123, 124], oligophenylene ethynylenes [122, 123], porphyrins [111, 126], oligo-anilines and thiophenes [116, 127], metal transition complexes [118, 128–132], carotenes [133], ferrocenes [134, 135], perylene tetracarboxylic bisimide [93, 136, 137] and redox-active proteins [138–143]), can be switched electrochemically. Such experiments, typically performed by STM on redox-active molecules tethered via Au–S bonds between a gold substrate and a tip under potential control, allow the possibility to examine directly the correlation between redox state and the conductance of individual molecules.

3 Large-Area Junctions

Even if, in the future, the dimensions of the metal contacts of electronic circuits will become compatible with those of organic molecule – therefore allowing for single-molecule based circuitry – the fabrication of single-molecule-based devices will always require sophisticated and expensive equipments. For commercial applications the processing cost of molecular devices should be as low as possible. Mesoscale devices as LAJs incorporating organic films and able to perform electronic functions can provide comparatively simpler and more convenient fabrication. LAJs, in addition, uniquely allow spectroscopic characterization "in situ" of the chemisorbed molecular layers [144].

3.1 Fabrication of Different LAJs

The fundamental requirement to obtain reliable electrical measurements in molecular junctions is the absence of electrical artefacts, such as formation of conducting filaments and short-circuits. The first, LAJs were fabricated by forming an organic film on top of a metal surface (bottom electrode) and depositing on top of the film a second electrode by standard metal-deposition techniques, such as electron-beam evaporation or sputtering. Nonetheless, even when SAMs are well-organized and well-packed on an atomically flat surfaces of noble metals (Au, Ag, Pd) [145], the presence of pinholes and defects at the grain boundaries and the fragility of the organic molecular monolayer represent a source of shorting events masking the electrical response. A number of strategies have been adopted to fabricate top electrodes that can make electrical contact with the organic SAM [19, 63, 146], each one presenting a number of advantages and disadvantages, as outlined in the following.

3.1.1 Gold Atoms Deposition

Vacuum deposition of metals by evaporation is one of the most common methods used in laboratories and industry for making electrical contacts. However, the vaporized metal atoms and clusters, by recondensing on top of the SAMs, can reach the substrate with high temperatures and kinetic energies, so that they can percolate along the molecules forming the organic layer [13, 144, 147, 148]. Deposition of low-melting metals on a cooled sample can in some cases circumvent thermal damage of the SAM [14]. In very early studies, Reed et al., in order to avoid short-circuits, reduced the area of the SAM exposed to the Au evaporation by fabricating nanopores in a film of silicon nitride membrane deposited on top of the SAM [28]. The top electrode was fabricated by Au deposition inside the nanopore. Alternatively, "indirect" evaporation can be adopted to reduce the damage and to provide more reproducible results; it is performed in an inert atmosphere on a cooled sample not directly exposed to the crucible [149]. Recently, McCreery and Bonifas presented a method of forming soft metallic contacts on molecular layers through surface-diffusion-mediated deposition (SDMD) [35]. The metal atoms, deposited remotely, diffuse onto the molecular layer, thus eliminating the problems of penetration and damage to the organic SAM.

3.1.2 Nano-Transfer Printing and Lift-Off, Float-On

The techniques of nano-transfer printing (nTP) and lift-off, float-on (LOFO) used to fabricate top electrodes in LAJs are based on transferring preformed electrodes onto the molecular-device substrate, thus eliminating metal penetration into monolayer defects. In particular, a number of authors [150–153] has shown that LAJs can be fabricated successfully by nTP, i.e. by transferring metal patterns from an elastomeric stamp onto a molecular layer chemically bonded to a substrate. By using the nTP approach it is possible to fabricate arrays of electrodes, suitable for testing multiple devices. However, the nTP technique requires a molecular layer that adheres to the top contact, usually via a thiol group, and therefore it severely restricts the materials that can be incorporated. This approach, when used to fabricate large-area top electrodes, produces wrinkled, problematic metal films.

The LOFO approach, based on capillary interactions induced by liquid–solid interfaces, is used for transferring prefabricated thin solid metal films onto molecularly modified solid substrates. In spite of the fact that the glass/metal pad during the lift-off process leaves a relatively rough (1 nm) surface, several types of device have been fabricated by LOFO [154–156].

The polymer-assisted lift-off (PALO) method, a particular application of LOFO, aims to mitigate energetic and kinetic phenomena that lead to wrinkling and tearing of the preformed electrode [78]. The key component is a hydrophobic polymer backing layer on the preformed electrodes that provides mechanical stability and a thermodynamic driving force to eliminate wrinkling. Through this technique high-quality metal-electrode devices can be fabricated in parallel over a wide range of electrode dimensions, and with lithographically defined spatial registry. Non-shorting molecular junctions with active areas up to 9 mm^2 were obtained with a very low number of short circuits ($<$10%). A PALO technique has recently been used to probe molecular junctions by using Surface Plasmon Resonance Spectroscopy [157].

3.1.3 Interlayers Based Junctions (IJs)

A different strategy in fabricating LAJs consists of depositing on top of self-assembled organic monolayers an organic or inorganic ultrathin film as protection. Polymer films, acting as a mechanical protective barrier, allow for simple metal deposition and prevent short circuits or molecular damage. De Boer et al. [80] have shown that a layer of PEDOT:PSS polymer allows one to evaporate a film of gold on top of the SAMs, preventing Au atom percolation. Rampi et al. [79] covered the organic SAMs by spin-coating a nanometer-thick layer of commercial PPV polymer. These junctions seem to be very stable and provide highly reproducible electrical measurements [80].

Recently, Melosh has obtained electrically stable LAJs as large as 9 mm^2 by atomic deposition of a nanometer-thick passivating layer of aluminium oxide on top of self-assembled organic monolayers with hydrophilic terminal groups [158, 159]. Obviously, interlayers based junctions limit electrical measurements only to organic SAMs less conductive than the protecting layer.

3.1.4 Liquid Metal Based Junctions (LMJs)

LAJs based on liquid metal electrodes have been extensively used in different geometries and modes to incorporate and study a large number of organic compounds [76, 85, 88, 106–108, 132, 160–171]. The wide use of Hg-based electrodes relies on the properties of this metal: (1) it is highly conductive, (2) it forms well-ordered SAMs in a few seconds [166], (3) its surface, as a liquid, is free of structural features that cause defects in adsorbed monolayers and (4) it can form a good conformal contact with molecular SAMs organized on metal electrodes (Fig. 5a, b).

The fabrication of these junctions is straightforward: in all cases, SAMs(1) are formed separately on the Hg drop extruded from a syringe, and SAMs(2) are formed on the solid metal surface.

The two metal surfaces covered by SAMs are brought into contact by the use of a micro-manipulator in the presence of a liquid medium, such as hexadecane: the presence of hexadecane transforms the defects of the SAMs into insulating sites. The use of a semitransparent solid surface (Au or Ag) allows (1) evaluation of the contact area by collecting the image of the contact area by a mirror and (2) electrical measurements under irradiation of the SAMs through the Au surface. The disadvantage of Hg-based junctions is related to the environmental unfriendly characteristics of Hg, which prevent any application. For this main reason, these junctions are valuable only as versatile test-beds for organic electronics.

Recently, Whitesides et al. [88, 164, 165] have replaced Hg with an In/Ga eutectic alloy (E-GaIn) (Fig. 5c). In/Ga alloy-based electrodes present few advantages related to (1) the lower affinity for the bottom Au or Ag electrode, so that the junction can be assembled in air, (2) low toxicity and (3) good processability and mouldability. These characteristics indicate E-GaIn electrodes as possible candidates for incorporation into functional devices. Some disadvantages are related to the surface of the In/Ga alloy: (1) unlike Hg, it is not atomically flat and (2) it forms in few minutes a discontinuous layer of oxide [81].

4 Results

4.1 Non-Active LAJs

4.1.1 Correlation Between Electrical Properties and Chemical Structure: The β Factor

In 1971 the results obtained by Kuhn [6], by incorporating LB films of fatty acids in LAJs based on Hg electrodes, showed a decrease in the measured current with increasing length of the fatty acid chains. In 1998 Rampi et al. assembled and characterized a junction based on two Hg drop electrodes (Fig. 5a) and opened the door to the use of a variety of different Hg-based LAJs [160–171]. Majda et al., by using this two-Hg-drop junction, extracted a β parameter for alkane chains of 0.89 $Å^{-1}$ [76]. By replacing of one of the Hg electrodes with a solid Au or Ag surface (Fig. 5b) it was possible to characterize the organization of the SAMs incorporated into the junction [167]. By measuring the current flowing through SAMs formed by alkanethiols, oligophenylene thiols and benzylic derivatives of oligophenylene thiols of different lengths (Fig. 6), the respective decay factors β of 0.87, 0.61 and 0.66 $Å^{-1}$ [77, 167, 168] have been calculated.

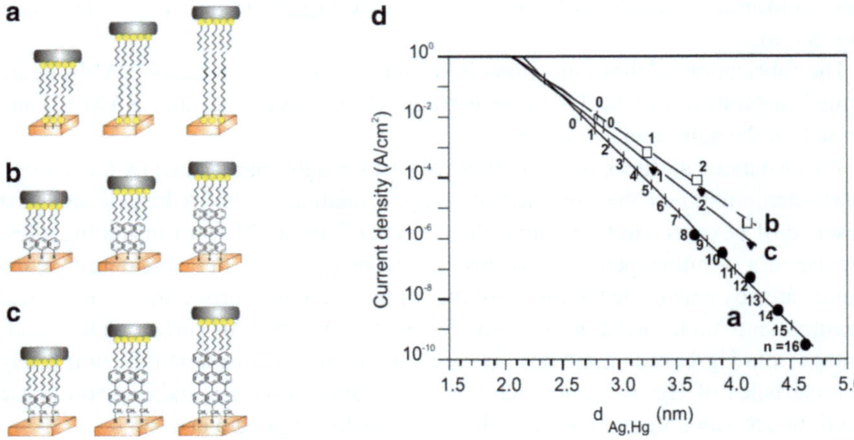

Fig. 6 Schematic representation of an Hg-drop LAJ incorporating SAMs of organic molecules of (**a**) alkanethiols, (**b**) oligophenylene thiols and (**c**) benzylic derivatives of oligophenylene thiols of different length formed on an Ag electrode. (**d**) Semi-logarithmic plot of measured current at applied bias $V = 0.5$ V vs electrode gap flowing through the **a**, **b**, **c** interfaces

Significantly, these values of β are in good agreement both with those extracted from measurements performed on the same compounds with c-AFM [83], and with those measured by transient spectroscopy on D-B-A supramolecular systems incorporating similar bridges, Bs (see Table 1). McCreery has used junctions based on Hg-drop electrodes to measure electrical properties of SAMs formed by monolayers of phenyl homologues covalently bonded to flat, graphitic carbon [169]. Cahen has extensively used Hg-based junctions to study charge transport through SAMs assembled on surfaces of semiconductors of different nature [64, 161, 162, 170]. In recent work, Whitesides et al. [164], using LAJ based on E-GaIn, reported for alkane chains a value of $\beta = 0.43$ Å^{-1} significantly lower than that reported in the literature [76, 77, 83, 84, 168].

LAJs incorporating polymer films on top of the organic SAMs showed that these junctions allow for studying the correlation between electrical properties and chemical structure. Both de Boer et al. [80] (Fig. 7) and Rampi et al. [79], by spin-coating respectively PEDOT:PSS and PPV on top of alkane SAMs, extracted from the electrical measurement $\beta = 0.57$ Å^{-1} and $\beta = 0.90$ Å^{-1}. Rampi et al. also showed that it is possible to measure electrical properties of polyphenyl chains with a $\beta = 0.61$ Å^{-1}, and that the PPV polymer layer is much more conductive that polyphenyl-based SAMs [79].

LAJs fabricated based using the PALO technique provided for alkane chains a β factor of 0.86 Å^{-1} (Fig. 8), a value in good agreement with literature [78].

By using Hg-electrode-based junctions, the electrical properties of the highly conjugated units of hexa-*peri*-hexabenzocoronene (HBC) have been characterized [171]. The HBC unit was anchored to the Au metal surface in an Hg-based junction by a decanonanethiol tether (C19), as depicted in Fig. 9. Comparison of the electrical behaviour of three different junctions, whose interfaces are schematized

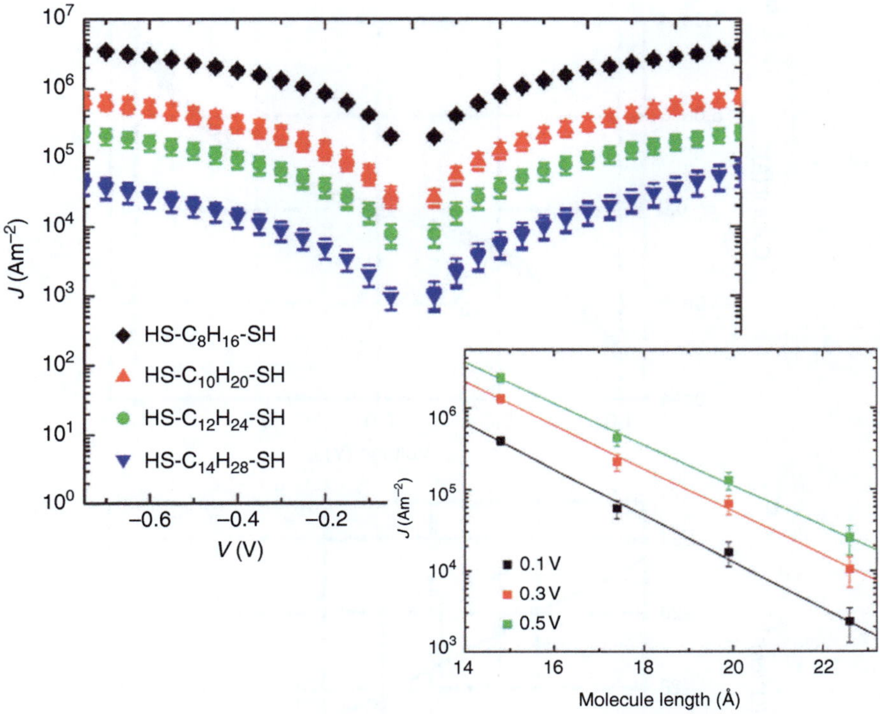

Fig. 7 Current densities *J* vs applied bias measured in a polymer-interlayer based LAJs for 1,8-octanedithiol, 1,10-decanedithiol, 1,12-dodecanedithiol and 1,14-tetradecanedithiol. *Inset*: lnJ as a function of the molecular length measured at 0.1 V, 0.3 V and 0.5 V bias. (Reprinted with permission from [80])

in Fig. 9a–c, indicates that the aliphatic chains are mostly responsible for the total barrier to electron transport of the monolayer, whereas the HBC cores are comparably electron-"transparent".

4.1.2 Electron Transfer Through Different Molecular Interactions

The mechanical mode used to assemble the Hg-drop-based junctions (see Sect. 3.1.4), has been used to measure and compare rates of electron flow through different molecular interactions. By bringing in contact the two Hg electrodes functionalized with SAMs carrying different functional groups, such as $-CH_3$, $-COOH$, and $-NH_2$ (see Fig. 10), it was possible to measure and compare the current flowing through (1) van der Waals interactions, (2) hydrogen bonds and (3) covalent bonds. By forming on the Au surface a SAM terminated with anhydride groups and a SAM terminated in amine groups on the Hg drop, it was also possible to bridge the two electrodes by a covalent amide C(O)–NH bond. Figure 10b shows the large difference in currents flowing through these

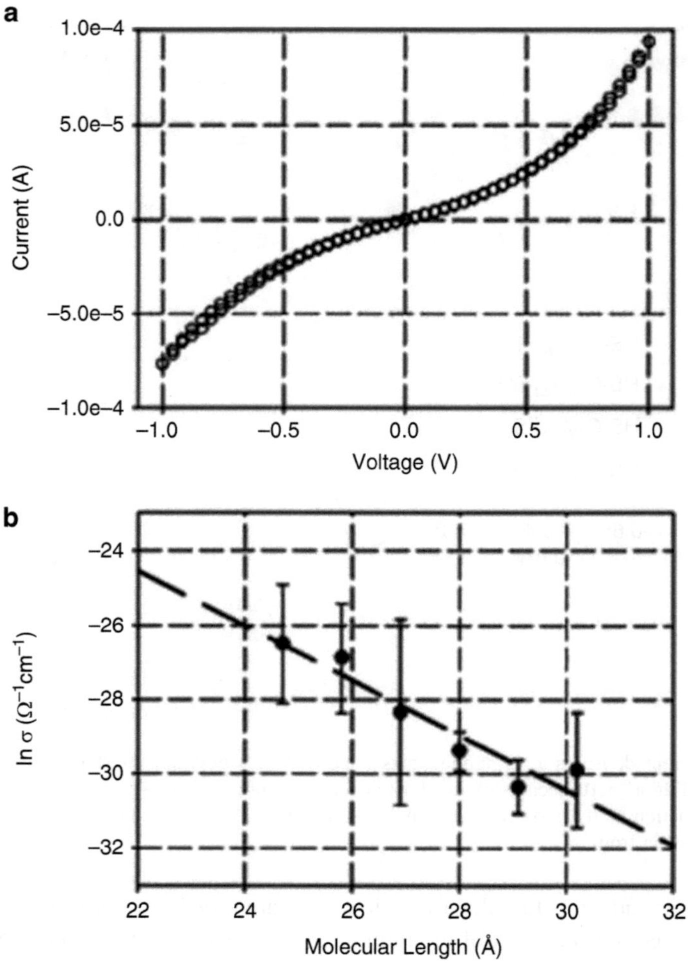

Fig. 8 (**a**) The *I–V* curve measured for a junction consisting of an LB monolayer of margaric acid formed on an Al bottom electrode and a top Au electrode fabricated via the PALO technique. (**b**) Semilogarithmic plot of current density vs the chain-length of fatty-acid monolayers. The fit to a straight line gives a slope of $\beta = 0.86\ \text{Å}^{-1}$. (Reprinted with permission from [78])

interactions. These results agree with the limited experimental and theoretical data obtained in D-B-A systems [172, 173].

4.1.3 Molecular Wires Based on Metal Centres (MCMWs)

Highly conductive and extra-long MWs are systems interesting both for gaining new insight into CT processes and for providing the possibility to bridge large-gap

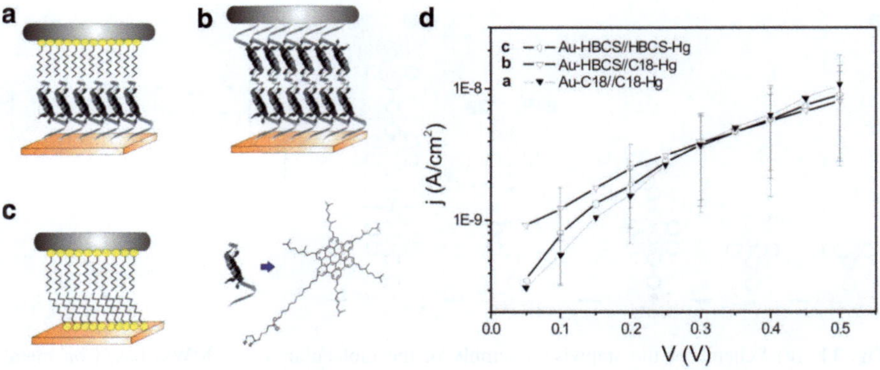

Fig. 9 Scheme of different Hg-drop LAJs incorporating: (**a**) SAMs of octadecanethiols (C18) on both electrodes; (**b**) SAMs of C18 on the Hg electrode and of HBCS on an Au electrode; (**c**) SAMs of HBCS on both electrodes; (**d**) *i–V* curves measured for the three a, b, c junctions

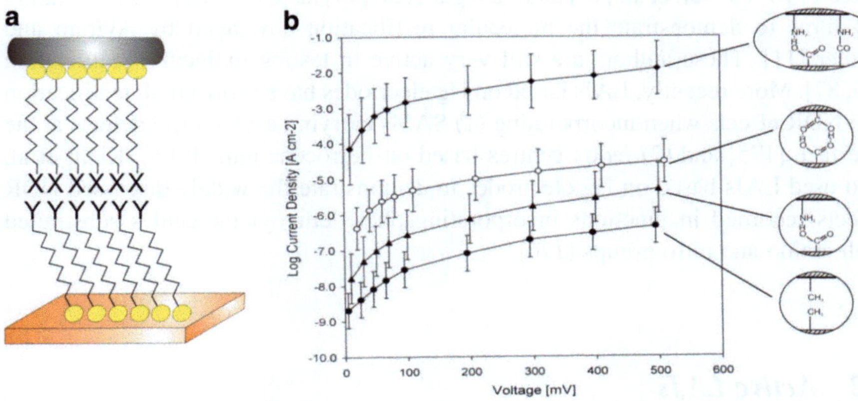

Fig. 10 (**a**) Scheme of the interfaces of Hg-drop junctions incorporating SAM with different X and Y terminal groups. (**b**) *i–V* curves for the junctions incorporating the schematized interactions

electrodes. MWs of different length, formed by chains of redox centres based on Fe (II) and Co(II) ions, were assembled as schematized in Fig. 11a, and have been incorporated in Hg-based LAJs [85]. The measured current values, plotted according to the equation $I = I_0 e^{-\beta d}$, yields β values which are extremely low, with respect to those obtained for other organic molecular wires, $\beta = 0.028$ and $\beta = 0.001$ Å^{-1} respectively for Fe(II)- and Co(II)-based MWs. While a tunnelling mechanism is inadequate for modelling the charge transfer process occurring across these MWs [174], a multistep electron/hole hopping mechanism between redox sites is suggested to be operative [85].

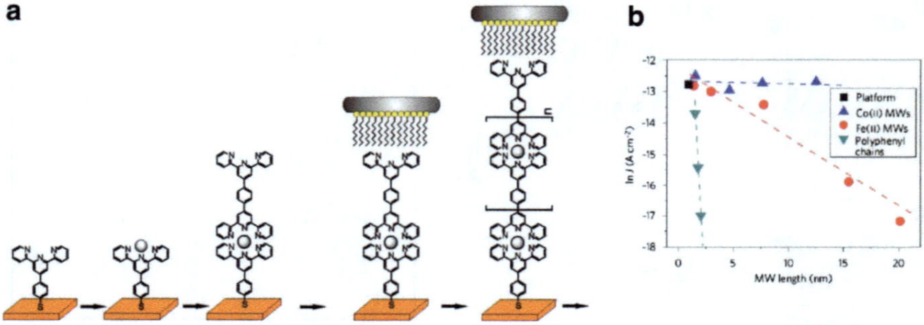

Fig. 11 (**a**) Scheme of the stepwise assembly of the molecular wires (MWs) based on metal centres incorporated into an Hg-drop junction. (**b**) lnJ vs length of the MWs containing Fe(II) (*circles*) or Co(II) (*blue triangles*) (at 0.5 V). For comparison the plot lnJ vs length of polyphenyl chains (*green triangles*) [85] is reported

4.1.4 Rectification and NDR Effects

Since 1990 Ashwell et al. [12] and Metzger et al. [14] have used LAJs incorporating LB films to demonstrate the molecular rectification envisaged by Aviram and Ratner [11]. These authors are still very active in testing molecular rectification [86, 87]. More recently, LAJs based on Hg electrodes have shown both rectification and NDR effects when incorporating (1) SAMs carrying asymmetric defects at the interface [175] and (2) redox centres based on Ferrocene units [165]. Kiehl et al. also used LAJs based on Hg electrodes to demonstrate the widely discussed NDR effects recorded in junctions incorporating phenylethynyl molecules substituted with amino and nitro groups [176].

4.2 *Active LAJs*

4.2.1 Photoactive LAJs

LAJs formed at least by one electrode consisting of a semitransparent Au surface allow for irradiation "in situ" of the incorporated molecular systems.

LAJs Incorporating Diarylethene SAMs

De Boer et al. [99], by using LAJs schematized in Fig. 1 incorporating SAMs of DAE, have shown that (1) the current flowing through SAMs of the open (O) and the closed (C) forms differ by one order of magnitude and (2) the currents measured under irradiation "in situ" differ of a factor of about seven (Fig. 12). In this junction

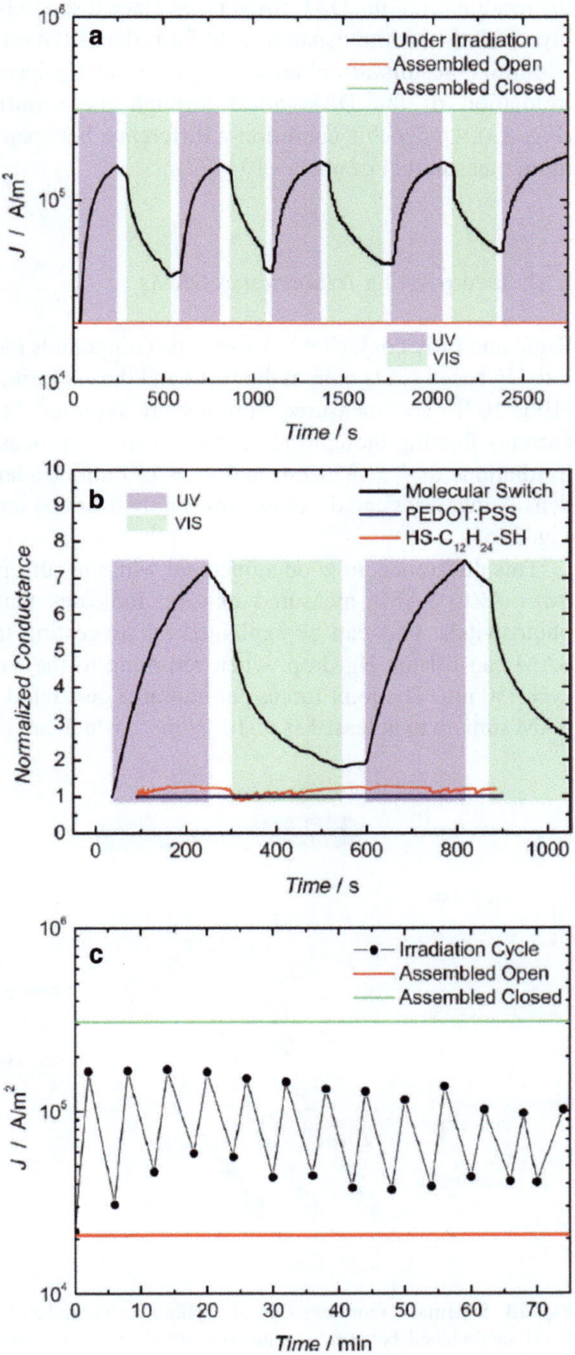

Fig. 12 In situ optical switching of the current flowing through a DAE monolayer assembled in a PEDOT:PSS interlayer-based LAJ. (**a**) Comparison of the current densities flowing through "as assembled" open and closed isomer and upon in-situ photoisomerization. (**b**) Comparison of the normalized conductance for DAE switches with in situ measurements for a non-switchable 1,12-dodecanedithiol and for PEDOT:PSS only. (**c**) Current density (at 0.5 V) vs time for the in situ optical switching of a monolayer of diarylethenes. Alternating UV and visible illumination yields a temporal modulation of the current flowing through the junctions. (Reprinted with permission from [99])

the irradiation of the DAE SAM takes place through both the PEDOT:PSS polymer layer and the semitransparent gold films deposited on top of the polymer layer.

Similar results are obtained by using an Hg-based junction under alternating irradiation of the DAE SAM through the semitransparent bottom electrode (Fig. 13). Under this condition a difference between the minimum and the maximum measured currents is ~10 [177].

LAJs Incorporating Azobenzenes SAMs

Rigid and fully conjugated azobenzene compounds (AZOs) have been incorporated into Hg based LAJs both in the *trans* and the *cis* form, as schematized in Fig. 14a, b [106, 107]. The measured currents are reported in Fig. 14c. Significantly, the currents flowing though the junction were also measured under in situ alternated irradiations at $\lambda = 370$ nm and $\lambda = 450$ nm. Under these condition the current density increases and decreases reversibly over one order of magnitude, as shown in Fig. 14d.

This difference, in good agreement with the current flowing through the *cis* and *trans* AZO SAMs measured ex situ, indicates that the junction behaves as a photoswitch. This can be explained by suggesting that the *cis* form of the AZO SAM can lift the Hg drop, when returning to the *trans* form under irradiation at $\lambda = 450$ nm. The joint forces per unit area generated by the molecules in the AZO SAM sum up to at least 1.0×10^5 N m^{-2}, which are large enough to lift or displace

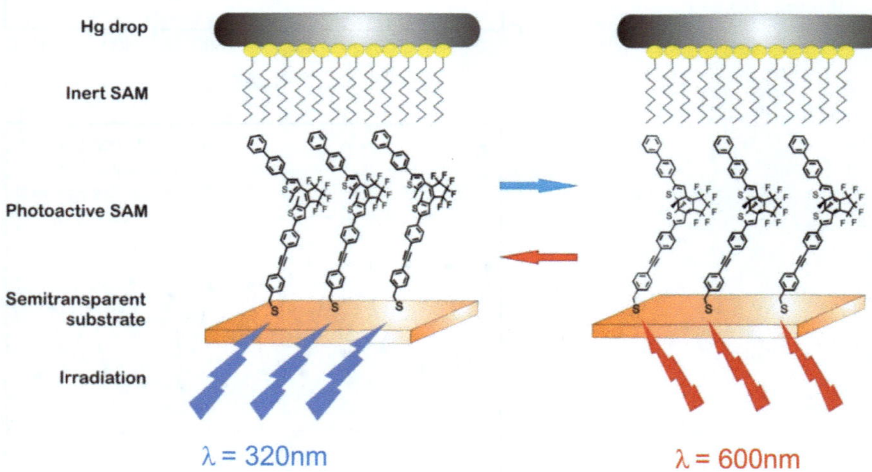

Fig. 13 Schematic representation of a photo-active molecular junction containing DAE-based SAM sandwiched between a semitransparent Au bottom electrode and an SAM/Hg-drop top electrode. The electrical response of the junction is measured upon irradiation at different wavelengths through the semitransparent bottom electrode

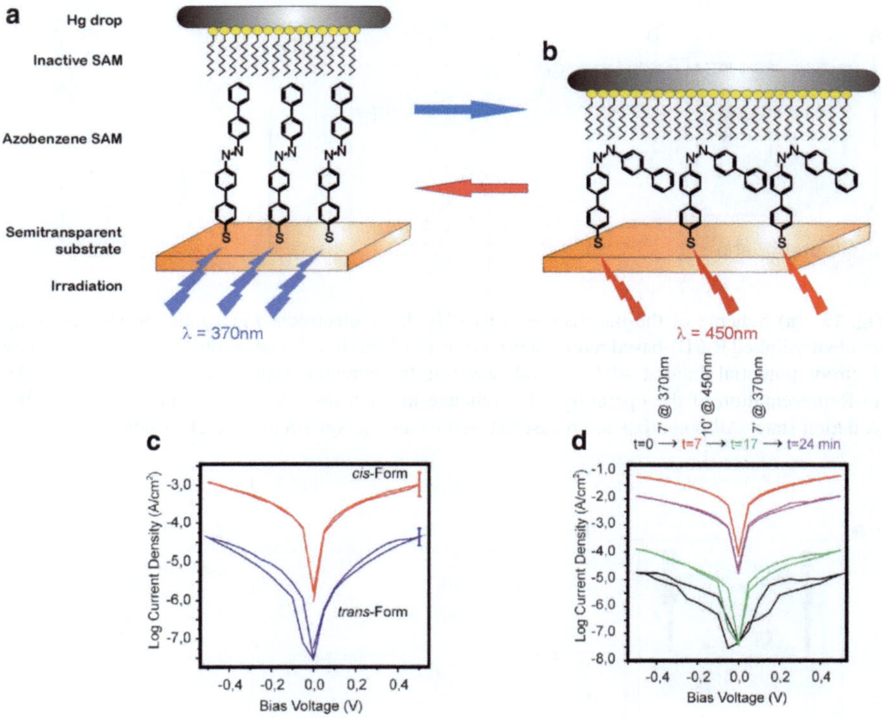

Fig. 14 Schemes of the Hg-drop LAJ incorporating AZO-based SAM: (**a**) in the *trans* form and (**b**) in the *cis* form. (**c**) *I*–*V* curves for junctions assembled by incorporating either the trans or the *cis* isomers. (**d**) *I*–*V* curves measured under "in-situ" photoisomerization by alternating irradiation at $\lambda = 370$ nm and $\lambda = 450$ nm through the semitransparent gold electrode

the Hg drop. Under the cooperative forces per unit area generated by the AZO SAM triggered by light, the junction behaves both as an electrical switch and as a "cargo lifter triggered by light" [107].

4.2.2 Electrochemical LAJs

Figure 5a schematizes LAJs based on two Hg electrodes. By bringing in contact the two Hg-drops inside an electrolyte solution and connecting them to a macroscopic reference electrode by a potentiostat (Figs. 15a and 16a), an electrochemical junction is created: this junction allows for independent control of the potentials applied to the two Hg electrodes, so that the cathode can act as electron donor (source) and the anode as electron acceptor (drain) with respect to the redox centre.

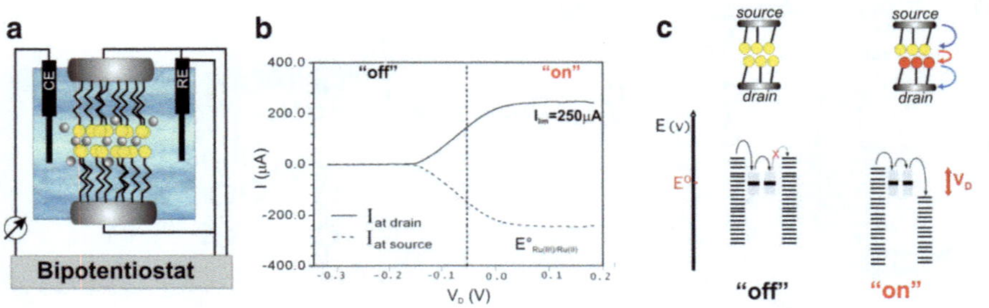

Fig. 15 (**a**) Scheme of the interface of a two Hg-drops electrochemical junction incorporating covalently linked Ru(II)-based redox sites (*yellow circles*). (**b**) *I–V* curves obtained by keeping one electrode potential fixed at −0.02 V and sweeping the potential applied at the second electrode. (**c**) Representation of the operating self-exchange mechanism: *red circles* represent the Ru(III) oxidation state. All potentials are measured against an Ag/AgCl reference electrode

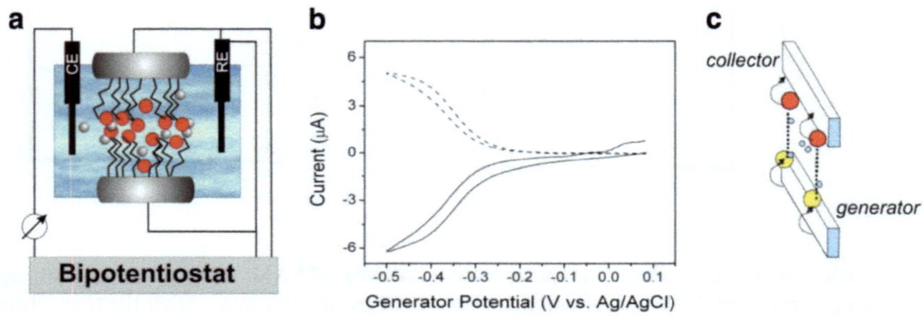

Fig. 16 (**a**) Scheme of the interface of a two Hg-drops electrochemical junction where $Ru(NH_3)_6^{3+}$ complexes (*red circles*) are electrostatically trapped in between the electrodes. (**b**) *I–V* curves obtained by keeping constant the potential of the collector and sweeping the potential of the generator (potentials are measured against an Ag/AgCl). (**c**) Scheme of the operating redox cycling mechanism

Redox Centres Covalently Linked to the Electrodes

Redox centres can be incorporated inside the junction by forming at each Hg electrode a SAM of HS-C10-Ru ($E^{o\prime}_{SAM} = -0.01$ V vs Ag/AgCl), as schematically shown in Fig. 15a [132, 184, 185]. The current flow is measured by fixing the potential of the source (V_S) at −0.20 V (the attached ruthenium is in its +2 oxidation state, yellow circles in Fig. 15a, c) and by sweeping the potential of the drain (V_D) across a half-wave potential of −0.04 V (that is near the formal potential $E^{o\prime}$ of the $Ru^{II/III}$ couple). The results reported in Fig. 15b show that the charge transport through the junction occurs as a result of (1) oxidation of Ru^{II} to Ru^{III} (red circles) at the drain, (2) electron exchange between Ru^{III} and Ru^{II} and (3) reduction of Ru^{III} at acceptor back to Ru^{II} ; these key steps are schematized in Fig. 15c. The electrochemical junction behaves as a switch (or a diode), where the current flow is switched from "off" to "on" by an external signal controlled by the reference electrode.

Redox Centres non-Covalently Linked to the Elelctrodes

Redox centres can be incorporated in a junction by electrostatic interactions. Figure 16 a represents the interface of such an electrochemical junction. The junction is assembled by immersing two Hg electrodes functionalised by $HSC_{10}COOH$ SAMs in an electrolyte solution at pH 9 containing $Ru(NH_3)_6^{3+}$. The positively charged Ru complexes remain trapped in the gap, as shown in Figure 16a, when the two electrodes are brought in contact by the micro-manipulator [186]. Figure 16b shows that the current flows through the junction only when the potential of the electrodes is controlled in such a way that one can reduce, and the other can oxidize the ruthenium species. In the present case, the current flowing at the electrodes is generated by the $Ru^{(III)}$ species diffusing between the electrodes in the nanometer wide gap and being alternatively oxidized and reduced by a redox cycling mechanism, as schematized in Fig. 16c [186–189].

5 Conclusions

The exponentially growing number of studies in the field of molecular electronics has provided encouraging results. Three major points emerge from the results presented in this chapter.

First, the comparison between the results provided by the kinetic and the electrical approach to study charge trasfer/transport precesses is very informative. In particular, we observe that electrical measurements performed by M-B-M junctions of different design (BJs, STMJs, c-AFMJs. LAJs) incorporating the same B, provide different conductance values but very often similar values of the β factor (see Table 1). In fact, while the measured conductivity incorporates a number of experimental parameters (geometry, electrical contacts, molecular environment) which are related to the type of the molecular junction, the β decay factor, calculated by scaling with length of the intrinsic conductivity of the MWs, is independent of these parameters. Consequently, the discrepancy among the conductance values represents a major source of information about the role of the electrical contacts and the effects of molecular environment on the electrical behaviour of a molecular device. From the data reported in Table 1 we can also observe that the β parameters extracted from electrical measurements performed in M–B–M junctions and from kinetic studies in D–B–A systems, by using respectively the relations $I = I_o e^{-\beta d}$ and $k_e = k_o e^{-\beta d}$, are very similar.

Each of these two experimental approaches are charaterized by respective advantages and disadvantages: (1) in D–B–A systems, the kinetic approach measures the average rate of the one-electron transfer occurring in each D–B–A supramolecular unit isolated in solution, while in M–B–M junctions the electrical approach measures the flux of many electrons reaching the electrodes; (2) while the errors affecting the ET rate measurements in D–B–A systems are very small, those affecting the electrical measurements in M–B–M junctions can be quite large (about one order of magnitude); (3) the calculation of the β value requires measurements on D–B–A or in M–B–M systems where the length of the molecular bridge, B is varied; it is clear that the kinetic approach is based on a serious effort in synthesizing a series of D–B–A systems with B of increasing length – in contrast, in M–B–M junctions Bs of different lengths are directly incorporated between the electrodes; (4) in M–B–M junctions, by tuning the applied bias, it is very easy to change the value of the barrier of the charge transfer process – in fact, several examples have shown that by tuning the applied bias, a change in the mechanism of the charge-transport process from tunnelling to resonant tunnelling/hopping can be induced; (5) in M–B–M junctions, by tuning the potential applied at the electrodes, one can explore the energy levels of the molecules. This possibility, (1) provides the condition for a new spectroscopy, and (2) produces interesting electronic effects as rectification and NDR.

Second, in designing new molecule-based electronic devices, one of the major goals is the precise control of the current flowing between the terminals. Electrochemical molecular junctions allow for control of the potentials of the electrodes with respect to the redox potential of incorporated redox-active molecules with well-defined, accessible, tunable energy states. These junctions represent unique systems able to predict precisely at which applied potential the current flow will take off. Even though the presence of a liquid electrolyte represents a detriment towards possible applications, they provide the concepts for designing molecular devices that mimic electronic functions and control electrical responses.

Three, even if molecular electronics remains still far from application, a number of attempts are underway for fabricating molecular devices. In contrast to single-molecule-based devices, LAJs are more suitable systems for applications in organic electronics. Nonetheless, the fabrication of a large part of mechanically and electrically stable LAJs requires nanometer-thick interlayers of a different nature: the presence of this layer certainly can represent an electron blocking barrier masking the results when measuring the electrical properties of highly conductive organic molecules.

The large number of results obtained by LAJs based on liquid metal electrodes (Hg, E-GaIn) indicates that these systems present a number of advantages: (1) they are very inexpensive and easy to assemble and to use; (2) they are highly versatile by supporting a range of organic structures on different substrates (Au, Ag, Pd, SiO_2, graphite); (3) by using semitransparent gold surfaces as electrodes, the junctions allow for electrical measurements under irradiation of the SAM trapped between the electrodes; (4) when used in the electrochemical mode they display diode-like characteristics analogous to those of solid-state devices. The Hg based junctions also present some disadvantages: (1) they are not compatible with measurements over a wide range of temperature and (2) they cannot be developed into practically useful microelectronic components. Nevertheless, these junctions can be considered as "horse-power test-beds" that can yield the electrical characteristics of a large number of molecular systems useful in future organic electronic devices. New sophisticated techniques have been suggested recently towards fabrication of convenient LAJs, which combine mechanical stability with a high level of electrical performance and considerable promise toward applications.

Even if many fundamental questions remain still open in molecular electronics, the answers to these issues are today the central topics in this field, and much of the driving force for understanding and developing molecular junctions derives from the original promise of possible applications in different fields, such as microelectronics, chemical and biological sensing, and in the realm of photonics and the photonic/electronic interface.

References

1. Heath JR, Ratner MA (2003) Phys Today Mol Electron 56:43–49
2. Reimers JR, Picconnatto CA, Ellenbogen JC, Shashidhar R (2003) Molecular electronics III. Ann New York Acad Sci 1006:1–330
3. Tour JM (2003) Molecular electronics: commercial insights, chemistry, devices, architecture and programming. World Scientific, Singapore
4. Brederode ME, Jones MR, Van Grondelle R (1997) Primary electron transfer kinetics in membrane-bound Rhodobacter sphaeroides reaction centers: a global and target analysis. Biochemistry 36:6855–6861
5. De Cola L (2005) (ed) Molecular wires: from design to properties. In: Topics in Current Chemistry 257. Springer, Heidelberg
6. Mann B, Kuhn H (1971) Tunneling through fatty acid salt monolayers. J Appl Phys 42:4398–4406
7. Kuhn H (1989) Present status and future prospects of Langmuir-Blodgett film research. Thin Solid Films 178:1–16
8. Lambe J, Jaklevic RC (1968) Molecular vibration spectra by inelastic electron tunnel. Phys Rev 165:821–832
9. Tredgold RH, Winter CS (1981) Tunnelling currents in Langmuir-Blodgett monolayers of stearic acid. J Phys D Appl Phys 14:L185
10. Tredgold RH (1987) The physics of Langmuir-Blodgett films. Rep Prog Phys 50:1609
11. Aviram A, Ratner MA (1974) Molecular rectifiers. Chem Phys Lett 29:277–283
12. Ashwell GJ, Sambles JR, Martin AS, Parker WG, Szablewski MJ (1990) Rectifying characteristics of Mg|($C_{16}H_{33}$-Q3CNQ LB film)|Pt structures. J Chem Soc Chem Commun 19:1374–1376
13. Martin AS, Sambles JR, Ashwell GJ (1993) Molecular rectifier. Phys Rev Lett 70:218–221
14. Metzger RM, Chen B, Höpfner U, Lakshmikantham MV, Vuillaume D, Kawai T, Wu X, Tachibana H, Hughes TV, Sakurai H, Baldwin JW, Hosch C, Cava MP, Brehmer L, Ashwell GJ (1997) Unimolecular electrical rectification in hexadecylquinolinium tricyanoquinodimethanide. J Am Chem Soc 119:10455–10466
15. Cui Z (2008) Nanofabrication: principles, capabilities and limits. Springer, Heidelberg
16. Bowker M, Dacie PR (eds) (2010) Scanning tunneling microscopy in surface science and catalysis. Wiley-VCH, Weinheim
17. Simeone FC, Albonetti C, Cavallini M (2009) Progress in micro- and nanopatterning via electrochemical lithography. J Phys Chem C 113:18987–18994

18. Heath JR (2009) Analytical chemistry in molecular electronics. Annu Rev Mater Res 39:1–23
19. McCreery RL, Bergren AJ (2009) Progress with molecular electronic junctions: meeting experimental challenges in design and fabrication. Adv Mater 21:4303–4322
20. Ulgut B, Abruña HD (2008) Electron transfer through molecules and assemblies at electrode surfaces. Chem Rev 108:2721–2739
21. Haick H, Cahen D (2008) Contacting organic molecules by soft methods: towards molecule-based electronic devices. Acc Chem Res 41:359–366
22. Salomon A, Cahen D, Kahn A (2007) What is the barrier for tunneling through alkyl monolayers? Results from n- and p-Si-C- alkyl/Hg junctions. Adv Mater 19:445–460
23. Beckman R, Boukai A, Bunimovich Y, Choi JW, DeIonno E, Green J, Johnston-Halperin E, Luo Y, Sheriff B, Stoddart JF, Heath JR (2006) Molecular mechanics and molecular electronics. Faraday Discuss 131:9–22
24. Chen F, Hihath J, Huang ZF, Li XL, Tao NJ (2007) Measurement of single-molecule conductance. Annu Rev Phys Chem 58:535–564
25. Zhang JD, Kuznetsov AM, Medvedev IG, Chi Q, Albrecht T, Jensen PS, Ulstrup J (2008) Single-molecule electron transfer in electrochemical environments. Chem Rev 108:2737–2791
26. Reed MA, Zhou C, Muller CJ, Burgin TP, Tour JM (1997) Conductance of a molecular junction. Science 278:252–254
27. Lörtscher E, Elbing M, Tschudy M, von Hänisch C, Weber HB, Mayor M, Riel H (1997) Charge transport through molecular rods with reduced π-conjugation. ChemPhysChem 9:2252–2258
28. Zhou C, Deshpande MR, Reed MA, Jones L, Tour JM (1997) Nanoscale metal/self-assembled monolayer/metal heterostructures. Appl Phys Lett 71:611–613
29. Xu B, Tao JN (2003) Measurement of single-molecule resistance by repeated formation of molecular junctions. Science 301:1221–1223
30. He J, Sankey O, Lee M, Tao N, Li X, Lindsay S (2006) Measuring single molecule conductance with break junctions. Faraday Discuss 131:145–154
31. Park YS, Widawsky JR, Kamenetska M, Steigerwald ML, Hybertsen MS, Nuckolls C, Venkataraman L (2009) Frustrated rotations in single-molecule junctions. J Am Chem Soc 131:10820–10821
32. Choi SH, Kim B, Frisbie CD (2008) Electrical resistance of long conjugated molecular wires. Science 320:1482–1486
33. Troisi A, Beebe JM, Picraux LB, van Zee RD, Stewart DR, Ratner MA, Kushmerick JG (2007) Tracing electronic pathways in molecules by using inelastic tunneling spectroscopy. Proc Natl Acad Sci USA 104:14255–14259
34. Elbing M, Blaszczyk A, von H€anisch C, Mayor M, Ferri V, Grave C, Rampi MA, Pace G, Samorı̀ P, Shaporenko A, Zharnikov M (2008) Single component self-assembled monolayers of aromatic azo-biphenyl: influence of the packing tightness on the SAM structure and light induced molecular movements. Adv Funct Mater 19:2972–2983
35. Bonifas AP, McCreery RL (2010) 'Soft' Au, Pt and Cu contacts for molecular junctions through surface-diffusion-mediated deposition. Nat Nanotechnol 5:612–6172
36. Paddon-Row MN (2003) Superexchange-mediated charge separation and charge recombination in covalently linked donor-bridge-acceptor systems. Aust J Chem 56:729–748
37. Davis WB, Svec WA, Ratner MA, Wasielewski MR (1998) Molecular wire behaviour in p-phenylenevinylene oligomers. Nature 396:60–63
38. Newton MD (1991) Quantum chemical probes of electron-transfer kinetics: the nature of donor-acceptor interactions. Chem Rev 91:767–792
39. Segal D, Nitzan A, Ratner MA, Davis WB (2000) Activated conduction in microscopic molecular junctions. J Phys Chem B 104:2790–2793
40. Simmons JG (1963) Generalized formula for the electrical tunnel effect between similar electrodes separated by a thin insulating film. J Appl Phys 34:1793–1803
41. Simmons JG (1964) Potential barriers and emission limited current flow between closely spaced parallel metal electrodes. J Appl Phys 35:2655–2658

42. Wang WY, Lee T, Reed MA (2005) Electron tunnelling in self-assembled monolayers. Rep Prog Phys 68:523–544
43. McConnell HM (1961) Intramolecular charge transfer in aromatic free radicals. J Chem Phys 35:508–515
44. Cuniberti G, Fagas G, Richter K (2005) (eds) Introducing molecular electronics. Lecture notes in physics. Springer, Heidelberg
45. Nitzan A (2001) Electron transmission through molecules and molecular interfaces. Annu Rev Phys Chem 52:681–750
46. Eng MP, Albinsson B (2006) Non-exponential distance dependence of bridge-mediated electronic coupling. Angew Chem Int Ed 45:5626–5629
47. Beebe JM, Kim B, Gadzuk JW, Frisbie CD, Kushmerick JG (2006) Transition from direct tunneling to field emission in metal-molecule-metal junctions. Phys Rev Lett 97:026801–026804
48. Beebe JM, Frisbie CD, Kushmerick JG (2008) Measuring relative barrier heights in molecular electronic junctions with transition voltage spectroscopy. ACS Nano 2:827–832
49. Huisman EH, Guedon CM, van Wees BJ, van der Molen SJ (2009) Interpretation of transition voltage spectroscopy. Nano Lett 11:3909–3913
50. Simeone FC, Kolb DM, Venkatachalam S, Timo J (2007) The Au(111)/electrolyte interface: a tunnel-spectroscopic and DFT investigation. Angew Chem Int Ed 46:8903–8906
51. Simeone FC, Kolb DM, Venkatachalam S, Timo J (2008) Tunneling behavior of electrified interfaces. Surf Sci 602:1401–1407
52. Berlin YA, Ratner MA (2005) Intra-molecular electron transfer and electric conductance via sequential hopping: unified theoretical description. Radiat Phys Chem 74:124–131
53. Andres RP, Bein T, Dorogi M, Feng S, Henderson JI, Kubiak JP, Mahony W, Osifchin RG, Reifenberger R (1996) "Coulomb Staircase" at room temperature in a self-assembled molecular nanostructure. Science 272:1323–1325
54. Akkerman HB, de Boer B (2008) Electrical conduction through single molecules and self-assembled monolayers. J Phys Condens Matter 20:013001
55. Salomon A, Cahen D, Lindsay S, Tomfohr J, Engelkes VB, Frisbie CD (2003) Comparison of electronic transport measurements on organic molecules. Adv Mater 15:1881–1890
56. Lindsay S, Ratner MA (2007) Molecular transport junctions: clearing mists. Adv Mater 19:23–31
57. Chen F, Li X, Hihath J, Huang Z, Tao N (2006) Effect of anchoring groups on single-molecule conductance: comparative study of thiol-, amine-, and carboxylic-acid-terminated molecules. J Am Chem Soc 128:15874–15881
58. Kim B, Beebe JM, Jun Y, Zhu XY, Frisbie CD (2006) Correlation between HOMO alignment and contact resistance in molecular junctions: aromatic thiols versus aromatic isocyanides. J Am Chem Soc 128:4970–4971
59. Yaliraki SN, Kemp M, Ratner MA (1999) Conductance of molecular wires: influence of molecule-electrode binding. J Am Chem Soc 121:3428–3434
60. Hipps KW (2001) It's all about contacts. Science 294:536–537
61. von Wrochem F, Gao D, Scholz F, Nothofer HG, Nelles G, Wessels JM (2010) Efficient electronic coupling and improved stability with dithiocarbamate-based molecular junctions. Nat Nanotechnol 5:618–624
62. Park YS, Whalley AC, Kamenetska M, Steigerwald ML, Hybertsen MS, Nuckolls C, Venkataraman L (2007) Single molecule conductance and link chemistry: a comparison of phosphines, methyl thiols and amines. J Am Chem Soc 129:15768–15769
63. Haick H, Cahen D (2008) Making contact: connecting molecules electrically to the macroscopic world. Prog Surf Sci 83:217–261
64. Nesher G, Shpaisman H, Cahen D (2007) Effect of chemical bond-type on electron transport in GaAs-chemical bond-alkyl/Hg junctions. J Am Chem Soc 129:734–735
65. Salomon A, Boecking T, Gooding JJ, Cahen D (2006) How important is the interfacial chemical bond for electron transport through alkyl chain monolayers. Nano Lett 6:2873–2876

66. Donhauser ZJ, Mantooth BA, Kelly KF, Bumm LA, Monnel JD, Stapleton JJ, Price DW, Rawett AM, Allara DL, Tour JM, Weiss PS (2001) Conductance switching in single molecules through conformational changes. Science 292:2303–2307
67. Cygan MT, Dunbar TD, Arnold JJ, Bumm LA, Shedlock NF, Burgin TP, Jones L II, Allara DL, Tour JM, Weiss PS (1998) Insertion, conductivity, and structures of conjugated organic oligomers in self-assembled alkanethiol monolayers on Au{111}. J Am Chem Soc 120:2721–2732
68. Lindsay SM (2004) Single molecule electronics. Interface 13:26–30
69. Ramachandran GK, Hopson TJ, Rawlett AM, Nagahara LA, Primak A, Lindsay SM (2003) A bond-fluctuation mechanism for stochastic switching in wired molecules. Science 300:1413–1416
70. Landau A, Kronik L, Nitzan A (2008) Cooperative effects in molecular conduction. J Comp Theor Nanosci 5:535–544
71. Selzer Y, Cai L, Cabassi MA, Yao Y, Tour JM, Mayer TS, Allara DL (2005) The effect of local environment on molecular conduction: isolated molecule versus self-assembled monolayer. Nano Lett 5:61–65
72. Paddon-Row MN (2001) Covalently linked systems based on organic components. In: Balzani V (ed) Electron transfer in chemistry, vol 3. Wiley-VCH, Weinheim, pp 179–271
73. Paddon-Row MN (1994) Investigating long-range electron-transfer processes with rigid, covalently linked donor-(norbornylogous bridge)-acceptor systems. Acc Chem Res 27:18–25
74. Scandola F, Argazzi R, Bignozzi CA, Chiorboli C, Indelli MT, Rampi MA (1992) In: Balzani V, De Cola L (eds) Supramolecular chemistry. Kluwer Academic, Dordrecht, The Netherlands, pp 235–248
75. Helms A, Heiler D, McLendon G (1992) Electron transfer in bis-porphyrin donor-acceptor compounds with polyphenylene spacers shows a weak distance dependence. J Am Chem Soc 114:6227–6238
76. Slowinski K, Fong HKY, Majda M (1999) Mercury–mercury tunneling junctions. 1. Electron tunneling across symmetric and asymmetric alkanethiolate bilayers. J Am Chem Soc 103:7257–7261
77. Holmlin RE, Haag R, Chabinyc ML, Ismagilov RF, Cohen AE, Terfort A, Rampi MA, Whitesides GM (2001) Electron transport through thin organic films in metal-insulator-metal junctions based on self-assembled monolayers. J Am Chem Soc 123:5075–5085
78. Shimizu KT, Fabbri JD, Jelincic JJ, Melosh NA (2006) Soft deposition of large-area metal contacts for molecular electronics. Adv Mater 18:1499–1504
79. Milani M, Grave C, Ferri V, Samorì P, Rampi MA (2007) Ultrathin π-conjugated polymer films for simple fabrication of large-area molecular junctions. ChemPhysChem 8:515–518
80. Akkerman HB, Blom PWM, de Leeuw DM, de Boer B (2006) Towards molecular electronics with large-area molecular junctions. Nature 441:69–72
81. Dickey MD, Chiechi RC, Larson RJ, Weiss EA, Weitz DA, Whitesides GM (2008) Eutectic gallium-indium (EGaIn): a liquid metal alloy for the formation of stable structures in microchannels at room temperature. Adv Funct Mater 18:1097–1104
82. Bumm LA, Arnold JJ, Dunbar TD, Allara DL, Weiss PS (1999) Electron transfer through organic molecules. J Phys Chem B 103:8122–8127
83. Wold DV, Haag R, Rampi MA, Frisbie CD (2002) Distance dependence of electron tunneling through self-assembled monolayers measured by conducting probe atomic force microscopy: unsaturated versus saturated molecular junctions. J Phys Chem B 106:2813–2816
84. Wang W, Lee T, Reed MA (2003) Elastic and inelastic electron tunneling in alkane self-assembled monolayers. Phys Rev B 68:035416
85. Tuccitto N, Ferri V, Cavazzini M, Quici S, Zhavnerko G, Licciardello A, Rampi MA (2009) Highly conductive 40-nm long molecular wires assembled by stepwise incorporation of metal centres. Nat Mater 8:41–46
86. Metzger RM (2009) Unimolecular electronics and rectifiers. Synth Met 159:2277–2281

87. Ashwell GJ, Urasinska-Wojcik B, Phillips J (2010) In situ stepwise synthesis of functional multijunction molecular wires on gold electrodes and gold nanoparticles. Angew Chem Int Ed 49:3508–3512
88. Nijhuis CA, Reus WF, Whitesides GM (2009) Organometallic molecular rectification in metal-SAM-metal junctions. J Am Chem Soc 131:17814–17827
89. Metzger RM (2003) Unimolecular electrical rectifiers. Chem Rev 103:3803–3834
90. Lortscher E, Riel H (2010) CHIMIA 64:376–382
91. Chen J, Reed MA, Rawlet AM, Tour JM (1999) Large on-off ratios and negative differential resistance in a molecular electronic device. Science 286:1550–1552
92. Horowitz G (1998) Organic field-effect transistors. Adv Mater 10:365–377
93. Xu B, Xiao X, Yang X, Zang L, Tao N (2005) Large gate current modulation in a room temperature single molecule transistor. J Am Chem Soc 127:2386–2387
94. Park J, Pasupathy AN, Goldsmith JI, Chang C, Yaish Y, Petta JR, Rinkoski M, Sethna JP, Abruña HD, McEuen PL, Ralph DC (2002) Coulomb blockade and the Kondo effect in single-atom transistors. Nature 417:722–725
95. Van der Molen SJ, Liljeroth P (2010) Charge transport through molecular switches. J Phys Condens Matter 22:133001
96. Browne WR, Feringa B (2009) Light switching of molecules on surfaces. Annu Rev Phys Chem 60:407–428
97. Kotiaho A, Lahtinen R, Efimov A, Metsberg HK, Sariola E, Lehtivuori H, Tkachenko NV, Lemmetyinen H (2001) Photoinduced charge and energy transfer in phthalocyanine-functionalized gold nanoparticles. J Phys Chem C 114:162–168
98. Kittredge KW, Fox MA, Whitesell JK (2001) Effect of alkyl chain length on the fluorescence of 9-alkylfluorenyl thiols as self-assembled monolayers on gold. J Phys Chem B 105:10594–10599
99. Kronemeijer AJ, Akkerman HB, Kudernac T, van Wees BJ, Feringa BL, Blom PWM, de Boer B (2008) Reversible conductance switching in molecular devices. Adv Mater 20: 1467–1473
100. Irie M (2000) Diarylethenes for memories and switches. Chem Rev 100:1685–1716
101. Tamai N, Miyasaka H (2000) Ultrafast dynamics of photochromic systems. Chem Rev 100:1875–1890
102. He J, Chen F, Liddel PA, Andreasson J, Straight SD, Gust D, Moore TA, Moore AL, Li J, Sankey OF, Lindsay SM (2005) Switching of a photochromic molecule on gold electrodes: single-molecule measurements. Nanotechnology 16:695–702
103. Dulic D, van der Molen SJ, Kudernac T, Jonkman HT, de Jong JJD, Bowden TN, van Esch J, Feringa BL, van Wees BJ (2003) One-way optoelectronic switching of photochromic molecules on gold. Phys Rev Lett 91:207402
104. Katsonis N, Kudernac T, Walko M, van der Molen SJ, van Wees BJ, Feringa BL (2006) Switching of single diarylethenes on a gold surface. Adv Mater 18:1397–1400
105. Griffiths J (1972) Photochemistry of azobenzene and its derivatives. Chem Soc Rev 1: 481–493
106. Pace G, Ferri V, Grave C, Elbing M, Zharnikov M, Major M, Rampi MA, Samorì P (2007) Cooperative light-induced molecular movements of highly ordered azobenzene self-assembled monolayers. Proc Natl Acad Sci USA 104:9937–9942
107. Ferri V, Elbing M, Pace G, Dickey MD, Zharnikov M, Samorì P, Mayor M, Rampi MA (2008) Light-powered electrical switch based on cargo-lifting azobenzene monolayers. Angew Chem Int Engl Ed 47:3407–3409
108. Mativetsky JM, Pace G, Elbing M, Rampi MA, Mayor M, Samorì P (2008) Azobenzenes as light-controlled molecular electronic switches in nanoscale metal–molecule–metal junctions. J Am Chem Soc 130:9192–9193
109. Mc Creery RL (2009) Electron transport and redox reactions in molecular electronic junctions. Chem Phys Chem 10:2387–2391

110. McCreery RL (2010) Electrochemical concepts in functional materials. Electrochemistry – The Electrochemical Society of Japan 78:103
111. Tao NJ (1996) Probing potential-tuned resonant tunneling through redox molecules with scanning tunneling microscopy. Phys Rev Lett 76:4066–4069
112. Xiao X, Nagahara LA, Rawlett AM, Tao NJ (2005) Electrochemical gate-controlled conductance of single oligo(phenylene ethynylene)s. J Am Chem Soc 127:9235–9240
113. Wong EW, Collier CP, Běhloradský M, Raymo FM, Stoddart JF, Heath JR (2000) Fabrication and transport properties of single-molecule-thick electrochemical junctions. J Am Chem Soc 122:5831–5840
114. Li Z, Pobelov I, Han B, Wandlowski T, Błaszczyk A, Mayor M (2007) Conductance of redox-active single molecular junctions: an electrochemical approach. Nanotechnology 18:044018
115. Gittins DI, Bethell D, Schiffrin DJ, Nichols RJ (2000) A nanometre-scale electronic switch consisting of a metal cluster and redox-addressable groups. Nature 408:67–69
116. Chen F, He J, Nuckolls C, Roberts T, Klare JE, Lindsay S (2005) A molecular switch based on potential-induced changes of oxidation state. Nano Lett 5:503–506
117. He J, Chen F, Lindsay S, Nuckolls C (2007) Length dependence of charge transport in oligoanilines. Appl Phys Lett 90:072112
118. Chen F, Nuckolls C, Lindsay S (2006) In situ measurements of oligoaniline conductance: linking electrochemistry and molecular electronics. Chem Phys 324:236–243
119. Pobelov I, Li Z, Wandlowski T (2008) Electrolyte gating in redox-active tunneling junctions – an electrochemical STM approach. J Am Chem Soc 130:16045–16054
120. Leary E, Higgins SJ, van Zalinge H, Haiss W, Nichols RJ, Nygaard S, Jeppesen JO, Ulstrup J (2008) Structure–property relationships in redox-gated single molecule junctions–a comparison of pyrrolo-tetrathiafulvalene and viologen redox groups. J Am Chem Soc 130:12204–12205
121. Li Z, Liu Y, Mertens SFL, Pobelov IV, Wandlowski T (2010) From redox gating to quantized charging. J Am Chem Soc 132:8187–8193
122. McCreery RL (2009) Electron transport and redox reactions in molecular electronic junctions. ChemPhysChem 10:2387–2391
123. Haiss W, van Zalinge H, Higgins SJ, Bethell D, Hobenreich H, Schiffrin DJ, Nichols RJ (2003) Redox state dependence of single molecule conductivity. J Am Chem Soc 125:15294–15295
124. Haiss W, Albrecht T, van Zalinge H, Higgins SJ, Bethell D, Hobenreich H, Schiffrin DJ, Nichols RJ, Kuznetsov AM, Zhang J, Chi Q, Ulstrup J (2007) Single-molecule conductance of redox molecules in electrochemical scanning tunneling microscopy. J Phys Chem B 111:6703–6712
125. He J, Fu Q, Lindsay S, Ciszek JW, Tour JM (2006) Electrochemical origin of voltage-controlled molecular conductance switching. J Am Chem Soc 128:14828–14835
126. Han W, Durantini EN, Moore TA, Moore AL, Gust DRP, Leatherman G, Seely GR, Tao N, Lindsay S (1997) STM contrast, electron-transfer chemistry, and conduction in molecules. J Phys Chem B 101:10719–10725
127. He H, Zhu J, Tao NJ, Nagahara LA, Amlani I, Tsui R (2001) A conducting polymer nanojunction switch. J Am Chem Soc 123:7730–7731
128. Albrecht T, Guckian A, Ulstrup J, Vos JG (2005) Transistor-like behavior of transition metal complexes. Nano Lett 5:1451–1455
129. Albrecht T, Guckian A, Kuznetsov AM, Vos JG, Ulstrup J (2006) Mechanism of electrochemical charge transport in individual transition metal complexes. J Am Chem Soc 128:17132–17138
130. Albrecht T, Moth-Poulsen K, Christensen JB, Hjelm J, Bjørnholm T, Ulstrup J (2006) Scanning tunneling spectroscopy in an ionic liquid. J Am Chem Soc 128:6574–6575
131. Albrecht T, Moth-Poulsen K, Christensen JB, Guckian A, Bjørnholm T, Vos JG, Ulstrup J (2006) In situ scanning tunnelling spectroscopy of inorganic transition metal complexes. Faraday Discuss 131:265–279

132. Tran E, Duati M, Whitesides GM, Rampi MA (2006) Gating current flowing through molecules in metal–molecules–metal junctions. Faraday Discuss 131:197–203
133. Visoly-Fisher I, Daie K, Terazono Y, Herrero C, Fungo F, Otero L, Durantini E, Silber JJ, Sereno L, Gust D, Moore TA, Moore AL, Lindsay SM (2006) Conductance of a biomolecular wire. Proc Natl Acad Sci USA 103:8686–8690
134. Wassel RA, Credo GM, Fuierer RR, Feldheim DL, Gorman CB (2004) Attenuating negative differential resistance in an electroactive self-assembled monolayer-based junction. J Am Chem Soc 126:295–300
135. Xiao X, Brune D, He J, Lindsay S, Gorman CB, Tao N (2006) Redox-gated electron transport in electrically wired ferrocene molecules. Chem Phys 326:138–143
136. Li X, Hihath J, Chen F, Masuda T, Zang L, Tao N (2007) Thermally activated electron transport in single redox molecules. J Am Chem Soc 129:11535–11542
137. Li C, Mishchenko A, Li Z, Pobelov I, Wandlowski T, Li XQ, Wurthner F, Bagrets A, Evers F (2008) Electrochemical gate-controlled electron transport of redox-active single perylene bisimide molecular junctions. J Phys Condens Matter 20:374122
138. Alessandrini A, Salerno M, Frabboni S, Facci P (2005) Single-metalloprotein wet biotransistor. Appl Phys Lett 86:133902
139. Chi Q, Farver O, Ulstrup J (2005) Long-range protein electron transfer observed at the single-molecule level: in situ mapping of redox-gated tunneling resonance. Proc Natl Acad Sci USA 102:16203–16208
140. Alessandrini A, Corni S, Facci P (2006) Unravelling single metalloprotein electron transfer by scanning probe techniques. Phys Chem Chem Phys 8:4383–4397
141. Chi Q, Zhang J, Jensen PS, Christensen HEM, Ulstrup J (2006) Long-range interfacial electron transfer of metalloproteins based on molecular wiring assemblies. Faraday Discuss 131:181–195
142. Liu Y, Offenhausser A, Mayer D (2010) Rectified tunneling current response of bio-functionalized metal–bridge–metal junctions. Biosens Bioelectron 25:1173–1178
143. Davis JJ, Peters B, Xi W (2008) Force modulation and electrochemical gating of conductance in a cytochrome. J Phys Condens Matter 20:374123
144. Nowak AM, McCreery RL (2004) In situ Raman spectroscopy of bias-induced structural changes in nitroazobenzene molecular electronic junctions. J Am Chem Soc 126:16621–16631
145. Love JC, Estroff LA, Kriebel JK, Nuzzo RG, Whitesides GM (2005) Self-assembled monolayers of thiolates on metals as a form of nanotechnology. Chem Rev 105:1103–1170
146. McCreery RL (2004) Review molecular electronic junctions. Chem Mater 16:4477–4496
147. Chang SC, Li ZY, Lau CN, Larade B, Williams RS (2003) Investigation of a model molecular-electronic rectifier with an evaporated Ti–metal top contact. Appl Phys Lett 83:3198
148. DeIonno E, Tseng HR, Harvey DD, Stoddard JF, Heat JR (2006) Infrared spectroscopic characterization of [2]rotaxane molecular switch tunnel junction devices. J Phys Chem B 110:7609–7612
149. Haick H, Niitsoo O, Ghabboun J, Cahen D (2007) Electrical contacts to organic molecular films by metal evaporation: effect of contacting details. J Phys Chem C 111:2318–2329
150. Hsu JWP, Loo YL, Lang DV, Rodger J (2003) Nature of electrical contacts in a metal–molecule–semiconductor system. J Vac Sci Technol B 21:1928
151. Hsu JWP (2005) Soft lithography contacts to organics. Mater Today 8:42–54
152. Loo YL, Lang DV, Rogers JA, Hsu JWP (2003) Electrical contacts to molecular layers by nanotransfer printing. Nano Lett 3:913–917
153. Guerin D, Merckling C, Lenfant S, Wallart X, Pleutin S, Vuillaume D (2007) Silicon–molecules–metal junctions by transfer printing: chemical synthesis and electrical properties. J Phys Chem C 111:7947–7956
154. Moons E, Bruening M, Shanzer A, Beier J, Cahen D (1996) Electron transfer in hybrid molecular solid-state devices. Synth Met 76:245–248

155. Vilan A, Shanzer A, Cahen D (2000) Molecular control over Au/GaAs diodes. Nature 404:166–168
156. Vilan A, Cahen D (2002) Soft contact deposition onto molecularly modified GaAs. Thin metal film flotation: principles and electrical effects. Adv Funct Mater 12:795–807
157. Shimizu KT, Pala RA, Fabbri JD, Brongersma ML, Melosh NA (2006) Probing molecular junctions using surface plasmon resonance spectroscopy. Nano Lett 6:2797–2803
158. Preiner MJ, Melosh NA (2008) Creating large area molecular electronic junctions using atomic layer deposition. Appl Phys Lett 92:213301
159. Preiner MJ, Melosh NA (2009) Identification and passivation of defects in self-assembled monolayers. Langmuir 25:2585–2587
160. Ranganathan S, Steidel I, Anariba F, McCreery RL (2001) Covalently bonded organic monolayers on a carbon substrate: a new paradigm for molecular electronics. Nano Lett 1:491–494
161. Thieblemont F, Seitz O, Vilan A, Cohen H, Salomon E, Kahn A, Cahen D (2008) Electronic current transport through molecular monolayers: comparison between Hg/alkoxy and alkyl monolayer/Si(100) junctions. Adv Mater 20:3931–3936
162. Shpaisman H, Salomon E, Nesher G, Vilan A, Cohen H, Kahn A, Cahen D (2009) Electrical transport and photoemission experiments of alkylphosphonate monolayers on GaAs. J Phys Chem C 113:3313–3321
163. Guérin D, Lenfant S, Godey S, Vuillaume D (2010) Synthesis and electrical properties of fullerene-based molecular junctions on silicon substrate. J Mater Chem 20:2680–2690
164. Chiechi RC, Weiss EA, Dickey MD, Whitesides GM (2008) Eutectic gallium–indium (EGaIn): a moldable liquid metal for electrical characterization of self-assembled monolayers. Angew Chem Int Ed 47:142–144
165. Nijhuis CA, Reus WF, Barber JR, Dickey MD, Whitesides GM (2010) Charge transport and rectification in arrays of SAM-based tunneling junctions. Nano Lett 10:3611–3619
166. Rampi MA, Schueller OJ, Whitesides GM (1998) Alkanethiol self-assembled monolayers as the dielectric of capacitors with nanoscale thickness. Appl Phys Lett 72:1781–1783
167. Rampi MA, Whitesides GM (2002) A versatile experimental approach for understanding electron transport through organic materials. Chem Phys 281:373–391
168. Holmlin RE, Ismagilov RF, Haag R, Mujica V, Ratner MA, Rampi MA, Whitesides GM (2001) Correlating electron transport and molecular structure in organic thin films. Angew Chem Int Ed 40:2316–2320
169. Anariba F, McCreery RL (2002) Electronic conductance behavior of carbon-based molecular junctions with conjugated structures. J Phys Chem B 106:10355–10362
170. Hossam H, Cahen D (2008) Contacting organic molecules by soft methods: towards molecule-based electronic devices. Acc Chem Res 41:359–366
171. Duati M, Grave C, Tcbeborateva N, Wu J, Müllen K, Shaporenko A, Zharnikov M, Kriebel JK, Whitesides GM, Rampi MA (2006) Electron transport across hexa-peri-hexabenzocoronene units in a metal–self-assembled monolayer–metal junction. Adv Mater 18:329–333
172. De Redge PJF, Williams SA, Therien MJ (1995) Direct evaluation of electronic coupling mediated by hydrogen bonds: implications for biological electron transfer. Science 269:1409–1413
173. Beratan DN, Betts JN, Onuchic JN (1991) Protein electron transfer rates set by the bridging secondary and tertiary structure. Science 252:1285–1288
174. Perrine TM, Dunietz BD (2008) Conductance of a cobalt(II) terpyridine complex based molecular transistor: a computational analysis. J Phys Chem A 112:2043–2048
175. Chabinyc ML, Chen X, Holmlin RE, Jacobs H, Skulason H, Frisbie CD, Mujica V, Ratner MA, Rampi MA, Whitesides GM (2002) Molecular rectification in a metal–insulator–metal junction based on self-assembled monolayers. J Am Chem Soc 124:11730–11736
176. Le JD, He Y, Hoye TR, Mead CC, Kiehl RA (2003) Negative differential resistance in a bilayer molecular junction. Appl Phys Lett 83:5518
177. Duati M, Cattabriga M, Adamo V, Belser P, Shaporenko A, Zharnikov M , Rampi M A. to be submitted

178. Young RL, Nguyen PT, Slowinski K (2003) Long-range electron transfer through monolayers and bilayers of alkanethiols in electrochemically controlled Hg-Hg tunneling junctions. J Am Chem Soc 125:5948–5953
179. Selzer Y, Salomon A, Cahen D (2002) The importance of chemical bonding to the contact for tunneling through alkyl chains. J Phys Chem B 106:10432–10439
180. Selzer Y, Salomon A, Cahen D (2002) Effect of molecule-metal electronic coupling on through-bond hole tunneling across metal-organic monolayer-semiconductor junctions. J Am Chem Soc 124:2886–2887
181. Haag R, Rampi MA, Holmlin RE, Whitesides GM (1999) Electrical breakdown of aliphatic and aromatic self-assembled monolayers used as nanometer-thick organic dielectrics. J Am Chem Soc 121:7895–7906
182. York RL, Nguyen PT, Slowinski K (2003) Long-range electron transfer through monolayers and bilayers of alkanethiols in electrochemically controlled Hg-Hg tunneling junctions. J Am Chem Soc 125:5948–5953
183. Yu HZ, Liu YJ (2003) Alkyl monolayer passivated metal–semiconductor diodes: 2: Comparison with native silicon oxide. ChemPhysChem 4:335–342
184. Tran E, Rampi MA, Whitesides GM (2004) Electron transfer in a Hg-SAM//SAM-Hg junction mediated by redox centers. Angew Chem Int Ed 43:3835–3839
185. Schmickler W, Rampi MA, Tran E, Whitesides GM (2004) Electron exchange between two electrodes mediated by two electroactive adsorbate. Faraday Discuss 125:171–177
186. Tran E, Cohen A, Murray R, Rampi MA, Whitesides GM (2009) Redox site-mediated charge transport in a Hg–SAM//Ru(NH3)63+/2+//SAM–Hg junction with a dynamic interelectrode separation: compatibility with redox cycling and electron hopping mechanisms. J Am Chem Soc 131:2141–2150
187. Bard AJ, Crayston JA, Kittlesen GP, Shea TV, Wrighton MS (1986) Digital simulation of the measured electrochemical response of reversible redox couples at microelectrode arrays: consequences arising from closely spaced ultramicroelectrodes. Anal Chem 58:2321–2331
188. Pickup PG, Murray RW (1983) Redox conduction in mixed-valent polymers. J Am Chem Soc 105:4510–4514
189. Wittstock G, Burchardt M, Pust SE, Shen Y, Zhao C (2007) Scanning electrochemical microscopy for direct imaging of reaction rates. Angew Chem Int Ed 46:1584–1617

Top Curr Chem (2012) 313: 121–188
DOI: 10.1007/128_2011_238

Published online: 3 November 2011

Charge Transport in Single Molecular Junctions at the Solid/Liquid Interface

Chen Li, Artem Mishchenko, and Thomas Wandlowski

Abstract Charge transport characteristics in metal–metal nanocontacts and single molecular junctions were studied at electrified solid–liquid interfaces employing a scanning tunneling microscope-based break junction technique, in combination with macroscopic electrochemical methods, in non-conducting solvents and in an electrochemical environment. We aim to demonstrate recent attempts in developing fundamental relationships between molecular structure, charge transport characteristics, and nanoscale electrochemical concepts. After an introduction and brief description of the experimental methodology, a case study on the electrical and mechanical properties of gold atomic contacts in aqueous electrolytes is presented. In experiments with alkanedithiol and α,ω-biphenyldithiol molecular junctions the role of sulfur–gold couplings and molecular conformation, such as gauche defects in alkyl chains and the torsion angle between two phenyl rings, are addressed. The combination with quantum chemistry calculations enabled a detailed molecular-level understanding of the electronic structure and transport characteristics of both systems. Employing the concept of "electrolyte gating" to 4,4′-bipyridine and redox-active molecules, such as perylene bisimide derivatives, the construction of "active" symmetric and asymmetric molecular junctions with transistor- and diode-like behavior upon polarization in an electrochemical environment will be demonstrated. The latter experimental data could be represented quantitatively by the Kutznetsov/Ulstrup model, assuming a two-step electron transfer with partial vibration relaxation. Finally, we show that (individual) surface-immobilized gold clusters within the quantum-confined size range exhibit features of locally addressable multistate electronic switching upon electrolyte gating, which appears to be reminiscent of a sequential charging through several redox

C. Li, A. Mishchenko, and T. Wandlowski (✉)
Institute of Chemistry and Biochemistry, University of Berne, Freiestrasse 3,
3012 Berne, Switzerland
e-mail: chen.li@dcb.unibe.ch; artem.mishchenko@iac.unibe.ch; thomas.wandlowski@dcb.unibe.ch

states. The examples addressed here demonstrate the uniqueness and capabilities of an electrochemical approach for the fundamental understanding and for potential applications in nano- and molecular electronics.

Keywords Break junction · Charge transport · Electrolyte gate · Metal nanocluster · Molecular junction · Scanning tunneling microscopy · Scanning tunneling spectroscopy

Contents

1 Introduction

The idea of developing a single molecule-based electronics was first proposed by Aviram and Ratner [1], which is nowadays termed *Molecular Electronics* [2, 3]. The molecule-based approach bears several unique opportunities: (1) custom-design of nanoscale units, (2) self-organization and recognition properties of molecular building blocks, and (3) implementation of localized functions such rectification, switching or amplification [4–16]. The ability to measure and control charge transport through (single-) molecule junctions is of considerable fundamental interest, and represents a key step towards the development of a molecule-based electronics.

The following fundamental aspects represent some of the key challenges in attempts to understand charge transport in (single-) molecule junctions in more detail:

1. How does a metal electrode behave at an atomic scale?
2. How does one reliably "wire" a single molecule to macroscopic leads?
3. What role does the contact between an electrode and a molecule play?
4. What is the relation between molecular structure and charge transport characteristics, and what is the mechanism of charge transport in such configurations?
5. How could one control/tune electron transport through single molecules?

Molecular junctions represent model configurations, in which a voltage bias, as imposed by an outside source, triggers an electric current between two electrodes and a single embedded molecule, which reflects the electronic characteristics of the molecular junctions (Fig. 1).

Three challenges must be overcome to realize such a configuration: (1) fabrication of atomically sharp electrodes, (2) creation of a molecule-sized gap between the enclosed electrodes, and (3) a reliable embedding of the molecule into the gap. The former two were realized in various experimental approaches [17–23]. These

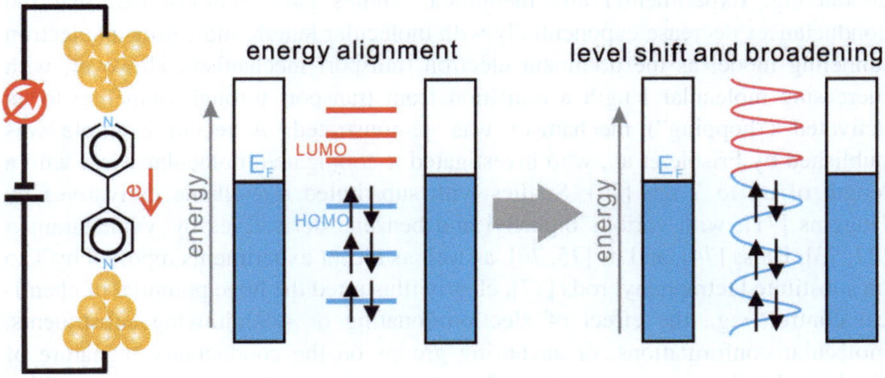

Fig. 1 Schematics of a typical "metal–molecule–metal" junction and energy level alignment

include scanning probe microscopies (SPM), scanning tunneling microscopy (STM) [24–27], conductive-probe atomic force microscopy (CP-AFM) [28–30], scanning tunneling spectroscopy (STS) [31–34]), nanoparticle (NP)-assisted junctions [35, 36], mechanical [37–41] and electromigration [42, 43] break junctions, crossed wires [44], nanopores [45], mercury drop electrodes [46], etc.

"Wiring" molecules to metal electrodes relies on the formation of specific chemical bonds. This is typically accomplished by modifying the core of the target molecules with "sticky" anchoring groups ("alligator clips" [47]) on both sides. In addition to the most prevalent thiol group (–SH), many other functional groups have been explored to establish a well-defined electronic coupling between molecular wires and metal contacts. These include amine ($–NH_2$) [48, 49], pyridine [32, 50, 51], isocyanide (–NC) [52], nitrile (–CN) [53, 54], isothiocyanate (–NCS) [55], selenol [56–58], carboxylic acid (–COOH) [59, 60], dimethyl phosphine ($–P(CH_3)_2$) [61], methyl sulfide ($–SCH_3$) [61], carbodithionate [62], *N,N'*-dimethyl amine [49], and fullerene [63] anchors. The different linker groups possess different coupling strengths and contact geometries, which affect considerably the charge transport characteristics of the various molecular junctions. For instance, covalent Au–S bonds exhibit strong geometry-dependent coupling properties [64, 65] while amine linkers were shown to prefer exclusively binding to undercoordinated gold atoms [66]. Complementary to the concept of covalent bonds, several groups recently demonstrated the formation of stable and conducting molecular junctions based on non-covalent intermolecular interactions. For example, Wu et al. for the first time verified the formation of an oligophenylene ethynylene (OPE) junction based on π–π interactions between adjacent molecular rods [67]. Chang et al. measured the tunneling conductance of hydrogen-bonded DNA bases [68].

Molecular wires, the "simplest" molecular targets, have attracted particularly high attention during the past decade. One may distinguish between two classes: (1) σ-bonded aliphatic molecules and (2) π-bonded aromatic molecules [9, 10]. α,ω-Alkanedithiols [28, 29, 64], oligophenylene vinylenes (OPV), and OPE rods [30, 35, 69] attached to gold electrodes or nanoclusters are prominent examples. Compared to the saturated aliphatic molecular wires, π-conjugated wires are more conducting. Experimental and theoretical studies have demonstrated that the conductances decrease exponentially with molecular length, indicating an electron tunneling model as the dominant electron transport mechanism. However, with increasing molecular length a transition from transport through tunneling to an activated ("hopping") mechanism was demonstrated. A recent example was published by Frisbie et al., who investigated π-conjugated molecular wires with a length of up to 7 nm [70]. Studies with substituted α,ω-alkane derivatives by Huggins [71], with various biphenyl and benzene derivatives by Venkataraman [72, 73], Haiss [74], and us [75, 76], as well as recent experiments reported by Tao on substituted tetraphenyl rods [77], clearly illustrated the huge potential of chemical control, e.g., the effect of electron-donating or -withdrawing substituents, molecular conformations, or anchoring groups on the conductance signature of single-molecule junctions at "solid/liquid" interfaces. However, there is still a considerable lack in understanding the basic relationships between molecular structure and conductance properties.

Complementary to "passive" molecular wires, several groups carried out proof-of-concept experiments on "active" solid-state "metal–molecule–metal" junctions, for instance with redox centers under ex-situ (ambient) and ultra-high vacuum (UHV) cryogenic conditions. They discovered various phenomena mimicking fundamental functions of silicon-based electronics, such as rectification [78, 79], negative differential resistance (NDR) [45, 80, 81], amplification [41, 42], and switching and memory characteristics [16, 82–86]. The experimental investigations are currently accompanied by powerful theoretical and computational approaches to interpret molecular signatures in these transport junctions. The realization of these functions requires external stimuli (temperature, applied potential, light, magnetic field, etc.) to modulate charge transport in a controlled way [42, 87–89]. Additional challenges appear upon introduction of a gate, i.e., the extension of two-terminal molecular junctions into three-terminal configurations. Working in an electrochemical environment offers the particular opportunity of introducing the concept of "electrolyte gating" [90–93]. The electrochemical approach is unique, as the measured current represents both the electrical contact to the external circuit and the functional state of the wired molecules or nanoclusters.

In the following sections we shall present selected examples of our own recent work on charge transport with tailored molecules and clusters at *electrified* "solid/liquid" interfaces. We will introduce in Sect. 2 a scanning tunneling microscope-based break junction (STM-BJ) technique and the concept of "Electrochemical Gating." Case studies on the electrical and mechanical properties of gold atomic contacts in an electrochemical environment will be discussed in Sect. 3. Sections 4 and 5 will focus on charge transport studies with α,ω-alkanedithiols and biphenyl-type molecular wires in non-polar organic solvents. The role of orientation gating upon application of an electrochemical gate field will be discussed with 4,4′-bipyridine as an example in Sect. 6. In Sect. 7 we will focus on single redox-active molecular nanojunctions in an electrochemical environment, as illustrated by perylene bisimide-type molecules bound either to one (asymmetric configuration) or to two (symmetrical configuration) adjacent electrodes. These molecular studies will be complemented with an example of quantized double-layer charging of gold nanoparticles. The chapter will end with Conclusions and Outlook.

2 Methodology

Experimental approaches for investigating charge transport characteristics in "metal–molecule–metal" junctions should fulfil the following requirements: (1) providing robust and reproducible contacts between the bridged molecule(s) and the outside world, (2) identification of molecular signatures, and (3) access to tuning molecular electronic signatures by external stimuli under well-defined conditions.

In the following we will focus on three "molecular electronics test beds" as developed and employed for applications at electrified "solid/liquid" interfaces: (1) STM and STS, (2) assemblies based on horizontal nanogap electrodes, and (3) mechanically-controlled break junction experiments. For a more detailed description of the methods we refer to several excellent reviews published recently [16–22]. We will also address specific aspects of "electrolyte gating" and of data analysis.

2.1 Scanning Probe Microscope

The power of STM for measuring electronic properties of single molecules lies in the combination of high-resolution imaging (STM) and spatially resolved current-sensing spectroscopy (STS). The latter is capable of providing the local density of states (LDOS) with atomic spatial resolution [94]. In an asymmetric tunneling junction "metal tip–molecule–conducting substrate," the resulting "apparent height" images contain coupled information on both the electronic and topological properties (true height). As an example, a conjugated molecule may appear higher than a non-conjugated one of the same geometric dimensions. Experiments at cryogenic temperatures and under UHV demonstrated access to well-resolved signatures in the differential conductance and vibrational excitations [95, 96], and enabled a wide range of manipulations with single atoms and molecules [97]. Studies under ambient conditions require the immobilization of the target molecules. This is accomplished by either matrix-isolation experiments, where the "active" molecule is inserted into an inert matrix of alkanethiols [98–100], the formation of highly ordered self-assembled monolayers [31, 101–105] or the chemical anchoring of the "active" molecules onto appropriate substrates [106–108]. These studies provide access to exploring electrical phenomena such as NDR, dielectric break-down, switching and diode-like responses at room temperature, and in solution from monolayers down to the level of a few and even single molecules and clusters embedded in nanoscale tunneling junctions [34, 109, 110]. All three approaches have also been applied to electrochemical systems with target molecules immobilized at "solid/liquid" interfaces [111, 112].

Tao et al. [32] pioneered a technique based on the formation of single molecular junctions between the tip of an STM and a metal substrate. The method was adapted by other groups, modified and applied to a large number of molecular conductance studies at (electrified) "solid/liquid" interfaces [33, 113–119]. For details we refer to Sect. 2.3.

Frisbie et al. [120] and Lindsay et al. [28] introduced CP-AFM for conductance measurements in (single) molecular junctions. CP-AFM allows precise control of the AFM probe, employing the feedback of the force signal and the simultaneous acquisition of the molecular current response. In order to achieve a better electric contact between the conducting AFM tip and the target molecules of interest, Cui et al. proposed contacting bifunctionalized target molecules embedded in inert alkanethiol-based matrices with metal nanoclusters. While the first attempt with cluster diameters ~1.5 nm led to artifacts due to Coulomb blockade, Lindsay et al. demonstrated later that larger gold nanoclusters (~4–5 nm diameter) provided direct access to clearly resolved molecular conductance signatures [121]. The cluster approach was also applied under electrochemical conditions [122]. In particular, Albrecht et al. resolved, for the first time, the intrinsic multistate switching of a single, monolayer-protected gold Au_{147} cluster through electrochemical gating [123].

2.2 Nanoscale Electrodes and Gaps

Conceptually perhaps the simplest method for wiring molecules in metallic junctions is the fabrication of constant-size metallic gaps with nanometer separation. The fabrication of electrodes with defined molecular-sized separation is technically challenging, but can be achieved by electron-beam lithography [124]. To trap molecules into the nanofabricated gap, as well as to tune gap dimensions, one uses electromigration [42], electrochemical dissolution or deposition strategies [125], molecular recognition and click-chemistry, employing acceptor-modified electrodes [126–128].

Due to the inherent size difference between nanofabricated gaps (>10 nm) and small organic molecules (~1 nm), many groups propose bridging these gaps by intermediate-size objects, such as carbon nanotubes [129, 130], nanoparticles [24, 35, 36, 131, 132], or oligomer strands [133]. For example, the Nuckolls group demonstrated how individual molecules can be trapped within a single-wall carbon nanotube, which was locally cut by oxidative etching [129, 130]. The nanotube itself is connected to two large nanofabricated metallic electrodes. Dadosh et al. have demonstrated that short dithiolates can connect with pairs of gold nanoparticles, giving rise to large particle–molecule–particle dumbbells, which could be successfully trapped and electrically addressed in larger gaps formed between two metallic leads [35].

2.3 Break Junction

The formation of molecule-sized gaps between electrodes embedding a few or even a single molecule comprises a major challenge. Gold appears to be one of the most suitable contact materials for application at electrified solid/liquid interfaces for several reasons: (1) it is malleable and ductile, (2) it forms covalent bonds with a large number of functional groups, (3) it could form atomically sharp contacts upon breaking a "simple" wire, and (4) it shows a rather large range of ideal polarizability in many electrolytes. Furthermore, using a piezoelectric transducer or other mechanical actuators (stepper motor, etc.) enables the formation of well-defined

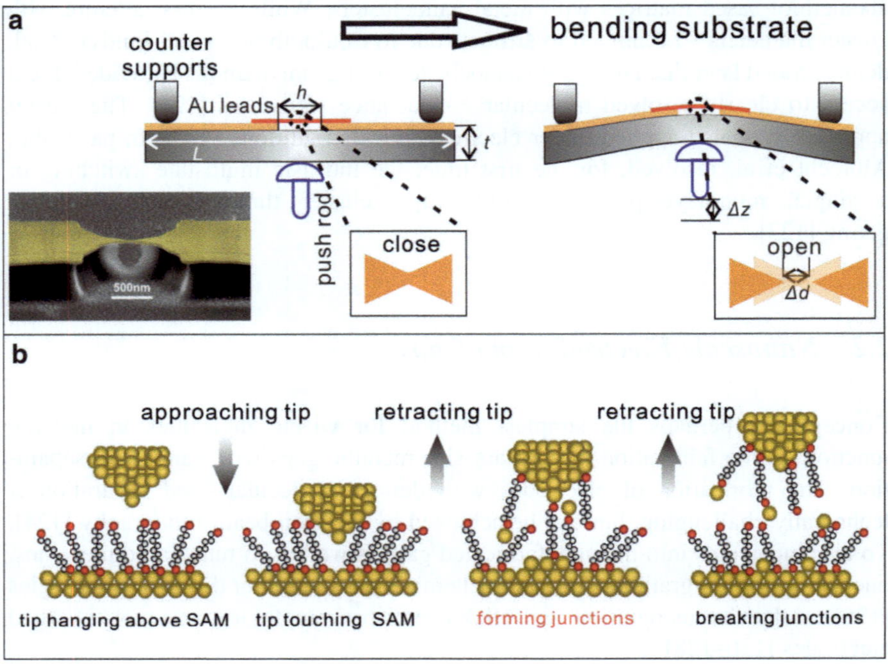

Fig. 2 (**a**) Schematic representation of a mechanically controlled break junction (MCBJ). The *inset* shows the SEM image of a nanofabricated gold bridge [40]. (**b**) Principle of an STM-based break junction experiment (STM-BJ)

gaps between two adjacent electrodes with high stability and sub-Ångstrom precision. These reasons led to the development of the so-called "break junction" techniques. We shall introduce next the two main types of break junction approaches: (1) mechanically controllable break junction (MCBJ) and (2) STM-based break junction (STM-BJ), (see Fig. 2).

2.3.1 Mechanically Controllable Break Junction

The basic working principle of an MCBJ is as follows. A flexible substrate is connected by two fixed beams on top and one pushing rod on the bottom. A mechanical actuator, such as a piezoelectric transducer or stepper motor, drives the pushing rod (moveable beam) to bend a substrate, leading to an elongation and subsequently the breaking of the metallic bridge on the substrate. The metallic bridge can be manually created by notching a metal wire [134], or by using micro-respective nanofabrication techniques [135] (see the inset of Fig. 2a). Applying a small bias voltage may lead to bridging the freshly created gap by bifunctional target molecules just after breaking of the metallic connection. Subsequently, (single) molecular junctions are formed. The reduction factor for the displacement

Δd of the notched wire or of a microfabricated metallic bridge with respect to the movement Δz of the pushing rod is $\Delta d/\Delta z = 1{:}100$ or $1{:}10^5$, respectively. This ensures a precise control of the gap width down to sub-nanometer dimensions, leading to a highly stable configuration.

The MCBJ technique was first introduced by Moreland et al. for studying the tunneling characteristics of Josephson junctions [136]. Müller et al. employed the approach to create metal quantum point contacts. These authors demonstrated in their studies conductance quantization phenomena [137]. The first (single) molecular junction employing an MCBJ set-up was reported by Reed et al., who measured the conductance of benzene-1,4-dithiol molecules (BDT) embedded between gold leads [37]. These authors used a BDT-coated gold wire before notching it under ambient conditions. Reichert et al. showed a superior stability of current–voltage characteristics measured with an MCBJ set-up under UHV conditions in a cryogenic environment [39]. Schönenberger et al. applied the MCBJ-approach to tunneling and conductance studies at solid/liquid interfaces [40, 138] and, rather recently, demonstrated a first application under electrochemical conditions [139].

2.3.2 Scanning Tunneling Microscopy Based Break Junction

Similar to the MCBJ technique, an STM configuration may also realize a break junction geometry. Different from the MCBJ, where two identical electrodes approach each other to form a *horizontal* molecular junction, the STM tip approaches and withdraws from the substrate in a vertical geometry. The STM-BJ was first applied to characterize electrical properties of metal nanocontacts [140]. Xu et al. extended the approach to form and to study single-molecule junctions [32]. The following sequence of steps was applied (see Fig. 2b). (1) The STM tip was driven to the initial position under feedback control by setting the appropriate tunneling conditions in a solution containing the target-molecule as solute. (2) Subsequently, the STM feedback was switched off, and the tip was moved by a piezoelectric transducer towards the adsorbate-modified substrate at constant x-y position, until a preset upper limit of the current (typically 1–20 μA for 0.1 V bias voltage) was reached. (3) After a short waiting time, long enough to form molecular junctions, the tip is pulled away from the substrate at a controlled rate. The current–distance curves are recorded while the tip is being retracted. The breaking of the previously formed molecular (single) junctions is indicated by abrupt steps following plateau-like regions of conductance.

In the present work, we applied two different strategies. (1) "Hard touch": the upper limit of the current is chosen to be higher than the quantum conductance. (i.e., the tip is gently driven into the substrate.) This strategy ensures "in situ" the fresh preparation of atomically-sharp electrodes before molecular junctions are being formed [32, 59, 76]. (2) "Gentle touch": this strategy avoids physical contact between tip and substrate, and thus keeps intact both the organic adlayer and the atomically flat substrate surface [33, 64].

Compared to the MCBJ, the STM-BJ has some unique advantages. (1) The STM-BJ facilitates a faster formation/breaking process, and gives access to a larger amount of data, thus yielding statistically more reliable data. (2) Due to the relative maturity of the STM technique, the approach can be directly applied to studies in solution [32], and even under electrochemical conditions [33, 34, 141]. (3) The STM configuration also enables the direct monitoring of the substrate surface before, during and after the transport measurements, leading to a superior structure control of the junctions. (4) The STM tip can be positioned precisely on the target area of an adsorbate-modified substrate, i.e., selectivity and flexibility in two dimensions (2D) during the measurements. A major drawback of the STM-BJ approach under ambient conditions and in solution is its limited stability as compared to the MCBJ approach.

Comparing the existing BJ techniques and their particular applications for studies at the solid/liquid interface, we emphasize the following concerns. (1) All technical realizations involve a fully "dynamic" process, in particular when comparing them to nano-fabricated chip assemblies. As a consequence, the lifetime of a molecular junction is often too short to be readily characterized by *I–V* measurement, local spectroscopy, and other structure-sensitive techniques. (2) The target molecules are always under tensile or compressive stress during the stretching process, which may lead to variations of the electronic properties of the respective molecular junctions. (3) When stretching the junctions, the microscopic geometry of the enclosing electrodes can vary from measurement to measurement, causing large variations of the interfacial geometric and electronic structure of the respective junctions.

2.3.3 Statistical Analysis

The formation and breaking of atomic contacts in BJ experiments is indicated by the existence of characteristic plateaus separated by steps in current–distance (i-Δz) traces with currents of the order of a few microamperes for typical bias voltages between 0.02 V and 0.30 V. In the case of molecular junctions one can observe additional plateaus and steps in the i-Δz traces, rather than a purely exponentially decaying tunneling current, or a sudden drop. However, they appear at significantly smaller currents, as compared to those in atomic metal contacts. The shape of individual traces varies considerably, due to variations in the structure of the nanojunctions, as well as because of mechanical instabilities of the assembly. The determination of most probable junction conductances, plateau lengths, stability, dynamics, etc., require the statistical analysis of a large number of individual curves recorded experimentally under a wide range of conditions, as well as their comparison with electronic structure and transport model calculations (simulations).

While initial attempts of quantification of i-Δz traces obtained in (single) molecule conductance studies suffered from rather poorly defined data selection strategies and limited current ranges [64, 116], most of the leading groups in the field focus now on the construction of all-data point conductance and plateau length histograms from a large number of individual traces (often >1,000), recorded in a wide current range, to evaluate the most probable junction characteristics. Linear histograms are based on the splitting of individual experimental traces into many equal-current respective conductance intervals (bins) and counting the number of data points in each interval. The statistical analysis of a complete experiment is based on adding up a large number of single-curve histograms. The most frequently observed conductance values appear as characteristic peaks in the conductance histograms.

An alternative strategy is based on the all-data point analysis on a logarithmic scale [40, 48]. This approach is equivalent to the construction of a linear histogram with a logarithmically increasing bin width. This analysis has two advantages. (1) The logarithmic-scale representation enables the direct visualization of experimental data in a wide current range, from several hundreds of microamps down to a few picoamps, covering the range of breaking of metal–metal contacts down to the molecular junctions and down to the tunneling regime. (2) The logarithmic representation also enables a straightforward distinction between molecular junction-related features and background [40, 54].

Complementary strategies for the statistical analysis of the experimental traces are based on *plateau data-point* histograms, *plateau count* histograms, and *two-dimensional (2D) conductance–distance* histograms [54]. The first two methods lead to the identifications of the plateaus in the conductance traces. A plateau is defined as a series of consecutive data points limited by two "abrupt changes," allowing a certain scatter in the current, which are found by computing the first derivative of the respective single traces. The number of data-points of the identified plateaus is subsequently collected to build the plateau data-point conductance histogram. Alternatively, in the plateau-count histogram each selected plateau adds one count to the histogram.

The extension of the data analysis to 2D histograms provides correlations between three variables, such as conductance, plateau length, and their probabilities, and therefore may lead to a wealth of additional information, as compared to 1D histograms. However, this is currently not yet fully exploited. Two examples of 2D histograms are the 2D conductance–distance histogram and the 2D covariance conductance histogram [54].

The 2D conductance–distance histogram shows the logarithm of the experimentally obtained conductance in units of G_0 plotted vs distance [54, 63, 142]. The distance axis is normalized with respect to the atomic contact rupture, to align the experimental traces to a common point. Examples are $z = 0$ at $G = 0.7\ G_0$ [54]. Individual conductance–distance traces obtained in this manner are binned in 2D space (usually 1,000 × 1,000 bins), and then all individual 2D histograms are added together. The result represents statistically significant conductance–distance trends.

The covariance analysis was inspired by work of Halbritter et al. [143]. The covariance provides a measure of the strength of a correlation between different parts of the 1D logarithmic conductance histograms. In other words, the covariance analysis enables evaluating whether different parts of a histogram are correlated to each other. The covariance is zero if the variations of counts in two different bins are independent. The covariance is positive if they are correlated, and negative if they are anti-correlated.

2.4 Electrolyte Gating

The addressing of nanoelectronic assemblies "metal–molecule (nanocluster)–metal" with device-like functions, such as rectifiers, switches, or transistors requires a source and a drain, and one or more localized electronic levels. The roles of source and drain (both as working electrodes WE1 and WE2) may be represented by the tip of an STM, combined with an appropriate substrate or, alternatively, a pair of nanoelectrodes; see Fig. 3.

Working in an electrochemical environment, such as an electrified solid/liquid interface, has the advantage that two potential differences can be controlled individually: the bias voltage, E_{bias}, between two working electrodes (WE1 and WE2) and the potential drop between each individual working electrode and the reference electrode (RE). The latter acts as an "electrochemical gate" and modulates the tunneling current between source (WE2) and drain (WE1) [144, 145]. The idea of an "electrochemical gate" to control charge transport in molecular electronics was introduced by Wrighton [90], Meulenkamp [91], Schönenberger [92], and McEuen [93], and further developed by Tao [31, 141, 146], Ulstrup et al. [144, 145, 147], Lindsay [113], Haiss et al. [33], Vanmaekelbergh [148], Mao [149], and us [34, 110, 150]. The effective "gate–molecule" distance is determined by the double layer thickness at the "electrode/electrolyte" interface, which is typically of the order of a few solvated ions. The "electrolyte gating" ensures a strong coupling, due to low contact resistances, thin adjacent double-layers (<1 nm) and the high mobility of the charge carriers (ions). The magnitude of the field in the electrochemical double layer (EDL) is close to the gate field required to modulate

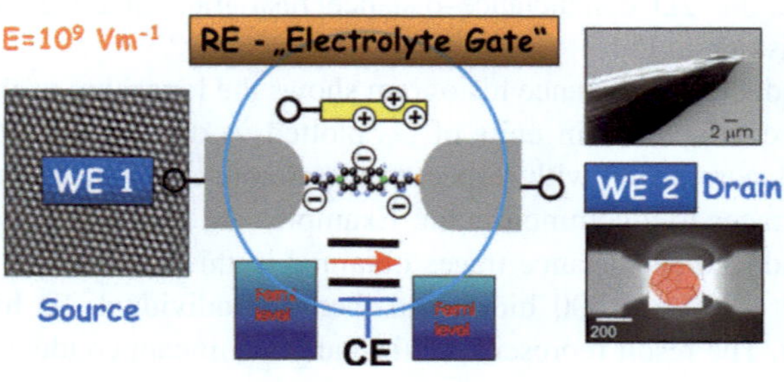

Fig. 3 Principle of "electrolyte gating." Tuning of the Fermi levels of WE1 and WE2 relative to the molecular levels enables measuring of current (i)–voltage (E) characteristics i vs ($E_{WE1}-E_{WE2}$) at fixed E_{WE1} or E_{WE2}, i vs E_{WE1} or E_{WE2} at fixed bias $E_{bias} = (E_{WE1}-E_{WE2})$ as well as barrier height profiles i vs distance z of tailored molecular junctions in a vertical SPM-based configuration respective horizontal nanoelectrode assembly

significantly the current through a molecule-sized nanostructure, according to first-principle calculation by di Ventra et al. [151, 152]. The electrochemical approach is powerful, as the measured current represents both electrical contact to the external circuit and the functional state of the nanostructures (single molecule, nanocluster).

Redox molecules are particularly interesting for an electrochemical approach, because they offer addressable (functional) energy states in an electrochemically accessible potential window, which can be tuned upon polarization between oxidized and reduced states. The difference in the junction conductance of the oxidized and the reduced forms of redox molecules may span several orders of magnitude. Examples of functional molecules used in these studies include porphyrins [31, 153], viologens [33, 34, 110, 114, 154, 155], aniline and thiophene oligomers [113, 146, 156, 157], metal–organic terpyridine complexes [46, 158–163], carotenes [164], nitro derivatives of OPE (OPV) [165, 166], ferrocene [150, 167, 168], perylene tetracarboxylic bisimide [141, 169, 170], tetrathiafulvalenes [155], fullerene derivatives [171], redox-active proteins [109, 172–174], and hydroxyquinones [175].

Employing an electrochemical surface-science-based strategy provides direct access to explore the properties of "metal–metal" atomic contacts (i.e., electrodes) as well as functional molecules and/or nanocluster under conditions of a strongly coupled gate field. The approach offers unique opportunities to develop a nanoscale electrochemistry, and to contribute to the better understanding of the basic knowledge on gating inorganic–organic hybrid nanostructures upon polarization in a well-controlled environment.

3 Electron Transport Through Gold Atomic Contacts in an Electrochemical Environment

3.1 Introduction

Landauer proposed in 1957 the first mesoscopic theoretical approach to charge transport [176]. Transport is treated as a scattering problem, ignoring initially all inelastic interactions. Phase coherence is assumed to be preserved within the entire conductor. Transport properties, such as the electrical conductance, are intimately related to the transmission probability for an electron to cross the system. Landauer considered the current as a consequence of the injection of electrons at one end of a sample, and the probability of the electrons reaching the other end. The total conductance is determined by the sum of all current-carrying eigenmodes and their transmission probability, which leads to the Landauer formula of a 1D system:

$$G = \frac{2e^2}{h}\sum_{n=1}^{N} T_n, \qquad (1)$$

where e is the electron charge, h is Plank's constant, and N is the number of channels with the transmission probability T_n.

In the case of a three-dimensional (3D) conductor, a more general formulation using a finite bias with a series of modes is given by

$$G = \frac{2e^2}{h} \sum_{i,j}^{N} T_{ij}, \tag{2}$$

where T_{ij} is the probability that an electron transmits from the ith mode of the left electrode to the jth mode at the right electrode. The quantum unit of the conductance G_0 is defined by

$$G_0 = \frac{2e^2}{h} \approx 77.5\ \mu\text{S} \approx (12.9\ \text{k}\Omega)^{-1}. \tag{3}$$

For an ideal conductor, no scattering occurs, and the transmission is given by $T = 1$. The quantum of conductance G_0 is obtained, indicating a maximum conductance. In other words, a perfect single-channel conductor between two electrodes has a finite, non-zero resistance. The exact interpretation of this result was provided by Imry [177], who associated the finite resistance with resistance arising at the interface between leads and the electrodes.

Systematic studies of I–V curves [178] and conductance histograms [134] of atomic contacts of different metals being characterized by different chemical valence states, supported by theoretical simulations [179], lead to the following discovery. The number of conducting channels of monatomic metal contacts is determined by the number of available valence orbitals [180]. In the case of gold, the conduction channel is only a single 6 s channel with an almost perfect transmission possibility. Thus, it possesses a total quantum conductance, which is quantized in units of G_0 (i.e., $G = N{\cdot}G_0$). These characteristics made gold an archetypal candidate to investigate properties of metal atomic contacts. Other s-metals such as Cu, Ag [181, 182], and the free electron metals Li, Na, K [183], exhibit a dominant conductance peak at 1 G_0, or slightly below that, in the respective conductance histograms. Metals like Al [184], Pb, Pt, Pd, Ir, Rh [185, 186], as well as magnetic metals, such as Fe, Co, Ni [187], do not have conductances with integer multiples of the quantum unit. For example, the first conductance peak for Nb was reported to be rather broad, in the range of 2.3–2.5 G_0 while Pt shows a pronounced feature at 1.6 G_0 in the conductance histogram.

The fabrication and characterization of atomic metal contacts have been based mainly on electro-deposition/dissolution [182] and break junction techniques (see review [134] and literatures cited therein). In particular, gold nanocontacts have been studied in great detail, due to the chemical inertness of the material, the malleability and ductility of gold. The processes of formation, evolution, and breaking of gold atomic contacts leads to step-like features in the current–distance curves [188, 189]. The abrupt changes in the current (conductance) response were attributed to sudden alternations of the atomic arrangement (i.e., plastic deformation), while the plateaus were assigned to relatively stable atomic configurations (i.e., elastic deformation) [190–192]. The plateau regions, often characterized by integers of the quantum conductance unit G_0 (e.g., 1 G_0, 2 G_0 and 3 G_0), represent specific atomic configurations, which are currently interpreted as one, two or three atoms bridging the gap between the enclosing metal leads. Other parameters, such as plateau length and the scattering level, are determined by the mechanical properties of the respective atomic configurations. We also note that a certain variation in each individual conductance trace can be found, due to a rather poor control of the structural arrangement during the elongation of the contacts.

In addition to the study of bare (clean) metal atomic contacts, adsorbates were explored, due to the unique reactivity properties of nanoscale junctions [193]. For example, a large Au surface is known to be completely inert to H_2, while single gold contacts form a strong chemical bond with H_2 [194]. Current experimental (ex situ and in situ, e.g., in solution) and theoretical studies on adsorbates could be separated into three groups: (1) small and inert gaseous molecules; examples are H_2 [194–196], O_2 [197, 198], CO [199], and H_2O [200]; (2) organic molecules [192, 201, 202]; (3) inorganic anions [203]. Molecular species can interact rather *strongly* with atomic contacts, which may lead to dissociation and local reactions [204]. The incorporation of molecules into atomic contacts often enhances the stability of metal atomic configurations, and significantly decreases the transmission probability of electrons through contacts, causing the appearance of additional fractional conductance peaks at $G \ll G_0$ in the conductance histograms.

To date, most experiments with Au atomic contacts have been carried out at cryogenic temperatures or at room temperature in UHV, at ambient conditions in the gas phase, or in solution. Very few studies were reported in an electrochemical environment [205–208]. Electrochemical polarization offers the unique opportunity of *tuning* both the electrical and the mechanical properties of the respective atomic contacts by variation of the electrode potential. The electrodes could be "charged" and the *local* concentration of adsorbates at the atomic contacts can be varied in a rather controlled matter.

In the following we present an experimental study of gold atomic contacts formed in various aqueous electrolytes under electrochemical potential control. We will focus in particular on the electrical and mechanical properties of the electrochemical nanojunctions, as well as on their interplay in the presence/absence of various adsorbates (e.g., H_2, ClO_4^-, SO_4^{2-}, Cl^-, Br^-, and I^-) [209, 210].

3.2 *Gold Atomic Contacts in 0.1 M H_2SO_4: A Case Study*

3.2.1 Characterization of Au(111) in H_2SO_4 Solution

Figure 4 shows a typical cyclic voltammogram (CV) and the corresponding in situ STM images of an Au(111) electrode in 0.1 M H_2SO_4. The CV (red) as recorded in the bias range $-0.35\ V < E_S < 1.35\ V$, reveals the characteristic oxidation peaks

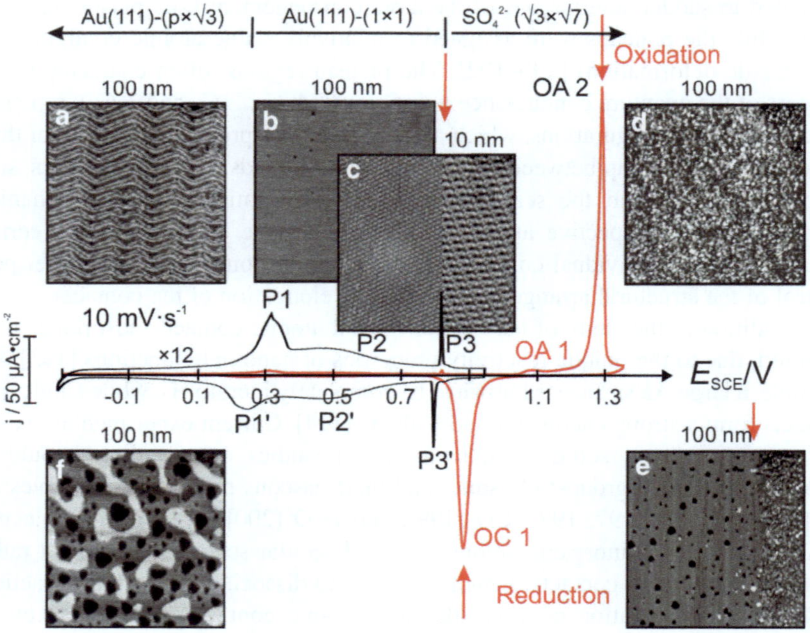

Fig. 4 Cyclic voltammograms (CV) of an Au(111) electrode in 0.05 M H_2SO_4 (*red*), scan rate 10 mV s^{-1}. The *black curve* represents the trace as recorded in the double layer region (current scale magnified by factor 12). The *red curve* extends the potential excursion into the surface oxidation/reduction potential region. The in situ STM images represent surface structures at different potentials: (**a**) thermally reconstructed Au(111)-(p × √3) surface, $E_S = -0.20$ V; (**b**) Au (111)-(1×1) surface after lifting of the reconstruction; (**c**) transition between a disordered and the ordered $(\sqrt{3} \times \sqrt{7})R19.1^0$ sulfate adlayer; (**d**) oxidized gold surface; (**e**) reduction of the oxidized gold surface during a negative-going potential scan; (**f**) gold surface with monatomic deep holes obtained after ten successive ox/red cycles

OA1 and OA2, attributed to the oxidation of step and terrace sites, respectively, the corresponding reduction peak OC1, and the onset of hydrogen evolution at the most negative potentials. The black trace in Fig. 4 represents the double-layer region, with three pairs of peaks (note the change in the current scale!). The pair of peaks labeled P1/P1' corresponds to the lifting/formation of the Au(111)−(22 × √3) reconstruction, which is close to the potential of zero charge (PZC) [211]. Figure 4a shows an STM image of the thermally reconstructed Au(111)−(p × √3) surface with the characteristic zig-zag pattern. The broad peak P2/P2' marks the potential region of disordered SO_4^{2-} adsorption. Adsorbed anions are rather mobile in this region. Figure 4b represents the unreconstructed Au (111)−(1 × 1) surface. The reversible sharp peaks P3/P3' indicate the transition between the disordered and the ordered $((\sqrt{3} \times \sqrt{7})R19.1^0)$ SO_4^{2-} overlayers, as illustrated in Fig. 4c [212, 213].

In consequence, we distinguish between the following characteristic potential regions: (1) hydrogen evolution at very negative potentials, (2) negatively charged gold surface with adsorbed, rather weakly hydrogen-bonded water [214], (3)

positively charged gold electrode with (hydrogen-) sulfate ions forming a disordered adlayer, (4) an ordered $((\sqrt{3} \times \sqrt{7})R19.1^0)$ adlayer at $E_S > P3$ until the onset of surface oxidation in region (5) at $E > OA1$.

3.2.2 Conductance of Gold Atomic Contacts at Various Potentials

Figure 5 shows representative 1D conductance histograms (upper panel) and the corresponding plateau length distributions (lower panel) for Au–Au nanocontacts formed at various electrode potentials in the double-layer region of Au(111) in 0.1 M H_2SO_4 solution. All histograms show characteristic conductance peaks at integer multiples of G_0 (1 G_0, 2 G_0, and 3 G_0). The following additional features were observed. (1) At the most negative potential (−0.35 V), where hydrogen evolution takes place, fractional conductance peaks evolve at 0.1 G_0, 0.5 G_0, and 1.5 G_0, respectively. (2) The fractional conductance peaks disappear with the onset of sulfate anion adsorption at the positively charged electrode. (3) The plateau length distributions appear to be narrower with higher potentials, where the nanoelectrodes bear positive charges.

The appearance of fractional conductance peaks at the negatively charged electrode are attributed to the presence of hydrogen molecules or atoms (see Sect. 3.2.3), as supported by a study of Csonka et al. [194, 195], and in agreement with recent observations reported by Kiguchi et al. [207]. The decreasing width of the plateau length distribution with more positive potentials, i.e., with the onset of

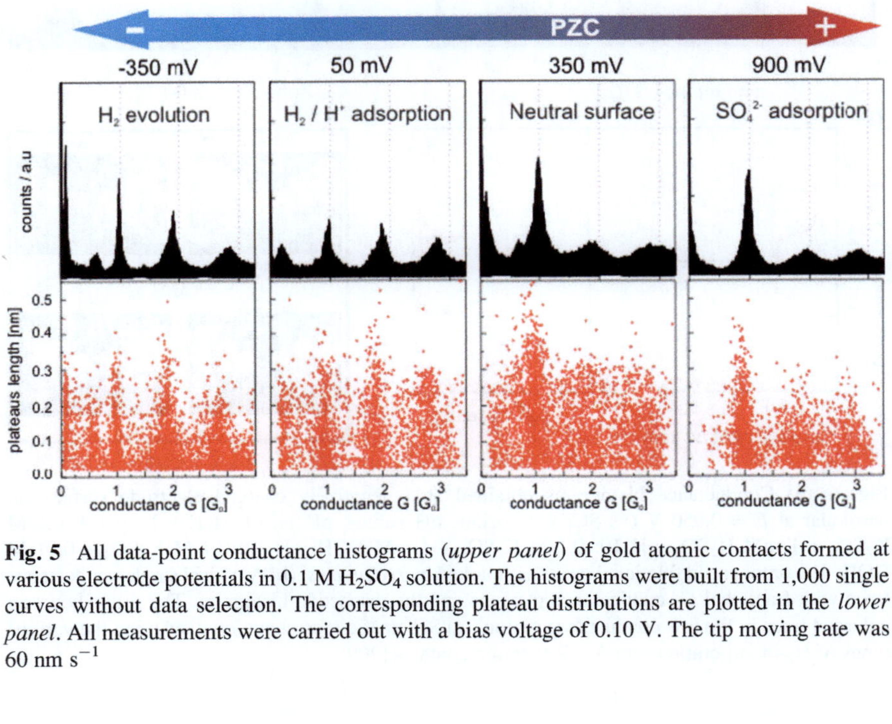

Fig. 5 All data-point conductance histograms (*upper panel*) of gold atomic contacts formed at various electrode potentials in 0.1 M H_2SO_4 solution. The histograms were built from 1,000 single curves without data selection. The corresponding plateau distributions are plotted in the *lower panel*. All measurements were carried out with a bias voltage of 0.10 V. The tip moving rate was 60 nm s^{-1}

sulfate ion adsorption (Fig. 5), is interpreted as an adsorbate-induced stabilization of certain atomic configurations [192, 201, 202, 215].

The exponential tail in the conductance histograms in regions $\ll G_o$ is attributed to an adsorbate-induced decrease in the adhesive interactions between Au atoms, and prevents the occurrence of a "jump-out-of-contact" characteristics. Similar observations were also reported under UHV conditions [195] and in air [216].

3.2.3 Effect of H_2 on Gold Atomic Contacts

Figure 6a illustrates conductance histograms recorded at $E = E_{pzc} - 0.30$ V at various solution pH values. These experiments demonstrate that the fractional conductance is sensitive to the concentration of hydronium ions or hydrogen in

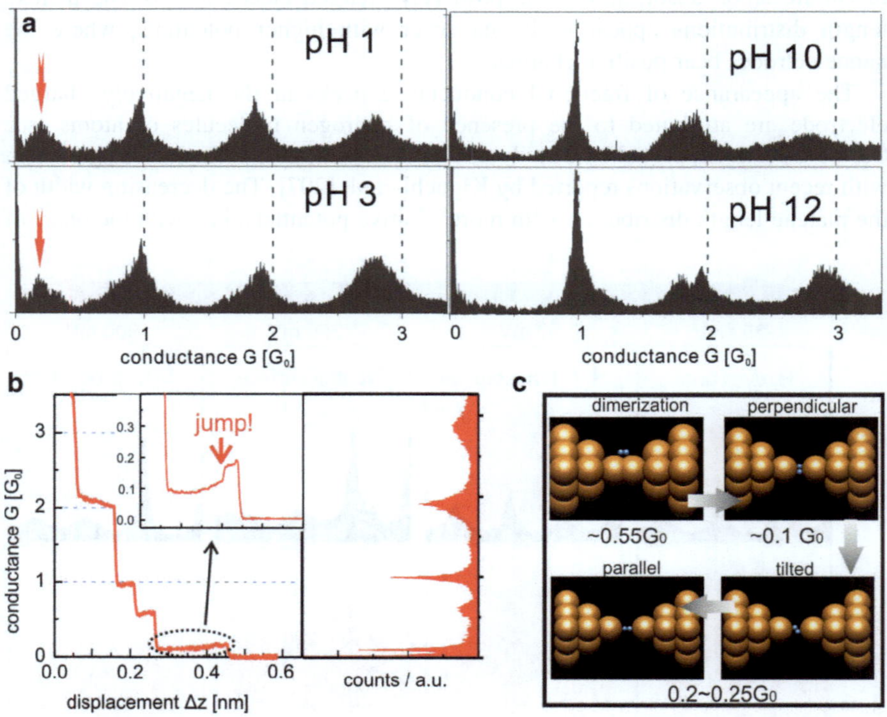

Fig. 6 (**a**) Conductance histograms obtained at a negatively charged electrode surface, in particular at $E = 0.050$ V (vs SCE) at various pH values: pH 1: 0.1 M H_2SO_4; pH 3: 0.1 M K_2SO_4 + 1 mM H_2SO_4; pH 10: 0.1 M K_2SO_4 + 1 mM KOH; pH 12: 0.1 M K_2SO_4 + 0.01 M KOH. The pronounced fractional peaks are marked by *red arrows*. (**b**) A typical single-conductance trace with fractional G_0 conductance plateaus and the respective histogram. The *inset* shows an enlarged trace in the low conductance region. (**c**) Schematic representation displaying the various types of H_2-incorporation into Au–Au atomic contacts [209]

either atomic or molecular forms, i.e., it is modulated chemically and not by the surface charge state.

Figure 6b shows a representative single-conductance trace, as well as the respective histograms. They exhibit well-developed plateaus at 0.1 G_0 and 0.5 G_0. The careful inspection of the low-conductance plateaus around 0.1 G_0 (see inset in Fig. 6b) reveals small jumps from ~0.1 G_0 to ~0.2 G_0, just before the conductance drops to values close to zero.

In an attempt to interpret these observations we refer to a recent ab initio density function theory (DFT) study by Barnett et al. [217]. These authors demonstrated that H_2 molecules form rather stable bridging structures with Au atomic contacts. They also report that a configuration of the H_2 molecules, with the H–H axis perpendicular to the axis of the contacting gold, leads to a conductance of ~0.1 G_0, while the conductances of tilted or parallel H_2 assemblies yield higher values, ranging between 0.20 and 0.25 G_0. We hypothesize that the breaking of Au–Au atomic contacts in the hydrogen evolution region of an electrochemical system might lead to a similar scenario, e.g., the relative positions of H_2 and gold leads are changing from a perpendicular one, via a tilted one, to a parallel one, just before the Au–Au nanocontacts breaks. Our tentative interpretation is supported by Au–Au pulling curves under cryogenic conditions [194].

The conductance peak near half-multiples of G_0 was found under both cryogenic [194, 195] and electrochemical [205, 207] conditions. Theoretical studies revealed that stretched single atomic chains have a strong tendency to form dimers spontaneously [218, 219]. DFT-based calculations suggest that the dimerization-induced conductances range from 0.58 G_0 [220] to 0.4 G_0 [218]. Adopting this result to our experimental observations, and supported by a recent work of Jelinek et al. [204], we suggest that the plateaus around 0.5 G_0 represent *H_2-assisted* dimerized gold atomic contacts (see Fig. 6c).

In consequence, the following scenario is suggested (c.f. Fig. 6c). During the elongation of an atomic contact, an H_2 molecule (which is electrochemically generated at the leads at potentials under consideration) couples weakly with gold atoms in the constriction region, assisting the formation of an Au dimer, which leads to a junction conductance of 0.5 G_0. Further stretching of the contact is followed by the incorporation of H_2 into the contact, initially in a perpendicular and subsequently into a parallel relative orientation of molecular and nanocontact axis, as represented by the small jump in the 0.1 G_0 plateau just before the contact breaks [209].

3.2.4 Statistical Analysis of the Plateau Lengths

The mechanical properties of Au atomic contacts were investigated by analyzing plateau lengths histograms. Figure 7a illustrates the most probable plateau lengths, obtained in a series of experiments with a stretching rate of 40 nm s^{-1}, as a function of the applied electrochemical potential for the first conductance peak developed in the interval between 0.75 G_0 and 1.25 G_0. The graph reveals a distinct maximum,

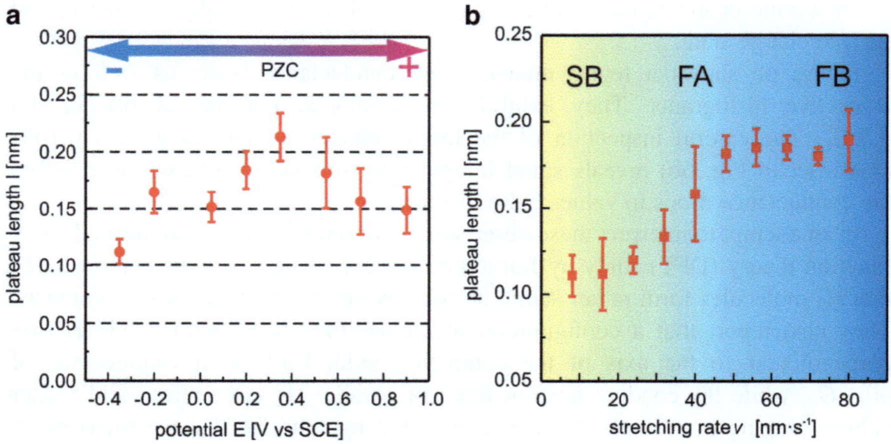

Fig. 7 (**a**) Plateau length of the first Au–Au nanocontact conductance peak (developed between 0.75 G_0 and 1.25 G_0) in function of the electrode potential in 0.1 M H_2SO_4, stretching rate $v = 40$ nm s^{-1}. (**b**) Dependence of the plateau length l of Au atomic contacts on the stretching rate v recorded at $E_S = -0.200$ V (vs SCE). Three characteristic regions are indicated as *SB*, *FA*, and *FB* (see text for more details)

in the vicinity of the PZC $E_{pzc} = 0.25$ V, on the Au(111) electrode under the current experimental conditions, which correlates with the position of the maximum in surface energy [221].

Every atom of the nanocontact constriction represents an active "surface" atom. The tensile force in the plateau region increases upon stretching atomic contacts elastically [188, 189]. The rupture of the gold monatomic contact takes place when the applied tensile force reaches a critical value, sufficient to break a single Au–Au bond. As a consequence, the atomic contact with the most "sticky" atoms at the E_{pzc} is the most difficult one to be break. The stronger adhesion eventually requires a larger load, which seems to be reflected in a longer stretching distance until the contact breaks. This interpretation is supported by the observation, that the plateau length of Au–Au atomic contacts before breaking is longer in the presence of weakly adsorbed ClO_4^- ions, as compared to the much stronger adsorbed SO_4^{2-} ions. The latter enhance the mobility (i.e., weaken adhesion) of surface Au atoms.

We extracted the dependence of the plateau length on the stretching rate in an attempt to explore the mechanical properties of Au–Au nanocontacts under electrochemical conditions. Figure 7b illustrates a typical example, obtained for the main conductance peak around 1 G_0 at $E_S = -0.200$ V. One may distinguish between three regions. At low stretching rates ($v < 16$ nm s^{-1}) the plateau length maintains a constant value of ~0.12 nm. At stretching rates higher than $v \sim 48$ nm s^{-1} the plateau length reaches a maximum of ~0.18 nm and remains rather constant. In the intermediate regions a monotonic increase of the plateau length with the stretching rate is observed. These observations illustrate that the Au–Au bond breaking process depends on the applied stretching rate.

In this context we note those force-dependent experiments with biological molecules [222–226] and the very few case studies reported for non-biological systems [224, 227, 228] demonstrate that bond rupture forces depend on the applied load. Specifically, barriers in the energy landscape are lowered, and bond lifetimes are shortened under external force. Evans et al. [229, 230] developed a thermodynamic model of the bond-breaking process. The theory predicts the spontaneous dissociation of a chemical bond in the absence of external forces, due to thermal fluctuations (spontaneous breakdown). At faster stretching rates the external force increases quickly enough to lower the dissociation energy barrier E_b before spontaneous breakdown takes place. In this regime, the breakdown force is finite and increases linearly with the logarithm of the force loading rate [229]. Finally, at very fast stretching rates the external force lowers E_b to zero, and a maximum of the breakdown force is reached, which leads to the adiabatic regime.

This approach is adopted to discuss the load-dependence of breaking Au–Au nanocontacts at different stretching rates (Fig. 7b). We benefited from a discovery of Tao et al. [227, 228], who found that the external force is linearly proportional to the plateau length. Thus, the breaking force and the loading rate of force can be replaced by plateau length and stretching rate. Considering the analogy between force and stretching rate in the current experiment, we assign the regions at low ($v < 16$ nm s^{-1}) stretching rate to a self-breaking (SB) regime, the intermediate (16 nm s^{-1} $< v <$ 48 nm s^{-1}) stretching rate to a force-assisting (FA) regime, and the high ($v >$ 48 nm s^{-1}) stretching rate to a force-breaking (FB) regime [231]. The three regimes observed are also consistent with the theoretical simulations by Leiva et al. [232] and Liu et al. [233].

3.3 Role of Anion Adsorption on Gold Atomic Contacts

3.3.1 Electrochemical Characterization

The adsorption strength of anions on Au electrodes follows the sequence of ClO_4^- $< SO_4^{2-} < Cl^- < Br^- < I^-$. The strong specific adsorption of halide ions leads to a partial charge transfer between the adsorbate and the metal electrode [234].

The CVs of Au(111) in 0.1 M $HClO_4$, 0.1 M H_2SO_4, 0.1 M $HClO_4$ + 1 mM KCl, 0.1 M $HClO_4$ + 1 mM KBr, and 0.1 M $HClO_4$ + 1 mM KI are displayed in Fig. 8. All curves, except that for 0.1 M $HClO_4$,, exhibit common features. However, they are located in different potential regions. The most prominent peaks reveal the lifting/formation of reconstruction of the Au(111) surface, coinciding approximately with the respective potentials of zero charge E_{pzc}, followed by a potential region of disordered (mobile) anion adsorption. At the most positive potentials displayed, various ordered overlayers of the specifically adsorbed anions are formed (for a detailed review we refer to [235]). In addition, we would like to emphasize two trends: (1) the potentials of zero charge shift with increasing strength of specific adsorption towards more negative values ($I^- < Br^- < Cl^- <$ $SO_4^{2-} < ClO_4^-$) [234] and (2) the regions of disordered anion adsorption shrink with increasing adsorption strength.

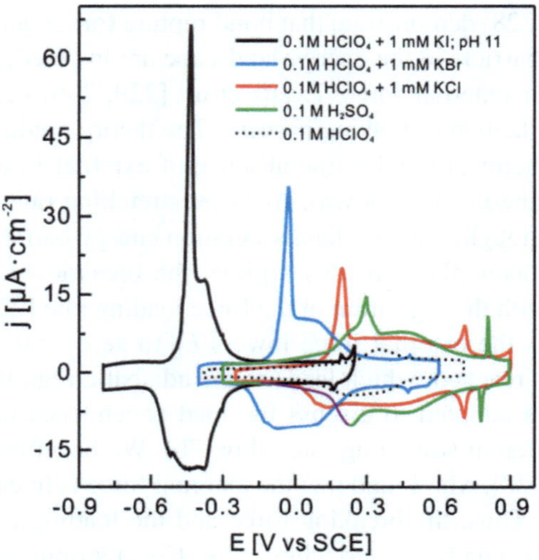

Fig. 8 Cyclic voltammograms of Au(111) in 0.1 M $HClO_4$ (*dotted line*), 0.1 M H_2SO_4 (*green*), 0.1 M $HClO_4$ + 1 mM KCl (*red*), 0.1 M $HClO_4$ + 1 mM KBr (*blue*), and 0.1 M $HClO_4$ + 1 mM KI (*black*) in the double layer region. Scan rate 50 mV s^{-1}

3.3.2 Single Conductance Traces and Conductance Histograms

Figure 9a displays selected conductance–distance traces and Fig. 9b, c the corresponding conductance histograms of Au atomic contacts formed in the presence of SO_4^{2-} and Cl^- ions at various potentials.

At negative potentials and charge densities, i.e., in the absence of specific anion adsorption, all conductance histograms display distinct maxima, reflecting the well-developed, flat, and long steps at integer multiples of G_0 in the individual conductance–distance traces. Examples are displayed in Fig. 9a as *curves A*. Conductance histograms constructed in the potential regions of the ordered sulfate and halide ion adlayers also exhibit pronounced peaks. However, one observes that the corresponding single traces exhibit rather unstable and fluctuating plateaus (*curves B* in Fig. 9a). In potential regions of the disordered anion adsorption we found mostly individual conductance–distance traces, with rather abrupt steps resulting in a completely broken junction (*curves C* in Fig. 9a). The temporarily created atomic junctions appear to be very fragile, which does not allow a “step-by-step” evolution during the elongation of contacts. This leads to a rather small data base for the quantitative analysis, as reflected in the low number of counts in the histograms at potentials of disordered anion adsorption.

These experimental results demonstrate that the specific adsorption of anions leads to a significant decrease in junction stability, as illustrated in particular in the potential regions of disordered anion adsorption, as well as in the fluctuations in the plateau currents at potentials of the 2D-ordered anion adlayers. The instability of the conductance plateaus were statistically analyzed, by determining the standard

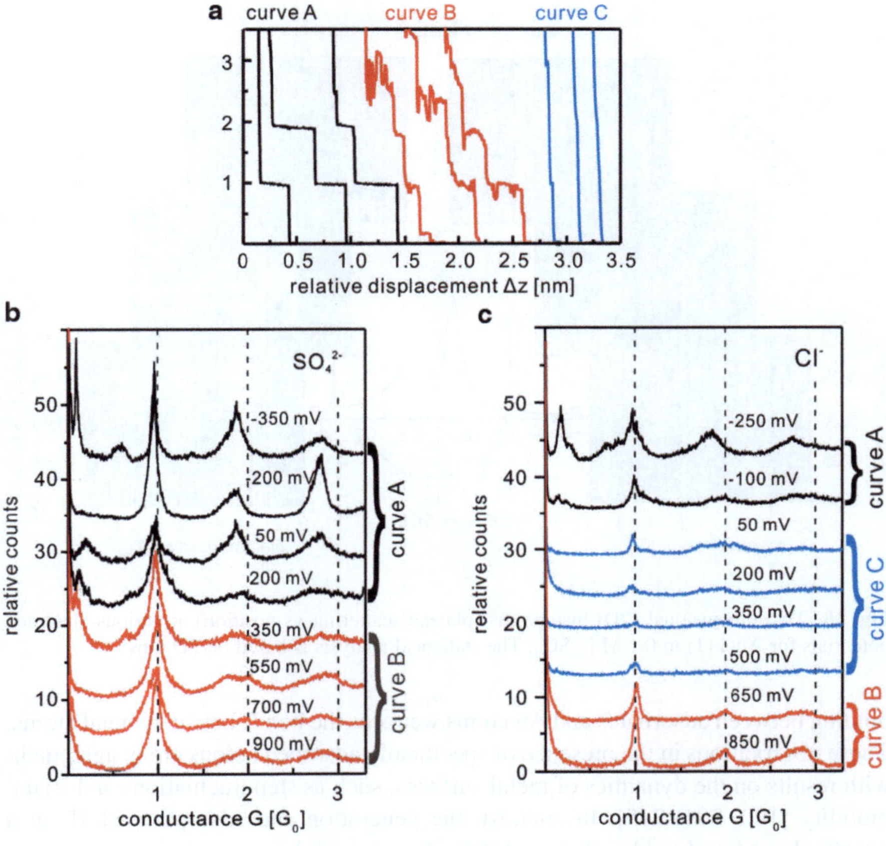

Fig. 9 (**a**) Three types of single conductance traces for Au atomic contacts as observed in the presence of specifically adsorbed anions characterized by well-developed plateaus (*curves A*), noisy plateaus (*curves B*), and abrupt steps with no plateaus (*curves C*), respectively. The stretching rate was 60 nm s^{-1}, and bias voltage applied was 0.100 V. (**b,c**) All data-point conductance histograms at various electrode potentials in the presence of specifically adsorbed SO_4^{2-} (**b**) and Cl^- (**c**), respectively

deviations of the plateau currents in individual traces, and subsequently summing over all families of plateaus, as recorded for a large number of individual traces. The standard deviation basically reflects the population of data points of a specific plateau, and thus provides information on the noise level of the respective plateau.

Figure 10 shows a 2D plot of the plateau position vs scattering level, as constructed from 1,000 single traces at various potentials in the double-layer region for Au atomic contacts in 0.1 M H_2SO_4. The scattering levels ($\pm 0.075\ G_0$) of plateaus formed at a positively-charged surface are larger, approximately by a factor of 3, as compared to those at a negatively charged electrode ($\pm 0.025\ G_0$). This result correlates with the shape of the families of conductance–distance traces displayed in Fig. 9a as *curves A* and *curves B*, respectively. The trend is more pronounced for the more strongly specifically adsorbed halide ions. The strong

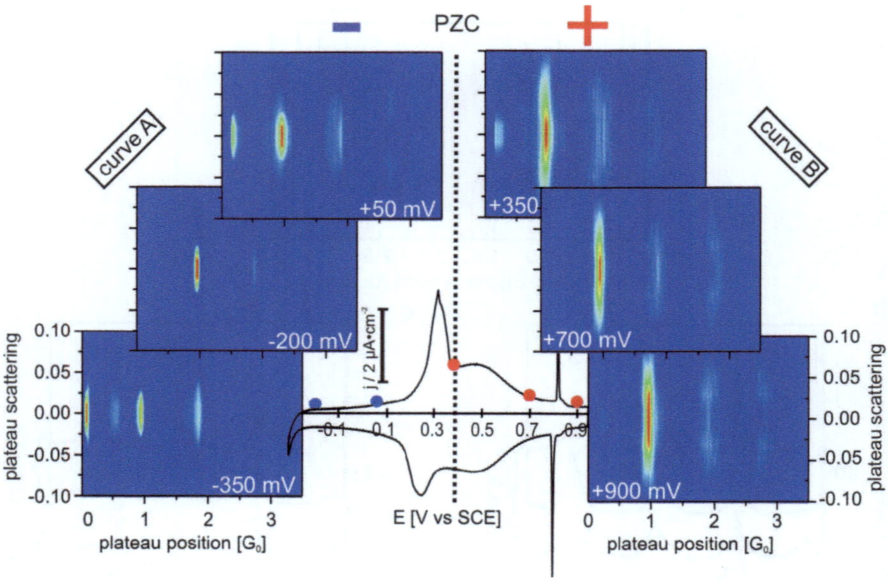

Fig. 10 Two-dimensional (2D) histograms (plateau scattering vs position) at various electrode potentials for Au(111) in 0.1 M H_2SO_4. The statistical analysis is based on 2D bins

binding between adsorbates and Au atoms weakens the bonds between metal atoms. These observations in the presence of specifically adsorbed anions are in agreement with results on the dynamics of metal surfaces, such as step fluctuations and island mobility [192, 201, 215]. In contrast, the generation and adsorption of H_2 at a negatively charged gold surface stabilizes the created Au–Au nanocontacts.

3.3.3 Shift of the 1 G_0 Conductance Peak

A closer inspection of the predominant peak in the conductance histogram at ~ G_0 (=77.5 μS) reveals that its position and magnitude depend on the applied electrode potential, as well as on the strength of anion adsorption (Fig. 11). The peak position shifts in the presence of weakly specifically adsorbed ions (e.g., ClO_4^-, SO_4^{2-}) to value smaller than G_0.

The down-shift of the 1 G_0 peak in the conductance histogram has been observed experimentally and theoretically [236, 237]. In addition, the above study revealed that the down-shifts are larger for higher-order conductance peaks. The down-shift observed for the 1 G_0 peak of the Au atomic contacts cover a range between 100 and 500 Ω. Theoretical model studies suggest that this downshift could be attributed to impurities or lattice defects, disorder or boundary corrugations, which generate backscattering in the atomic contact region [238, 239].

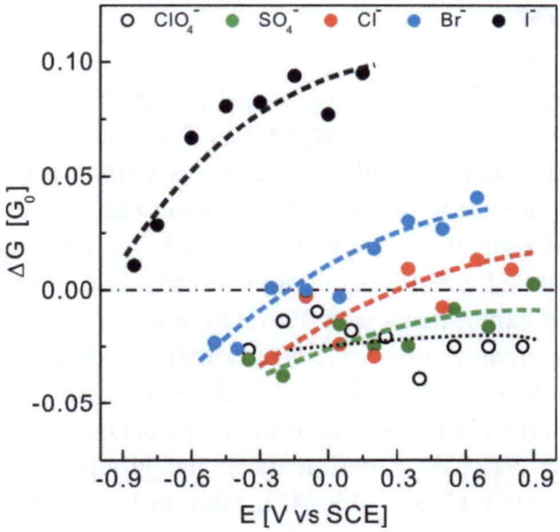

Fig. 11 Dependence of $\Delta G_0 = G_P - G_0$ on the electrode potential upon adsorption of different anions. G_P is defined by a Gaussian fit to the conductance peak closest to G_0 in the respective conductance histograms

In the presence of specifically adsorbed ions, e.g., for halides at potentials $E > E_{pzc}$, the trend changes and the position of the 1 G_0 peaks shifts to higher values than the quantum conductance unit. The up-shifts follow the order of adsorption strength, $Cl^- < Br^- < I^-$.

We note that the measured conductances are the total conductance, which is determined by the number of conducting channels and the transmission probability of the respective channels. The number of channels of monatomic contacts is determined by the number of available valence orbitals [180]. The transmissions of these channels vary from zero to one, and are affected by the atomic configuration and the state of strain [240]. The strong adsorption of halides on the gold surface shows a partially covalent character, which increases in the order $Cl^- < Br^- < I^-$. The mixture (hybridization) of electronic states of absorbed anions and metal atoms may lead to a non-negligible alteration in the electronic state and atomic structures. In other words, the number of conductance channels contributing to the experimentally measured signal might increase upon specific anion adsorption.

The enhancement of the ballistic transmission through atomic wires upon strong chemical adsorption is supported by ab initio DFT-based calculations [220], and was also observed in the presence of molecular adsorbates [201].

4 Charge Transport in "Au–α,ω-Alkanedithiol–Au" Junctions

"Au–α,ω-alkanedithiol–Au" junctions represent one of the simplest molecular junctions which have been studied since 1971 [241]. These alkanedithiol-based junctions were developed as a model system and test bed for various experimental approaches, as well as for theoretical studies for exploring fundamental concepts of charge transport in molecular junctions.

4.1 Introduction

Alkane(di)thiols possess a large energy gap between the HOMO and LUMO of about 7–8 eV and strongly localized σ-orbitals. Thus, these molecules are considered as insulators, and, consequently, the tunneling current is expected to decrease exponentially with increasing molecule length and to be temperature-independent.

Charge transport properties of alkane(di)thiol-based molecular junctions have been extensively studied, employing various techniques (c.f. reviews in [9, 10, 20, 23]). Most transport experiments were carried out under ambient conditions in air and/or in non-polar organic solvents. 1,8-Octanedithiol (ODT) embedded between two gold leads developed as a particularly well-studied system, employing a wide range of experimental techniques. Single-molecule conductance data of ODT have been obtained from STM-BJ experiments in the groups of Tao [32, 59, 242], Venkataraman [61, 116], Haiss [243–245], Fujihira [115], Segalman [117], Chen [55], and us [64, 114], by CP-AFM in the groups of Lindsay [28, 171, 246, 247], Frisbie [29, 248, 249], Xu [250], and Scoles [251], and by MCBJ in the groups of Schoenenberger [40, 252] and Ruitenbeek [253]. Dropping mercury electrode [254, 255], nanopores [256], and crossed wire assemblies [257] have also been used to measure the conductance of alkane(di)thiol molecular junctions composed of a few or a larger number of contacted target molecules.

Apart from the more conventional transport measurements of molecular junctions at constant bias voltage, alkane(di)thiols-based molecular junctions were also characterized by transition voltage spectroscopy [258, 259], AC voltage modulation [260], and inelastic electron tunneling spectroscopies [261].

The conductance values of ODT single-molecular junctions can be categorized in three distinctly different families around $2.5 \times 10^{-4}\, G_0$ (~20 nS), $0.5 \times 10^{-4}\, G_0$ (~4 nS), and $0.1 \times 10^{-4}\, G_0$ (~1 nS). The mid-conductance value was reported by almost all research groups, while the high and the low values were strongly dependent on the chosen experimental conditions, as well as on strategies of data analysis [19, 242, 245, 252]. Systematic studies on the distance respective molecular length dependence have led to decay parameters β_N ranging between 0.5 and 1.2 [9, 10, 28, 32, 40, 64, 242, 244].

The nature of the conductance families reported for single alkanedithiol junctions is still under debate. Current interpretations are based on different contact geometries [64, 115, 242], different molecular conformations [64, 244], substrate roughness [245], or the control of tip movement [252].

The discrepancies in the reported conductance data of "Au–alkanedithiol–Au" junctions attracted our attention, and we decided to carry out an in-depth experimental study of the charge transport properties of "Au–α,ω-alkanedithiol–Au" molecule junctions in a non-conducting solvent. The combination with quantum chemistry ab initio simulations yielded a detailed view of this archetype of molecular junctions, and helped to resolve the puzzle on the role of microscopic geometries at the contacts and in the molecular conformation.

4.2 Single-Molecule Conductance Measurements

Charge transport characteristics in "gold–alkanedithiol–gold" junctions were recorded with an STM-based contact junction approach ("gentle touch") with 0.1 mM of the respective alkanedithiol in 1,3,5-trimethylbenzene [64]. The i-Δz retraction curves were either exponentially decaying (50%), noisy (10%), or showed characteristic plateaus with a typical length of 0.04–0.15 nm, separated by abrupt steps (40%). Figure 12 illustrates a set of representative i-Δz curves of the latter type for the breaking of nonanedithiol (NDT) junctions. The observed currents are three to four orders of magnitude smaller as compared to metal nanocontacts. These conductance steps are ascribed to the breaking of individual respective multi-molecule junctions formed between the gold STM tip and the substrate. Control experiments in the absence of NDT displayed almost exclusively (98%) exponentially decaying traces.

The statistical analysis of the plateau current yields histograms with characteristic peaks (Fig. 13). The careful analysis of these current peaks allows the identification of three distinctly different sequences of equally spaced maxima, which are attributed to low (L), medium (M), and high (H) conductance molecular junctions. The current within each series scales approximately linearly with the number of peaks. The peak of each sequence is attributed to a *single* molecular junction. The corresponding currents also depend linearly on the applied bias voltage up to ± 0.2 V (see insets in Fig. 13a, b). By fitting the current maxima with a Gaussian function, the following conductance values of the three specific single-molecule "Au-NDT-Au" junctions were obtained: 0.47 ± 0.03 nS (L), 2.0 ± 0.2 nS (M) and 9.9 ± 0.9 nS (H).

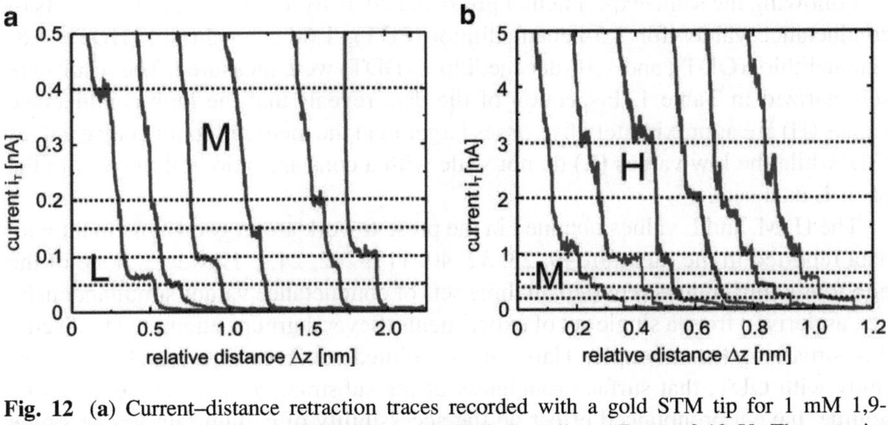

Fig. 12 (**a**) Current–distance retraction traces recorded with a gold STM tip for 1 mM 1,9-nonanedithiol in 1,3,5-trimethylbenzene on Au(111)-(1 × 1), at $E_{bias} = 0.10$ V. The setpoint current before disabling the feedback was chosen at $i_0 = 0.1$ nA. The retraction rate was 4 nm s^{-1}. (**b**) Same conditions as in (**a**), except that the preamplifier limit was chosen at 10 nA. The *dotted lines* represent characteristic regions of the low, mid, and high conductances

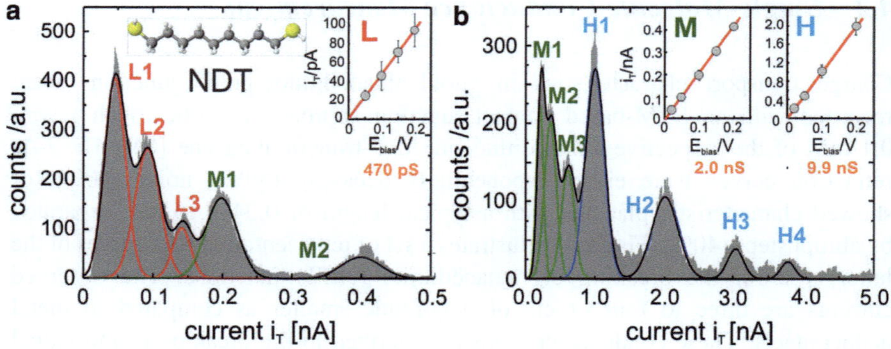

Fig. 13 Plateau count conductance histograms constructed from the plateaus found in the step-like conductance–distance traces for Au–1,9-nonanedithiol–Au junctions. (**a**) 1,600 selected out of 4,300 traces employing a 1 nA (max) preamplifier; (**b**) 1,100 selected out of 4,300 traces recorded with the 10 nA (max) preamplifier. All other conditions are identical to those in Fig. 12. The *insets* in (**a**) and (**b**) show that the currents within each series scale approximately linearly with the number of peaks [64]

Table 1 Summary of the conductance values of Au–α,ω-alkanedithiols–Au molecular junctions recorded in 1,3,5-trimethylbenzene [64]

α,ω-Alkanedithiol	Conductance values (nS)		
	L	M	H
PDT ($N = 5$)	1.9 ± 0.05	64 ± 5	–
HDT ($N = 6$)	2.45 ± 0.06	20 ± 2	95 ± 10
ODT ($N = 8$)	0.89 ± 0.08	4.4 ± 0.4	21 ± 2
NDT ($N = 9$)	0.47 ± 0.03	2.0 ± 0.2	9.9 ± 0.9
DDT ($N = 10$)	0.22 ± 0.02	0.45 ± 0.04	1.68 ± 0.03

Following the same experimental protocol and analysis strategy, multiple sets of conductance values for 1,5-pentanedithiol (PDT), 1,6-hexanedithiol (HDT), 1,8-octanedithiol (ODT), and 1,10-decanedithiol (DDT) were measured. The results are summarized in Table 1. Inspection of the data reveals that the high-conductance values (H) are approximately five times larger than the medium-conductance values (M), while the low values (L) do not scale with a constant ratio with respect to the M or L data sets.

The H, M, and L values obtained in the present work are in good agreement with data reported in the literature [9, 23, 32, 40, 117, 242, 243]. However, none of the previous contributions reported all three sets of conductance values simultaneously, i.e., as derived from a single set of experiments. Several groups attempted to discuss this situation. For example, Haiss et al. pointed out recently, based on a case study with ODT, that surface roughness of the substrate, as well as the approach regime, have a pronounced effect on the accessibility of certain families of single molecule conductances [245]. Gonzalez et al. put forward an interpretation based on multimolecular junctions in certain most probable microscopic arrangements of lead-molecule contact geometries [252].

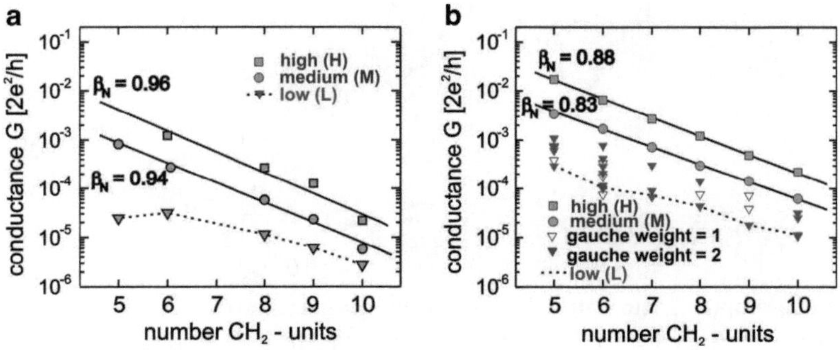

Fig. 14 Experimental (**a**) and calculated (**b**) conductance values of Au–n-alkanedithiol–Au junctions vs number n of methylene units in a semilogarithmic representation. The three sets of conductance values – high (H), medium (M), and low (L) – are shown as *squares*, *circles*, and *triangles*. The *straight lines* were obtained from a linear regression analysis with decay constants β_N defined per methylene (CH_2) unit. The conductances of many different, nonequivalent gauche isomers cover the window below the medium values in (**b**) [64]

Figure 14 shows the semi-logarithmic plots of the various single junction conductances vs molecular length (Fig. 14a – experimental data; Fig. 14b – calculated conductances). The length is expressed as the number of CH_2 units. The H and M conductance values follow a simple tunneling model given by $G = G_C\exp(-\beta_N N)$ with decay constants β_N of 0.96 ± 0.15 for H and 0.94 ± 0.05 for M. These values of β_N are in a good agreement with the literature on single junctions "Au–alkanedithiol–Au" data [32, 118], as well as the electron transfer through compact and aligned monolayers of alkanethiols using nanopores [262], mercury contacts [254, 263, 264], CP-AFM [247], or redox probes [265, 266].

On the other hand, the low-conductance values (L) give a poor linear correlation of the molecular length with an approximate decay constant $\beta_N \sim 0.45 \pm 0.09$, distinctively different from the H and M sequences. The estimated value of β_N(L) is rather close to results reported by Cui [28] and Haiss [243]. Haiss et al. [244] found a pronounced temperature dependence of these L values, which scales logarithmically with T^{-1} in the temperature range 293–353 K, indicating a transport mechanism different from a simple tunneling model.

4.3 Interpretation of Multiple Conduction States

To provide further insight into the nature of multiple conduction states observed experimentally, DFT-based calculations of alkanedithiols coupled to Au electrodes were carried out. Calculations were performed for different configurations of an "extended molecule" composed of an n-alkanedithiol with variable chain length ($n = 4\ldots10$) bridged between two pyramids of 45–55 Au atoms (Fig. 15a–c). These clusters mimic the contact region of the gold electrodes. Molecular

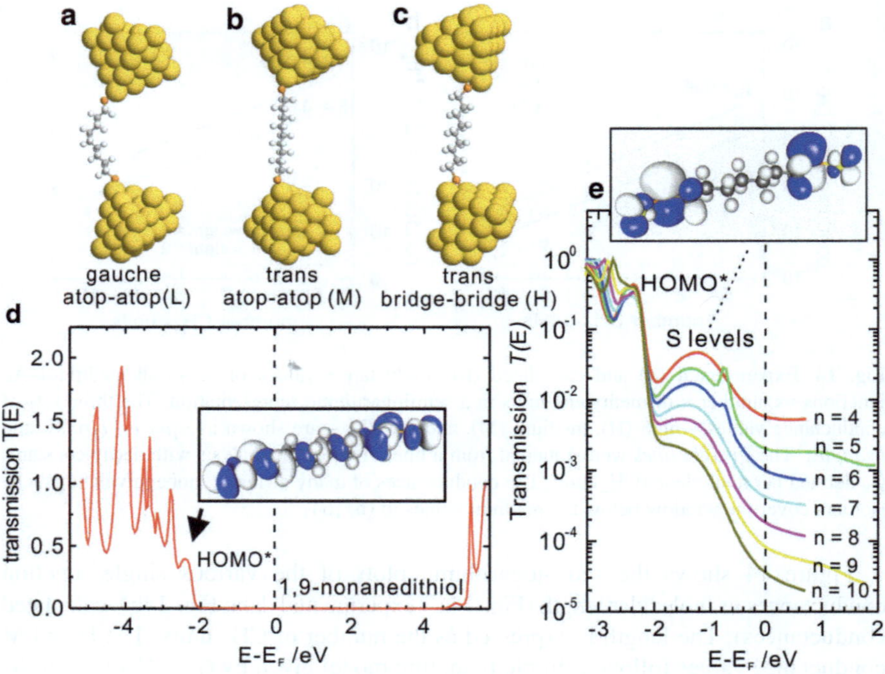

Fig. 15 Three typical arrangements of a single alkanedithiol molecule bridged between Au electrodes as used for the conductance calculations. (**a**) 1,9-Nonanedithiol (NDT) with one gauche defect and both terminal sulfur atoms coordinated in atop position (low, L). (**b**) NDT in all-trans conformation and atop–atop coordination (medium, M). (**c**) All-trans NDT in a bridge–bridge coordination (high, H). (**d**) Calculated energy-dependent transmission *T*(E) for a single NDT molecule bridged between gold electrodes in an all-trans atop–atop arrangement. The *inset* shows a molecular orbital of the Au–S–$(CH_2)_9$–S–Au cluster, formed by the HOMO of nonane hybridized with the sulfur and the Au electronic levels. This state determines the tunneling barrier. (**e**) Energy-dependent transmissions of *n*-alkanedithiols of various lengths *n* around the Fermi level. The *inset* displays the alkanedithiol's HOMO, which is an antisymmetric superposition of atomic wave functions localized on both sulfur atoms. The energy difference within its symmetric part, HOMO-1, is exponentially decaying with *n*

conformation and contact geometries have been varied. Examples are *gauche* and *trans* conformation of the alkyl chain, with the sulfur atoms bridged to one (*atop*), two (*bridge*) or three (*hollow*) gold atoms and/or combinations of them. The electron charge flow was described within the Landauer approach, formulated in terms of Green's functions, as implemented in a custom-made simulation package [64, 267].

Figure 15d illustrates a typical energy-dependent transmission curve, as calculated for *n*-nonanedithiol molecular junctions. The transmission represents the probability for electrons injected with the energy *E* from one electrode to be transmitted through the molecular junction. The conductance is defined by the transmission $T(E_F)$, evaluated at the Fermi energy E_F in units of the quantum of conductance $G_0 = 2e^2/h$. Molecular states of alkanedithiols appear as resonance peaks in the transmission spectrum (see Fig. 15d).

The Fermi energy E_F is situated in the alkane HOMO–LUMO gap (7.5 eV), giving rise to an effective barrier $\Phi_B = E_F - E_{HOMO*} = 2.14$ eV. The current flow involves mainly the HOMO (HOMO* of the alkanedithiol) depicted in the inset of Fig. 15d. Accordingly, the transmission $T(E)$ drops rapidly around −2.2 eV below E_F, and a nearly insulating gap spreads up all the way until 5.3 eV above E_F (alkane-LUMO, Fig. 15d, inset). Furthermore, the conductance through the n-alkanedithiol junction of the all-trans isomers with the sulfur coupled to single gold atoms at each electrode (atop–atop geometry) decays exponentially according to $G(N) = G_C\exp(-\beta_N N)$ with a decay constant $\beta_N = 0.83$ per CH_2 unit $G_C = 0.24$, $G = 18.5$ nS as contact conductance (Fig. 14d, circle). We note that the thiol linker-groups introduce two evanescent gap states, built of symmetric and anti-symmetric combination of wave functions, and localized at the sulfur atoms (Fig. 15e, inset). These states appear as broad resonances centered around 1.4 eV below E_F in the transmission spectrum (see Fig. 15e). The tails of such broad peaks, when approaching the Fermi energy (Fig. 6e), define the amplitude of the molecular conductance $G(n) \approx 2G_0\Gamma_S\gamma_0\exp[-\beta(E_F)n]/(E_F - E_S)^2$ and, therefore, predict an exponential distance dependence. Γ_S denotes an on-resonance probability amplitude (inverse lifetime) for an electron to hop from the sulfur atom to the nearest electrode. The other probability $\gamma_n(E) \ll \Gamma_S$ reflects tunneling through the molecule and is exponentially small, $\gamma_n(E) = \gamma_0\exp[-\beta(E)n]$, where $\beta(E) \propto [E - E^*_{HOMO}]^{1/2} d_0$ with d_0 as the unit length of the alkane chain, $d_0 = 1.28$ Å. Note that an evanescent state localized close to an Au surface does not change the tunneling asymptotics.

The model calculations revealed that the *bridge* contact geometry increases the conductance twice, as compared to the *atop* one. Due to the increase of the coordination number of the sulfur to the Au atoms (factor 2), the opening angle for the incoming electron wave is enlarged, providing an unchanged conformation of the alkyl chain. In other words, when two Au atoms are placed near one sulfur atom, the hopping probability increases by a factor of 2. On the other hand, this trend does not continue with a threefold *hollow* contact geometry. The contributions of three simultaneously contacted gold atoms to the overall junction conductance are overcompensated by a significant increase of the Au–S bond length.

Independent of the contact geometry, the calculations also demonstrated that the introduction of *gauche* defects resulted in a decrease of the bridge conductance by a factor of 10, as compared to an *all-trans* alkanedithiol chain (see Fig. 14b, triangles). Due to variations in the number and positions of gauche defects, as well as various contact geometries, the molecular junctions can exhibit conductance values up to two orders of magnitude below the conductance values of an *all-trans* conformation of the alkyl chain.

Comparison of the experimental data with the trends predicted from quantum chemistry ab initio simulations demonstrated that the multiple conductance values of "Au–alkanedithiol–Au" junctions could be attributed to different Au–sulfur coordination geometries and to different conformations of the alkyl chain. In particular, the medium conductance corresponds to an *all-trans* conformation of the alkyl chain, with each sulfur atom coordinated in *atop* position to a single

Au atom (see Fig. 15b). The "high" conductance values H represent an all-trans alkyl chain in combination with both sulfurs coordinated to two Au atoms in a *bridge* geometry (Fig. 15c). The theoretically predicted fourfold increase of the conductance, as compared to an *atop–atop* all-trans bridge, is close to the experimentally observed increase by a factor of 5. Other possible configurations with *all-trans* conformations (e.g., *atop–bridge*, multiple molecular junctions) are assumed to contribute as part of the distribution to the less-dominant peaks in the conductance histograms (e.g., M2, H2, and H3 in Fig. 13).

The sequence of low conductance values L is attributed to isomers of alkanedithiols with *gauche* defects. This interpretation is also supported by the experimentally observed temperature dependence of L conductance values. The alteration of gauche states appears to be an activated process [244].

5 Chemically Controlled Conductance: Torsion-Angle Dependence in Single-Molecule Biphenyl Junctions

5.1 Introduction

Biphenyl derivatives (Fig. 16), consisting of two aromatic rings interconnected by a single C–C bond, gained considerable interest as subunits for single-molecule rectifying or switching systems, as the two π systems can be either in the same plane, or perpendicular to each other, representing "on" and "off" states [74, 268–272]. π–π coupling dominates the electron transport in aromatic bridges, and transport can be manipulated and tailored by controlling the degree of electron delocalization. Examples are the triplet energy transfer in a series of mixed

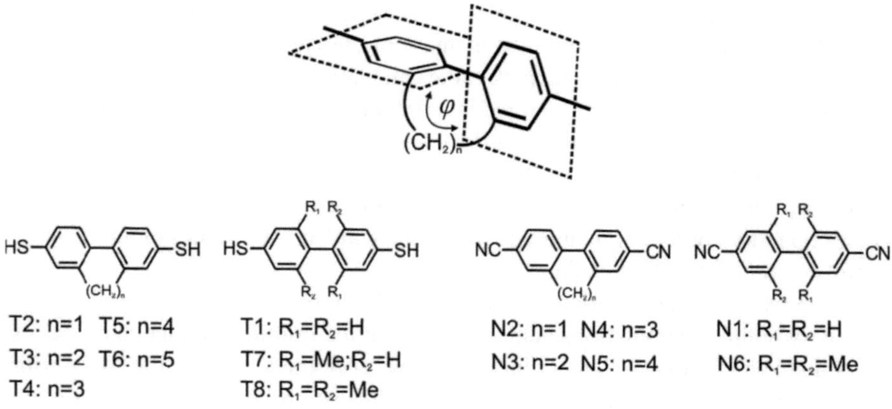

Fig. 16 Molecular structures of the biphenyldithiols T1–T8 and of the biphenyldinitrile N1–N6. The interplanar torsion angle φ of the biphenyl backbone is adjusted by methyl-group substitution or the length of the inter-ring alkyl chain [54, 75]

Ru–Os–bis(2,2′:6′,2″) terpyridine complexes attached to biphenyl spacer groups [273] and the electron transport characteristics of ferrocene covalently bound to OPE rods [274] or oligoaniline junctions [275].

The effect of rotating the middle ring of 1,4-bis-phenylethynyl-benzene on the conductance was estimated by Thomfohr and Sankey using the Landauer formalism in combination with a complex band structure analysis [276]. Venkataraman et al. demonstrated, in a series of terminally amine-functionalized biphenyls comprising electron-donating and -withdrawing substituents, that the junction conductances scale approximately linearly with the square of the cosine of the torsion angle φ between the planes of the two rings [48].

Despite the strong interest in the correlation between torsion angle and transport properties, suitable model compounds enabling the systematic variation of the torsion angle in biphenyl systems have not been realized so far, and the role of geometric and electric effects of the substituents is still being actively discussed [48, 72, 74, 271, 277].

As displayed in Fig. 16, an alkyl chain of varying length connected at the 2,2'-positions of the biphenyl system can adjust the resulting torsion angles φ. For both families of compounds, T and N, the length of the inter-ring alkyl chain is the only structural variation, keeping the electronic structure of the biphenyl systems as uniform as possible. Furthermore, the biphenyl conformation is locked by an intra-molecular bridge, with the number of CH_2 units dictating the torsion angle φ, lowering the expected motion and conformational variation of each molecule immobilized in the junction. The torsion angles φ have been estimated, based on crystal structure data and DFT calculations of the isolated molecules in the gas phase [54, 75].

The choice of the thiol anchoring groups ensures a strong chemical bonding to the leads, i.e., a stable junction configuration, with the current flow dominated by the molecular HOMO level [271].

We note that, although the thiol anchoring groups have been used most often, this choice is far from ideal. It shows a large variation in binding geometries, and does not allow an optimal electronic coupling between target molecules and contacting leads [64].

Venkataraman et al. [49] recently introduced the amine group ($–NH_2$) as a promising alternative, characterized by a more uniform contact geometry [73]. Other anchoring groups that have been explored in (single) molecule conductance studies are pyridine [32, 51], isonitrile (–NC) [49, 52], isothiocyanate (–NdCdS) [55], methyl selenide ($–SeCH_3$) [56], methyl thiol ($–SCH_3$) [56], dithiocarboxylic acid [–CS–SH] [62], dimethylphosphine [$–P(CH_3)_2$] [56], carboxylic acid (–COOH) [59], nitro ($–NO_2$) [278], and even fullerene [60]. The search for the "best" anchoring group for the formation of stable and energetically well-aligned metal-molecule contacts still represents a major challenge for fundamental and applied work in molecular-based electronics.

In this paragraph we complement experimental conductance data of a family of structurally well-tuned biphenyldithiols (BPDT) T1 to T8 [75, 76] with a family of biphenyl-dinitriles (BPDN) N1 to N6 [54] having the same molecular backbone, but different anchoring groups. The electron-withdrawing nitrile group causes a LUMO-controlled electron transport in BPDN molecular junctions [54], which are distinctively different from BPDT. Conductance properties of the –CN anchoring group have been explored in only two previous studies. Kiguchi et al. [53] reported preliminary data with 1,4-dicyanobenzene, but did not observe any clear molecular conductance signatures. Zotti et al. [278] performed current–voltage measurements of 4,4'-dicyanotolane molecules attached to gold leads using an MCBJ technique, but did not report a statistical analysis of their experimental traces, such as based on 1D or 2D conductance histograms.

5.2 *Single-Molecule Junctions of Thiol-Terminated Biphenyls*

The conductances of single-molecule "Au–BPDT–Au" junctions were extracted from STM-BJ experiments based on the repeated formation and breaking of atomic contacts between a gold STM tip and an Au(111) substrate in 0.25 mM solutions of the respective BPDT in a 1:4 (v/v) mixture of tetrahydrofuran (THF) and mesitylene ("hard touch") [75, 76].

Figure 17a shows, as an example, the plateau data-point histograms of T3 at three bias voltages. Each histogram, constructed from more than 1,000 individual traces, reveals a distinct maximum. The peak positions from individual experiments are very reproducible for low bias voltages $|V_{bias} < 0.30\ V|$. The broad asymmetric tail region toward higher conductance values is attributed to contributions from

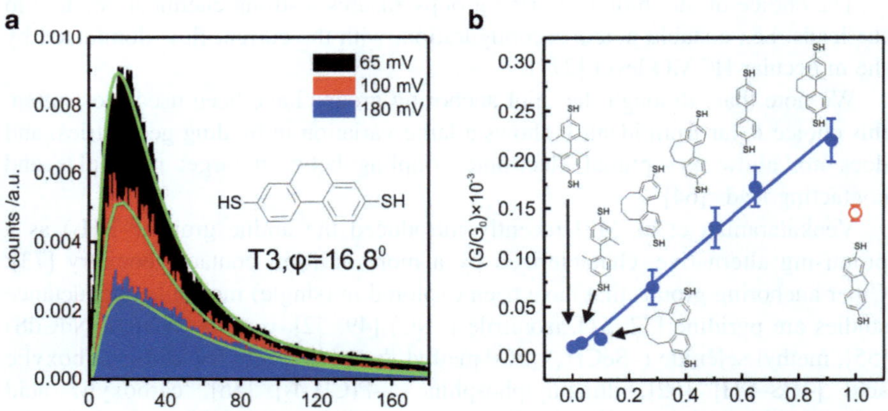

Fig. 17 (**a**) Plateau data-point histograms of the $(CH_2)_2$-bridged biphenyldithiol derivate T3, as recorded from individual current–distance traces in an STM-based break junction experiment (gold tip, Au (111) substrate) at three different bias voltages in mesitylene; $E_{bias} = 0.065$ V (*black*), 0.10 V (*red*), and 0.18 V (*blue*). (**b**) Experimentally determined single-molecule junction conductances of a series of biphenyldithiol molecules vs $\cos^2\varphi$, where φ is the torsion angle between the two phenyl rings as given by the X-ray data [76]

minority junctions with multiple molecules, a modification in substrate-adsorbate coordination, atomic rearrangements upon stretching, or local surface roughness [64, 66, 116, 169, 245].

Introducing a bridging alkyl chain $(CH_2)_n$, $n = 0$–5, in the 2,2′-position (T1 to T6) or $-CH_3$ substitution in 2,2′- (T7) and 6,6′- (T8) hinders the free rotation and enables tuning of the torsion angles φ between the two phenyl rings from 0° to 90°. The values of φ were estimated experimentally from the X-ray structure analysis of the acetyl-protected derivatives [75, 279, 280]. As the angle between the two phenyl rings increases, the conductance drops from 2.2×10^{-4} G_0 (T2) to 9.0×10^{-6} G_0 (T8). The lowest conductance was found for 2,2′,6,6′-tetramethyl-biphenyl-4,4′-dithiol (T8) with $\varphi = 89^\circ$.

Figure 17b displays a linear correlation between the experimentally obtained single junction conductances of the eight BPDT derivatives and $\cos^2\varphi$. The fluorine derivative T2 appears to be an exception. Using $G = b + a \cos^2\varphi$ and excluding T2 from the data fit, we estimate the slope and the intersection as $a_{ex} = (2.44 \pm 0.097) \times 10^{-4}$ G_0 and $b_{ex} = (3.4 \pm 4.7) \times 10^{-6}$ G_0, respectively. The residual conductance b_{ex} at $\varphi = 90^\circ$ stems mostly from σ–π couplings [276, 281]. However, this contribution is rather small. Clearly, the π–π overlap between the phenyl rings dominates the junction conductance [48, 271, 272, 277]. We conclude that twisting the BPDT systems from flat ($\varphi = 0^\circ$) to perpendicular ($\varphi = 90^\circ$) decreased the conductance by a factor of 30, following a linear $\cos^2\varphi$ dependence.

A HOMO-controlled transport mechanism and the experimentally observed linear relation between the junction conductance G and the square of the cosine of the torsion angle φ were also predicted by DFT-based model calculations [282–284] as well as by a simple Two-Level Model (TLM) [267, 285]. In these calculations, the molecular junctions were represented by an "extended molecule" composed of the BPDTs bridging two pyramids of 14–120 gold atoms. Figure 18a exhibits the two most probable and optimized structures *atop–atop* and *bridge–bridge*, which have been subsequently employed in the transport calculations. Figure 18b displays the resulting conductances of T1–T8. The fits of $G = a_{DFT} \cos^2\varphi$ to both series of data reveal a nearly linear behavior (Fig. 17b). However, one should note that the slopes $a_{DFT} = 0.094$ G_0 (atop) and $a_{DFT} = 0.13$ G_0 (bridge) are approximately three orders of magnitude higher, as compared to the experimental values: this is attributed to insufficient accuracy in the HOMO–LUMO level alignment and level broadening in DFT, a still unresolved problem [2, 286].

The main result of the transport measurement and DFT calculations is understood in terms of a TLM [76]. The model as parameterized by the DFT calculations demonstrated that the transmission through the molecular junction factorizes under "off-resonance" conditions, which enables the separate treatment of the "ring-to-ring" and "electrode-ring" transmission components. The electron path through the "extended molecule" is shown to be represented by three barriers, two of them representing the coupling to the leads, and the third defined by the torsion angle φ between the two phenyl rings. The theoretical analysis revealed unambiguously that the angular dependence of the overlap of the π-orbitals dominates the junction

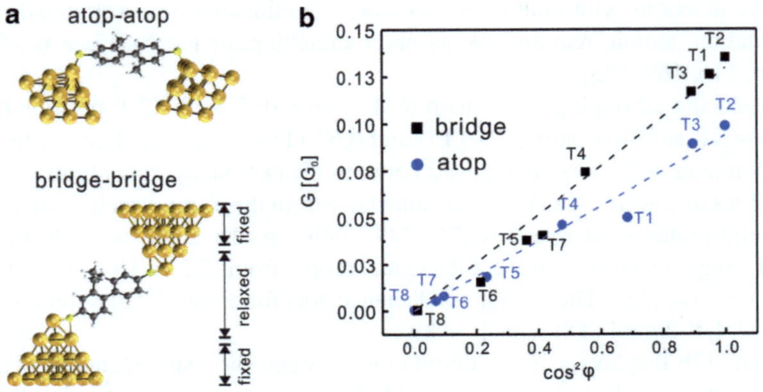

Fig. 18 (**a**) Examples of junction geometries of biphenyldithioles. *Upper panel*: T7 attached to gold pyramids in atop–atop position; *lower panel*: T4 attached in a bridge-bonded position. (**b**) Conductances as obtained from DFT-based transport calculations as a function of $\cos^2\varphi$: The *dashed lines* represent linear fits $G = a_{\text{DFT}} \cdot \cos^2\varphi$ with slopes $a_{\text{DFT}} = 0.094\ G_0$ (atop–atop) and $a_{\text{DFT}} = 0.130\ G_0$ (bridge–bridge) [76]

conductance for torsion angles in the range $0^{\circ} \leq \varphi < 80^{\circ}$. Contributions from degenerate σ–π and π–σ channels were only observed for biphenyl derivatives with completely broken conjugation, i.e., for $\varphi = 90^{\circ}$.

5.3 Single Molecule Junctions with Nitrile-Terminated Biphenyls

Figure 19a shows, as an example, single stretching traces and the 1D histogram of molecular junctions created between the dinitrile derivative N4, torsion angle $\varphi = 44.8^{\circ}$, and two gold leads. The predominant peak with a maximum at $G_{\max} = 4 \times 10^{-5}\ G_0$ is attributed to the most probable conductance [54].

Figure 19b displays the 2D histogram of the experimentally obtained conductance of N4 plotted vs distance [63]. The distance scale z' is normalized with respect to $z' = 0$ at $G = 0.7\ G_0$, to a common point. The chosen procedure is justified, because of the steep decay of the tunneling current after breaking of the last atomic contact. The histogram counts the occurrence of $[\log(G/G_0), z']$ pairs in a 2D field. Figure 19b exhibits the features of gold quantum contacts at $G \geq G_0$, and a second cloud-like pattern in $[10^{-5} \sim 10^{-4}\ G_0, 0 \sim 0.5\ \text{nm}]$. We attribute the latter to the formation of single-molecule junctions of only one type. The center of the cloud is located at $G = 3.5 \sim 4.5 \times 10^{-5}\ G_0$, close to the peak position in the 1D histogram (Fig. 19a). The extension of the cloud along the distance scale is around 0.5 nm, close to the typical length of the plateaus (the inset of Fig. 19a).

The dependence of the single-molecular conductances on the torsion angle φ between the two phenyl rings was addressed by choosing a series of five additional BPDN derivatives (Figs. 16 and 20a) [287]. All conductance histograms exhibited only one dominant single molecular junction-related feature, which is rather sharp

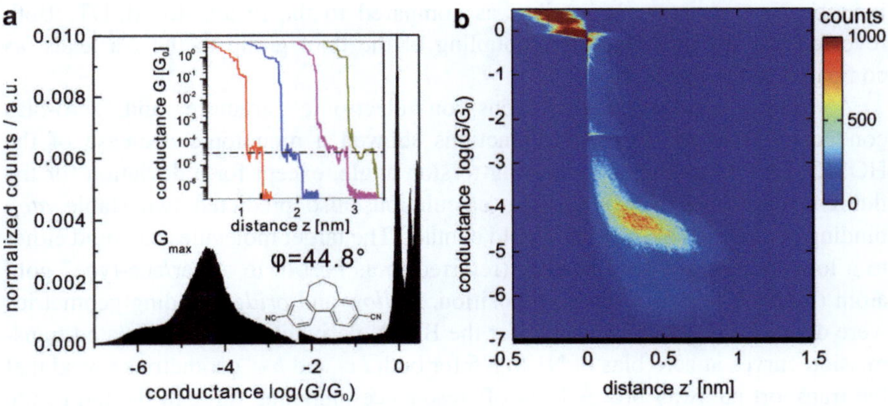

Fig. 19 (**a**) All data-point logarithmic conductance histogram as constructed from ~2,000 individual conductance–distance traces of N4 without any data selection. The experiment was carried out with 0.1 mM N4 solution in mesitylene: THF (4:1, v/v) at $E_{bias} = 0.1$ V. The *inset* shows selected individual stretching traces in a logarithmic conductance scale. (**b**) Two-dimensional (2D) conductance–distance histograms for N4, displaying the range of Au–Au contacts between 1 G_0 and 10 G_0, and a cloud-like feature between 10^{-3}·G_0 and 10^{-5}·G_0 representing the region of molecular junctions [54]

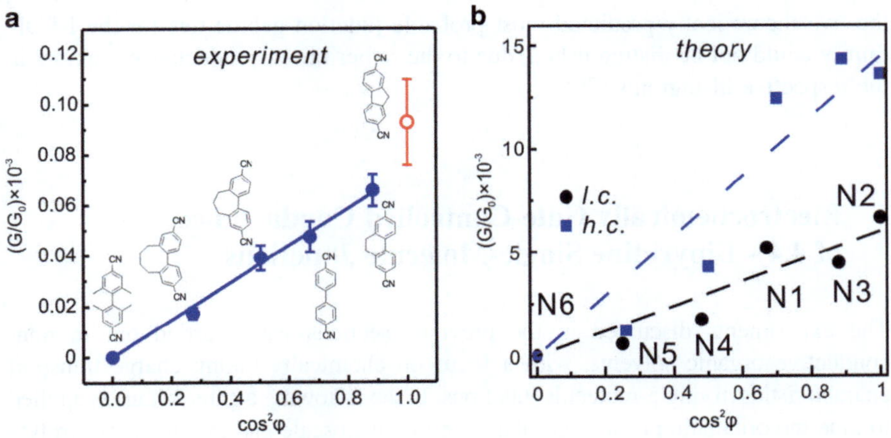

Fig. 20 (**a**) Experimentally measured conductances G/G_0 of the six biphenyl dinitriles N1–N6 as a function of $\cos^2\varphi$. (**b**) Computed conductances of N1–N6 as a function of $\cos^2\varphi$ for the low-coordination (*l.c.*, *circles*) contact geometry and the high-coordination (*h.c.*, *squares*) contact geometry [54]

and well-defined [54]. The result reflects the more uniform nitrile-gold bond, as compared to the family of BPDT derivatives T1 to T8 [64, 76, 242, 243]. The conductance values of all six BPDNs were plotted in Fig. 20a as a function of $\cos^2\varphi$, which demonstrates a linear dependence with the slope of $(7.3 \pm 0.2) \times 10^{-5}$ G_0. This slope, as well as the absolute magnitude of the BPDN molecular junction conductances, is smaller, as compared to the results for BPDT. Both revealed the stronger electronic coupling of the thiol group to the Au leads, as compared to the nitrile group.

Ab initio transport calculations on electronic structure and zero-bias conductances of the molecular junctions showed a monotonic decrease of the HOMO–LUMO gap with decreasing torsion angle, except for a deviation for the fluorene derivative N2 [285]. The calculations also predicted two stable *atop* binding geometries of the nitrile-gold contact. The target molecules can bind either to a low-coordinated gold adatom (referred to as *l.c.*) or to a "terrace-type" gold atom (referred to as *h.c.*) in *atop* position. *Hollow* and *bridge* binding geometries were demonstrated to be unstable for the BPDN derivatives. The calculated transmission curves at zero bias of N1 to N6 for both *l.c.* and *h.c.* geometries reveal that the transport pathway proceeds by off-resonance tunneling through the tail of the LUMO. This is consistent with the results of previous calculations [278, 288] and with thermopower measurements [289].

The computed conductance values of the BPDN junctions are plotted in Fig. 20b as function of $\cos^2\varphi$. The correlation is linear, with a higher slope for the *h.c.* geometry ($a = 1.44 \times 10^{-2}\ G_0$ and $a = 5.9 \times 10^{-3}\ G_0$, respectively). The higher slope for the *h.c.* geometry reflects the better coupling (broader resonance) between the molecular π-system and the metal states, as caused by the tilt of the molecular backbone and the presence of additional gold atoms on the surface. Experimentally, the two theoretically predicted most probable junction geometries for the BPDN family could not be distinguished, due to the rather broad conductance maxima in the respective histograms [76].

6 Electrochemically Gate-Controlled Conductance of 4,4'- Bipyridine Single-Molecule Junctions

The experiments discussed in the previous sections were carried out in non-conductive organic solvents, with a focus on chemically tuning charge transport characteristics in single-molecule junctions. In the following a gate voltage is applied to tune the orientation of a molecular wire in a nanoscale gap in a polar electrolyte under full electrochemical potential control. The approach will be exemplified by 4,4′-bipyridine attached to two gold contacts in aqueous electrolyte [290].

6.1 Introduction

4,4′-Bipyridine (44-BP) is a bifunctional, non-chelating molecule, which consists of two planar pyridyl rings connected by a C–C bond in 4 and 4′ positions. It has been extensively employed as a bridging ligand in coordination chemistry [291].

The binding energy per N-site in 44-BP–metal coordination complexes amounts to 60–120 kJ mol^{-1}, i.e., it ranges between strong covalent bonding and weak bonds in biological systems [292].

The rotation angle between the two planar pyridyl-rings was found to vary between 18.2° (solid-state [293]) and 37.2° (gas-phase [294]). H-NMR experiments in several solvents of different dielectric constants revealed that 4,4′-BP appears either highly twisted, or as a free rotor. The barrier to internal rotation has been estimated to be 17.0 kJ mol^{-1}. The two rings are rotating almost freely in most liquid environments [295–297].

44-BP has two lone electron pairs at each nitrogen atom. These lone pairs enable an effective molecule-substrate coordination, that is, the electronic overlap of molecule and substrate states. The relatively low lying π^*-orbitals may act as good acceptors of *d*-orbital electron density in metal (electrode)-ligand back-bonding. However, the nitrogen lone-pair electrons are not delocalized. They are orthogonal to the molecular π-system, causing the conductance of a 44-BP-based molecular wire to be quite sensitive to the orientation of the Au–N bond, relative to its principal conducting orbital. Theoretical calculations have demonstrated that the LUMO lies relatively close to the FERMI level, i.e., the LUMO π-state provides the dominant contribution to charge transport through the molecule. The HOMO–LUMO gap is around 5 eV [298, 299].

Tao et al. reported the first single-junction conductance data of 44-BP attached to two gold leads in an aqueous environment as ~ 0.01 G_0 [32, 300] by using the STM-BJ technique. Multiple single-conductance values have recently been reported by Zhou et al.: 4.7×10^{-3} G_0 and 0.59×10^{-3} G_0 [301] and Wang et al.: $(1.34 \pm 0.17) \times 10^{-4}$ G_0, $(5.38 \pm 0.64) \times 10^{-4}$ G_0, and $(75.5 \pm 15.3) \times 10^{-4}$ G_0, which were obtained in aqueous electrolyte, as well as in nonconductive organic solvents. These experimental observations were attributed to different contact geometries between the pyridyl anchoring groups and the gold electrodes [50]. Venkataraman et al. demonstrated experimentally, and in combination with DFT calculations, that two conductance states of 44-BP molecular junctions could be rationalized by a tilted (high conductance: ~6×10^{-4} G_0) and an elongated perpendicular orientation (low conductance: ~1.6×10^{-4} G_0) of the molecule between the two gold leads. These authors demonstrated reversible binary switching between the two conducting states in a single molecular junction by mechanical control of the metal-molecule contact geometry [299]. This interpretation was recently shown to be more general, in a conductance study with an extended family of pyridine-terminated derivatives in 1,2,4-trichlorobenzene [51].

In the following we will present results of a single-junction conductance study with 44-BP under electrochemical potential control, enabling the precise tuning of the molecular orientation, relative to the substrate, upon application of an adjustable gate voltage [290] in a well-controlled environment [86, 302].

6.2 *Interfacial Electrochemical Characterization of 4,4'- Bipyridine on Au(111)*

Figure 21 shows the cyclic voltammogram of 3 mM 44-BP in an aqueous solution of 0.05 M $KClO_4$ on Au(111) [304]. Four different interfacial regions, labeled I to IV, can be distinguished, which are separated by distinct peaks. The double-layer region was limited between the onset of OH^- adsorption or the beginning of gold oxidation, and the desorption/reduction of 44-BP on Au (111).

Excursion of the electrode potential from region I to more negative values triggers two first-order phase transitions, characterized by the loss of the parallel stacking rows, and accompanied by an orientation change of 44-BP from a perpendicular to a more tilted one, simultaneously with the rearrangement of co-adsorbed interfacial water molecules. The adlayer coverage decreases from 6.8×10^{-10} mol cm^{-2} (densely packed phase in region I, Fig. 21a) via 5.7×10^{-10} mol cm^{-2} (striped phase in region II, Fig. 21b) to 3.4×10^{-10} mol cm^{-2} (rhombohedral phase in region III, Fig 20c). The corresponding molecular orientations, as shown in Fig. 21d–f, display a tilt angle, with respect to the normal axis of the electrode surface, changing from zero via ~25° to ~60°. The position of the neighboring molecules are determined by the competition between molecular and substrate coordination sites and intermolecular interactions of the π-systems. The variation of the 44-BP surface structures, upon decreasing the electrode potential, is accompanied by a decrease in surface

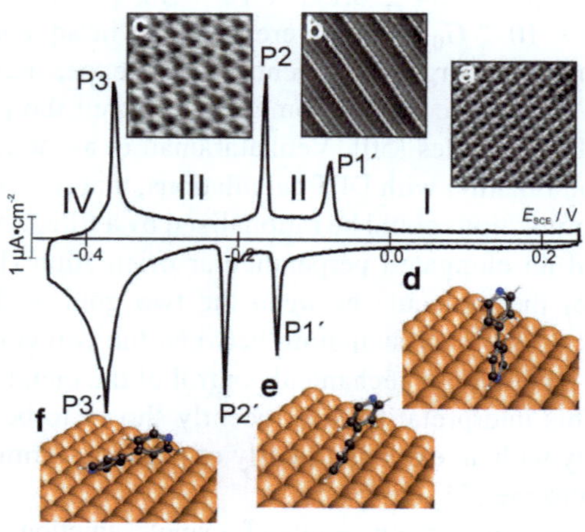

Fig. 21 Cyclic voltammogram and corresponding in situ STM images of 3 mM 4,4′-bipyridine (44-BP) on Au(111) in 0.05 M $KClO_4$, scan rate 10 mV s^{-1}. The sizes of the STM images are 10×10 nm^2. The following 44-BP adlayer structures have been observed in the potential regions I, II, and III: (**a**) high coverage densely packed phase, (**b**) striped structure, and (**c**) rhombohedral phase. The corresponding molecular orientations as derived from in situ IR studies on Au(111) are shown in panels (**d–f**) [303]. The pairs of peaks P1/P1′, P2/P2′, and P3/P3′ indicate first-order phase transitions between the respective adlayers [304]

coverage. Polarization at more negative potentials also leads to a decrease of the Au–N binding strength [303–306].

6.3 Single-Molecule Junction Conductance

STM-BJ experiments were performed to extract single-junction conductances of 44-BP bridged between a gold STM tip and an Au (111) substrate in non-polar solvents, such as mesitylene, isopropanol, and 1,2,4-trichlorobenzene, as well as under full electrochemical potential control in aqueous solutions of $HClO_4$ and $LiClO_4$ at various pH.

Figure 22 shows typical individual conductance traces of 44-BP molecular junctions, recorded over seven orders of magnitude with an STM-BJ technique, and the corresponding conductance histograms. The experiment started with the breaking of metal atomic contacts ("hard contact"), as indicated by the region labeled with G_0 in Fig. 22a. Several groups reported that, immediately after breaking an atomic Au–Au contact, the gap relaxes due to adhesion between Au atoms leading to an immediate separation between the leads of 3–5 Å [307, 192, 308]). Therefore, one can expect a large drop in conductance, which is observed experimentally. Simultaneously, adsorbed solution-based 44-BP molecules with a molecular length of ~7.2 Å will bridge the gap. The subsequent breaking of these (single) molecular junctions leads to plateaus and steps in the conductance–distance traces below 10^{-3} G_0 (c.f. Fig. 22a).

The statistical analysis of these conductance traces (an example is given in Fig. 22b for $E_S = 0.25$ V, which is close to the PZC [304]) reveals three distinct conductance peaks (high H, medium M, and low L) in the region between $10^{-2.5}$ G_0 and $10^{-4.5}$ G_0, with the most probable conductances evaluated by Gaussian fits as 14 nS (L), 38 nS (M), and 78 nS (H). The high conductance value H is twice as large as the medium one M, and the M value is three to four times larger than L. Similar

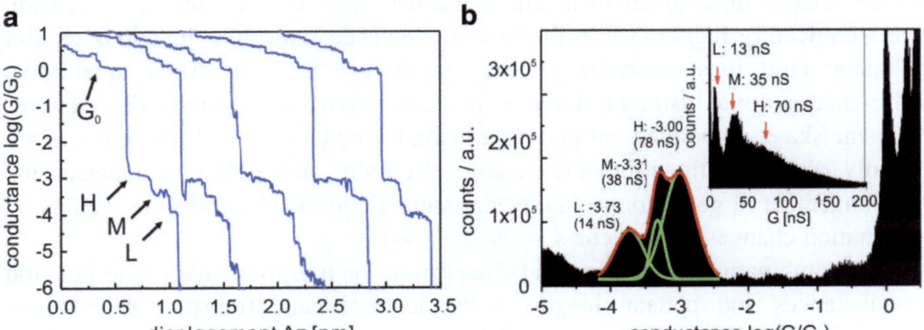

Fig. 22 (**a**) Conductance–distance traces for 3 mM 44-BP in 0.1 M $HClO_4$ at $E_S = 0.25$ V (vs SCE), stretching rate 60 nm s^{-1}. (**b**) Conductance histogram in logarithmic and linear (*inset*) scales built from ~2,000 single traces, created without any data selection

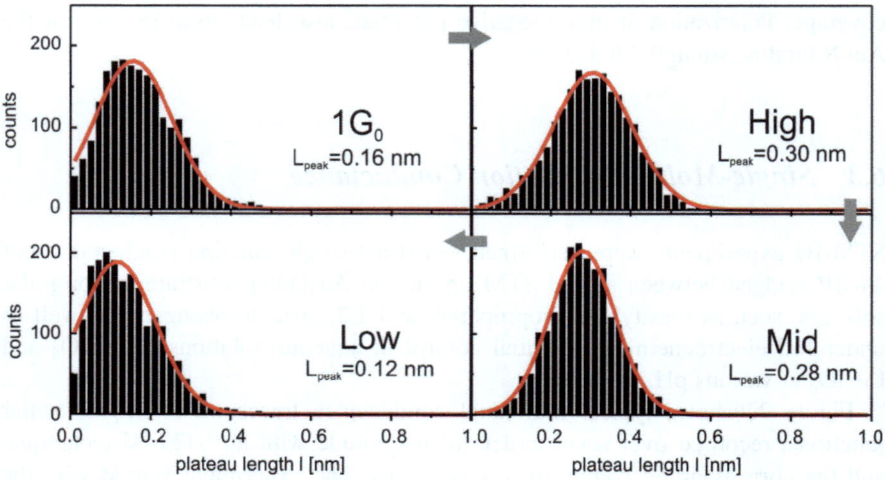

Fig. 23 Plateau length histograms [54] of the breaking of Au–Au nanocontacts in the presence of 44-BP, i.e., around 1 G_0, as well as of the molecular junctions labeled H, M, and L in Fig. 22. Gaussian fits (*solid red line*) revealed the most frequently observed plateau length in the individual conductance–distance curves

values of the three conductance states, as just reported for electrochemical conditions at the PZC ($E_S \sim 0.25$ V), were also found in non-conducting organic solvents, such as 1,3,5-trimethylbenzene, 1,2,4-trichlorobenzene, or isopropanol [290]. The M and L values are also in good agreement with results previously reported by Quek et al. [299] and Wang et al. [305].

The statistical analysis of the plateau-length histograms, c.f. Fig. 23, revealed that the most probable plateau length of an Au–Au atomic contact junction in the presence of 44-BP amounts to $l_{Au} \sim 0.16$ nm, which is slightly shorter than data obtained for Au nanocontacts in aqueous electrolytes containing weakly specifically adsorbed ions but no organic molecules (see also Sect. 3). The values for the 44-BP-related single-molecule junctions H and M, $l_H = 0.30$ nm and $l_M = 0.28$ nm, are significantly higher, while the low-conductance molecular junctions exhibit a rather short most probable plateau length $l_L = 0.12$ nm. These qualitative differences point to distinct differences in the respective single-molecule junctions. Kamenetska et al. pointed out that plateaus in the conductance–distance traces are usually caused by the stretching of gold electrodes and molecular contacts, the rearrangement of gold atoms, "molecular sliding" on the electrodes, as well as by orientation changes of molecules in the gap [66].

The combination of experimental observations on most probable single-junction conductances and plateau lengths with electronic and transport calculations employing the SMEAGOL code [309] suggests the following scenario [290]. A monatomic contact is initially formed, and subsequently elongated upon stretching, which leads to the plateaus in the conductance–distance traces around 1 G_0. Since the gap between the "pulled" electrodes formed immediately after relaxation of the

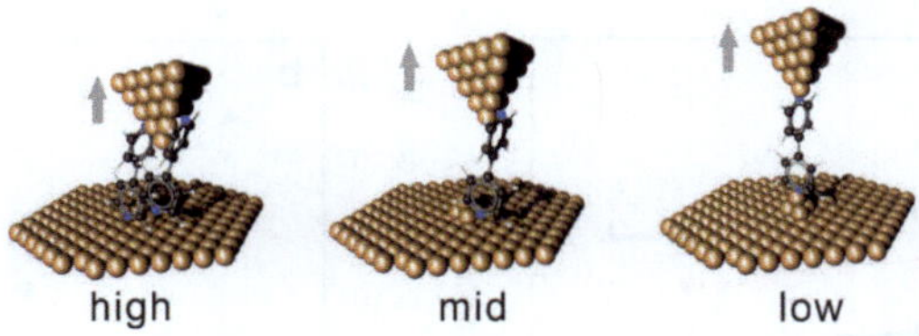

Fig. 24 (**a**) Suggested junction geometries representing a typical stretching trace: High (H) – two 44-BP molecules with the π-system of the pyridyl rings coupled to the gold electrodes; Mid (M) – single 44-BP coupled to sides of Au tip and substrate via adatom assemblies; Low (L) – single 44-BP molecule bound in upright orientation to two lower-coordinated gold adatoms [290]

Au atom positions ("snap back" [299]) is shorter than the molecular length of 44-BP, one or more 44-BP molecules can only bind sideways (c.f. Fig. 24a) with the conducting π-system being inclined with respect to the N–Au bond of the pyridyl nitrogen to an under-coordinated gold surface atom. This gives rise to an H-type high-conductance junction. As the junction is elongated, only one molecule is allowed to slide on the surface, due to the limited binding site near the apex of the electrode, causing the M plateaus. The further elongation leads finally to a binding geometry with the pyridyl nitrogen–gold bond to an under-coordinated Au atom being perpendicular to the molecular π-system, until the junction breaks, causing the L-type plateaus. Ideally, these four events are expected to take place sequentially. Figure 24 shows cartoons illustrating the various suggested junction geometries during the evolution of a "most probable" stretching trace [290].

6.4 *Electrochemical Gate Effect*

The conductance experiments with 44-BP junctions were extended to an electrochemical environment. This approach enables the application of an "electrolyte gate field" for tuning the 44-BP surface orientation on the Au(111) electrode between upright (in the potential region I) and strongly inclined with the π-system facing the substrate (in region III). The electrochemical experiments were performed in $LiClO_4$ or $HClO_4$ solutions containing 3 mM 44-BP. The solution pH was also varied to address the influence of 44-BP protonation, i.e., 44-BPH^+ and 44-BPH_2^{2+}, on the junction conductance characteristics ($pK_{a1} = 2.69$; $pK_{a2} = 4.77$) [310].

Figure 25a, as an example, shows the potential dependence of the single-junction conductances of 44-BP measured in 0.1 M $HClO_4$ solution (pH 1) in -0.10 V < E < 0.90 V in a semi-logarithmic representation. The values of L, M, and H decrease with more positive electrode potentials, and follow nearly the same trend for each family. The single-junction conductances decrease by a factor of 3–5 upon potential excursion towards positive values in the accessible potential region. A similar trend is also observed for electrolytes with variable pH ranging between 1 and 10, as

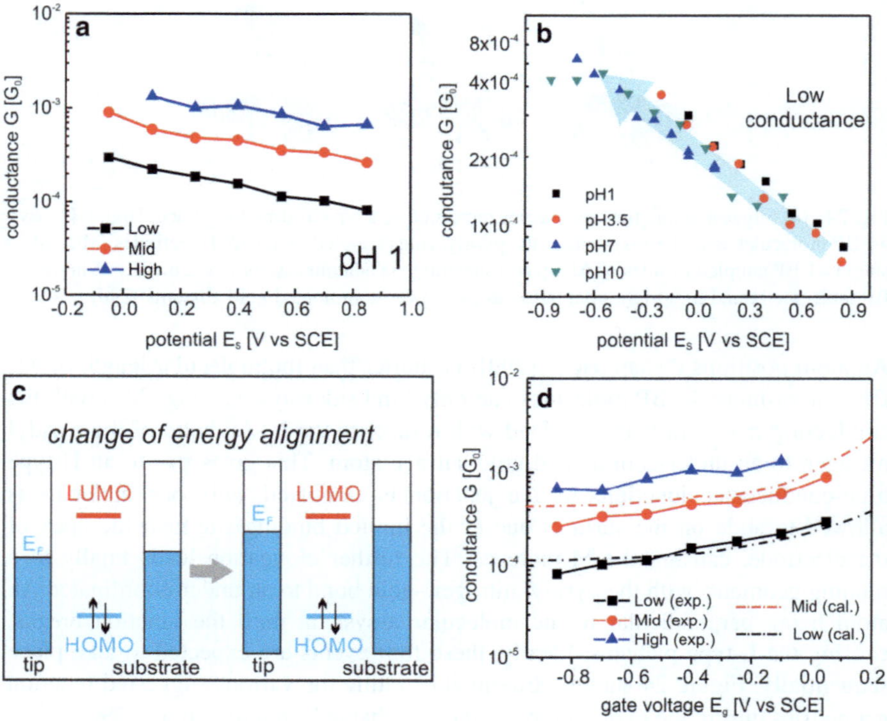

Fig. 25 (**a**) Potential dependence (E_S) of the single molecular junction conductances of 3 mM 44-BP in 0.1 M $HClO_4$. $E_{bias} = 0.10$ V; tip retraction rate 60 nm s^{-1} (see also Fig. 22). (**b**) Low conductance L plotted as a function of the applied electrode potential for different values of the solution pH. The pH was adjusted by $HClO_4$ or NaOH. (**c**) Schematic representation of the energy level alignment between the electrode's Fermi levels E_{tip} and E_S and the molecular frontier level upon sweeping the electrode potentials to more negative values. (**d**) Comparison between the experimentally measured conductances shown in (**a**) and computed conductances $\log (G(E_g)/G_0) = \log [T(E_F(E_G))] + A$ at various gate voltages [290]

exemplified for the case of low conductance values L in Fig. 25b. The L conductances increase monotonously with a decrease of the electrode potential, and do not depend on the solution pH. A deprotonation process seems to precede the formation of the molecule–metal contacts. We also note that the electrochemically-induced tuning of the junction conductance ("electrolyte gating") is fully reversible.

We attribute the enhancement of the junction conductance driven by an electrochemical potential to the alignment of the molecular frontier level with the Fermi levels of the gold leads. As illustrated in Fig. 25c, upon sweeping the electrode potential towards more negative values, the Fermi levels of the electrodes shift upwards and align better with the LUMO of 44-BP. We note that charge transport through 44-BP single-molecule junctions is dominated by the LUMO of the 44-BP extended molecule. Due to the relatively small HOMO–LUMO gap of 44-BP, the alignment of the 44-BP LUMO and the Au Fermi levels changes from off-resonance to a "partially" resonant state, leading to an increase in the transmission through the 44-BP single-molecule junctions. This trend is corroborated by a simultaneous change in orientation of the adsorbed 44-BP molecules on Au(111) from upright to tilted upon negative polarization of the electrode [303].

The above qualitative reasoning is supported by a theoretical study of the effect of electrochemical gating in molecular transport. The gate potential $E_G = -(E_S - E_{SCE})$ produces a shift in the Fermi energy, relative to the non-gated value. The gate-shifted Fermi level is expressed as $E_F(E_G) = \beta E_G + E_F(0)$, where $E_F(0)$ is the value of E_F in the absence of gating, and β is the gate control parameter [311]. Figure 25d shows the calculated gated conductances $\log(G/G_0) = \log T(E_F(E_G)) + A$, obtained from the tails of the LUMO resonance, where the optimum fit is obtained by choosing $\beta = 0.5$ and $E_F(0) = E_F{}^0 - 0.35$ V ($E_F{}^0$ is the bare Fermi energy predicted by DFT) [290]. The correction factor $A = -0.74$ is chosen to account for the slight discrepancies between the measured and the computed conductances, which arise from a range of factors, including the statistical distribution of atomic configurations and the choice of the basis set. Figure 25d shows that, over a wide range of gate potentials, the ratio G_{Mid}/G_{Low} is approximately 3.8, which is in good agreement with the experimentally observed ratio of M and L conductance values.

Finally, we note that the present observation is different from tunneling enhancement mediated by an oxidation/reduction process, which is accompanied by electronic structure changes of the junction ("inner-sphere reorganization energy") as well as reorientation of dipolar solvent molecules ("outer-sphere reorganization energy"). 44-BP keeps its neutral state, and therefore represents a universal case study, which will have analogs in many junctions of organic molecules exhibiting a relatively small HOMO–LUMO gap.

7 Electrochemically Gate-Controlled Charge Transport in Redox-Active Molecular Junctions

Redox-active molecular junctions represent a particularly unique class of systems for "electrolyte gating" and beyond. The electrochemical approach and a summary of systems explored have already been introduced in Sect. 2. In an attempt to contribute to a more advanced understanding of the phenomena involved, we have recently carried out several systematic experimental and theoretical studies with three families of molecules, i.e., derivatives of viologens [34, 110, 114], perylene bisimides [170], and ferrocenes [150]. In the following, several results as obtained with perylene bisimide derivatives (PBI) will be presented, to illustrate the uniqueness of the electrochemical approach in single-junction charge-transport experiments. In particular, the influence of anchoring group, chemical substitution at "bay positions", and electrochemical gating will be addressed.

7.1 Introduction

As a class of n-type organic semiconductors, PBI derivatives have received considerable attention for a variety of applications [312, 313], for example, for organic or polymer light-emitting diodes (OLEDs and PLEDs) [314, 315], thin-film organic field-effect transistors (OFETs) [316, 317], solar cells [318, 319], and liquid crystals [320]. They are also interesting candidates for single-molecule device applications, such as sensors [321], molecular wires [322], or transistors [141].

The PBI core exhibits a flat π-system, composed of two naphthalene half units, each of which is attached to an imide unit, and connected to the other naphthalene unit by two C sp^2–C sp^2 single bonds (Fig. 26a). The HOMO and LUMO exhibit nodes at the positions of the imide nitrogens, which provide unique opportunities for modifying the structure of the linker groups, without alterations of the electronic properties of the core π-system. On the other hand, the introduction of substituents at the 1,6,7,12 "bay positions" may significantly alter the electronic and optical properties (c.f. inset of Fig. 26a), from electron-poor to electron-rich, without changing the dimensions in the direction of the long molecular axis, which determines the length of a molecular junction. In addition, these substituents create steric strain in the bay area, leading to a propeller-like twisting of the two naphthalene half units [312, 323], which prevents π–π stacking between adjacent molecules, and therefore promotes the formation of single-molecule junctions. The twist angle of the π-system can be tuned from 0° (no substituent) up to 37° (tetra-chloro substitution). Voltammetric and spectro-electrochemical studies with PBI dissolved in organic solvents revealed two reversible one-electron reductions (formation of the stable radical anion $PBI^{\bullet-}$ and the dianion PBI^{2-}) and a single one-electron oxidation [324].

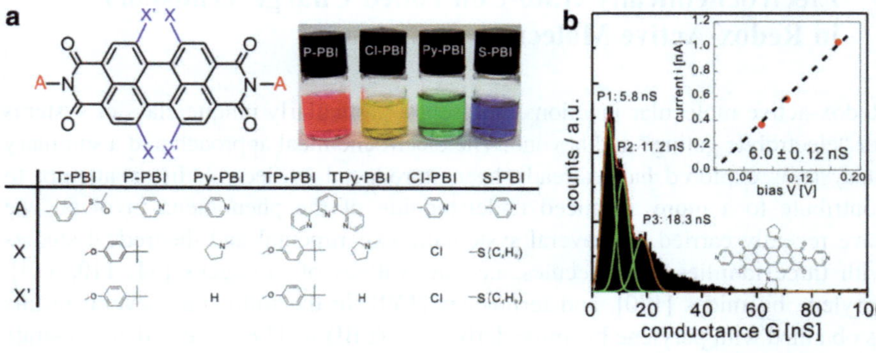

Fig. 26 (**a**) Structures of pyridine-, terpyridine-, and thiol-terminated PBI derivatives with different substituents at the "bay positions" X and X'. The *inset* illustrates the alternation of optical properties of the PBIs with different bay-area substituents. (**b**) Plateau data-point histogram of Py-PBI in a mixture of mesitylene/THF (4:1). $E_{\text{bias}} = 0.1$ V, tip retraction rate was 60 nm s^{-1}. The *inset* show the bias voltage dependence of the current through a molecular junction

Li et al. [169] reported that the single-junction conductance of PBI-type molecules can be modulated by two to three orders of magnitude by varying the electrode potential. These authors also found a temperature-dependent junction conductance, indicating a thermally activated process. Sun et al. predicted a LUMO-mediated electron transport process [325]. The experimentally observed temperature dependence was related to a modulation of hydrogen bonding between the four oxygen atoms of PBI and the surrounding water molecules [326].

The following section reports the charge-transport characteristics of symmetric PBI single-molecule junctions (where PBI is attached with two anchoring groups to two contacting leads) and also asymmetric PBI-mediated tunneling junctions (where the PBI derivative is bound only to one electrode), as a function of the anchoring groups and the bay-area substitution in two different environments, in non-conducting organic solvents, as well as under electrochemical conditions. Basic relationships between electron transport properties and molecular structure will be addressed [170].

7.2 Chemically Controlled Conductance of Perylene Bisimide Derivatives

The PBI molecules studied include pyridyl-(P-PBI), terpyridyl (TP-PBI), and thiol-terminated (T-PBI) perylene-3,4:9,10-tetra-carboxylic acid bisimides, that contain bulky *tert*-butylphenoxy substituents at 1,6,7,12 bay positions, and pyridyl-(Py-PBI) and terpyridyl- (TPy-PBI) terminated perylene-3,4:9,10-tetra-carboxylic acid bisimides, that contain bulky pyrrolidinyl substituents at 1,7 bay positions (Fig. 26a) [327–329].

Figure 26b shows, as an example, a plateau data-point histogram obtained for 0.1 mM Py-PBI at 0.1 V bias voltage in mesitylene/THF. The graph reveals three equally spaced maxima, separated by integer values of the conductance of the most prominent first peak at 6.0 ± 0.1 nS, which is assigned to the single-molecule conductance. The other peaks are interpreted as multi-molecule junctions. The conductance histograms were constructed from 10–25% of the experimentally recorded traces because of the low yield of successfully formed molecular junctions. Table 2 summarizes the results.

The low conductance of the π-conjugated system is due to the nodes at the imide nitrogen positions. Several additional trends have been observed. The conductance of pyridine-terminated PBI derivatives (P-PBI) is typically a factor of 2 larger than the thiol-terminated ones (T-PBI). Although the Au–S bond provides a stronger electronic coupling than the Au–N bond, the T-PBI has two extra methylene

Table 2 Summary of measured conductances of single PBI-type molecular junctions

Molecule	T-PBI	P-PBI	Py-PBI	TP-PBI	TPy-PBI
Conductance[nS]	2.5 ± 0.1	5.0 ± 0.2	6.0 ± 0.1	4.8 ± 0.3	3.8 ± 0.2

spacers at each side, which act as insulating units. The conductance of TP-PBI is slightly smaller than that of P-PBI, possibly caused by the less efficient bonding geometry (steric hindrance) and the longer bonding distance of the terpyridine group, as compared to pyridine.

The conductance of 1,7-pyrrolidinyl-substituted PBI (Py-PBI) is slightly larger than that of 1,6,7,12-butyl-phenoxy-substituted PBI (P-PBI). The bay-area substituents affect the transport properties of PBI molecular junctions in two ways: (1) the substituents alter the twist angle of the PBI core and (2) electron-withdrawing or -donating substituents shift the molecular frontier orbital relative to the electrode Fermi levels. According to a semiempirical AM1 calculation, the twist angles of the bay area unit for P-PBI and Py-PBI are 27° and 25°, respectively [330]. It is evident that a smaller twist angle leads to a higher degree of π-conjugation, and consequently also to a larger conductance. On the other hand, electron-donating substituents of the PBI core could confine the HOMO level. Oxygen or nitrogen lone-pair electrons delocalize into the PBI π-space, and thereby raise the HOMO level. The pyrrolidine has a stronger electron-donating effect, compared to butyl-phenoxy substituentens, causing a larger increase of the molecular frontier orbital energies. In conclusion, the changes of both the conformation and the energy levels of the frontier orbitals lead to the differences in single-junction conductances, as observed between P-PBI and Py-PBI.

Finally, we note that the idea of manipulating transport properties of molecular junctions by the modification of a "master rod" with electron-withdrawing or -donating groups was demonstrated before, experimentally [49] and theoretically [331–333].

7.3 Electrochemical Characterizations

Figure 27a shows a typical CV of 1 mM P-PBI in dicholoromethane (DCM) with 0.1 M tetrabutylammonium hexafluorophosphate ($TBAPF_6$) as the supporting electrolyte. One may distinguish three characteristic pairs of current peaks labeled as O^c/O^a, R_1^c/R_1^a, and R_2^c/R_2^a. The positions of the corresponding potentials are independent of scan rate v in the range $0.01\ \mathrm{V\ s^{-1}} < v < 1.00\ \mathrm{V\ s^{-1}}$. The peak heights scale linearly with the square root of the scan rate v. These observations indicate a reversible, bulk-diffusion-controlled process. O^c/O^a is assigned to a reversible one-electron oxidation process PBI/PBI$^+$, R_1^c/R_1^a and R_2^c/R_2^a to two one-electron reduction processes PBI/PBI$^{\bullet -}$ and PBI$^{\bullet -}$/PBI^{2-}, respectively [312, 329]. The reduction of Py-PBI with the electron donating dipyrrolidinyl-bay-area substituents requires more negative potentials. One also observes two reversible one-electron reduction peaks [328].

Figure 27b shows typical voltammograms of monolayers of three PBI derivatives on Au (111). The two reduction peaks merge into a single broad peak, due to specific solvent-PBI interactions, as well as the possibility of incorporating cations of the supporting electrolyte into the charged PBI adlayers. The peak-to-peak

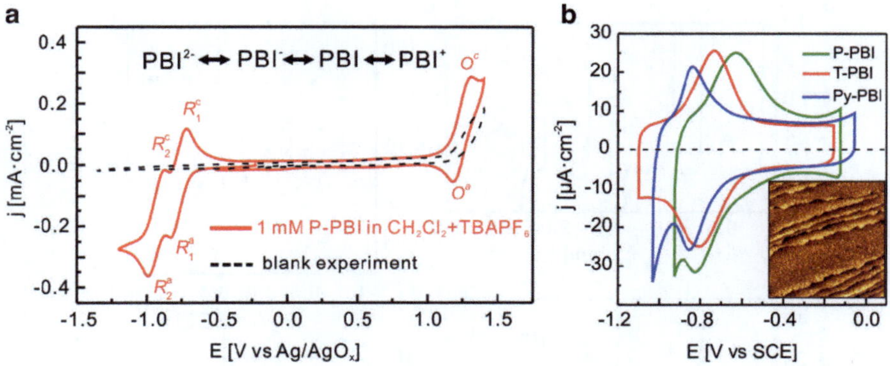

Fig. 27 (**a**) Cyclic voltammograms with (*red line*) and without (*dotted black line*) 1 mM P-PBI in CH_2Cl_2 + 0.1 M $TBAPF_6$ on Au (111), scan rate 0.1 V s^{-1}. (**b**) Cyclic voltammograms of monolayers of P-PBI (*green*), T-PBI (*red*), and Py-PBI (*blue*) immobilized on Au (111)-(1 × 1) in 0.05 M $LiClO_4$, pH ~ 12, scan rate 0.2 V s^{-1}. The *inset* shows a typical STM image of the disordered high coverage T-PBI adlayer on Au(111)-(1 × 1) in 0.05 M $LiClO_4$, pH ~ 12 (200 × 200 nm, $i_T = 0.1$ nA; $E_S = -0.02$ V(vs SCE), $E_{bias} = 0.1$ V) as recorded before or during the transport measurements [170]

separations are 0.07 ± 0.02 V for T-PBI, 0.15 ± 0.03 V for P-PBI, and 0.03 ± 0.01 V for Py-PBI. The peak positions are independent of scan rate v, and their heights scale linearly with v in the range of 0.01 V $s^{-1} < v <$ 1.00 V s^{-1}, indicating a reversible surface-confined redox process. The corresponding formal potentials were estimated as $E_{T\text{-}PBI}{}^{0} = 0.790 \pm 0.035$ V, $E_{P\text{-}PBI}{}^{0} = 0.725 \pm 0.095$ V, and $E_{Py\text{-}PBI}{}^{0} = 0.820 \pm 0.010$ V. Current integration, after correction of the double-layer contribution and assuming a two-electron transfer process, yields charges of ~20 μC cm^{-2} for all derivatives, which correspond to surface coverage and cross sectional areas of $\Gamma \sim 1.03 \times 10^{-10}$ mol cm^{-2} and A ~ 1.6 nm^2, respectively. These cross-section areas represent densely packed monolayers of the respective PBI derivatives, with the long molecular axis through the two imide nitrogens slightly tilted with respect to the surface normal. A parallel orientation could be ruled out, due to packing reasons.

7.4 *Electrochemically Gate-Controlled Conductance of Perylene Bisimide Derivatives*

Figure 28a shows a set of typical current–distance traces (i_T-Δz) recorded upon retracting a coated-gold STM tip away from a T-PBI covered Au(111) surface in 0.05 M $LiClO_4$ solution (pH 12). These curves exhibit characteristic single plateaus (dominant) or a series of plateaus separated by current steps. The steps are assigned to the breaking of individual (respectively multi-molecular) junctions of the PBI derivatives previously formed between the Au STM tip and the substrate surface.

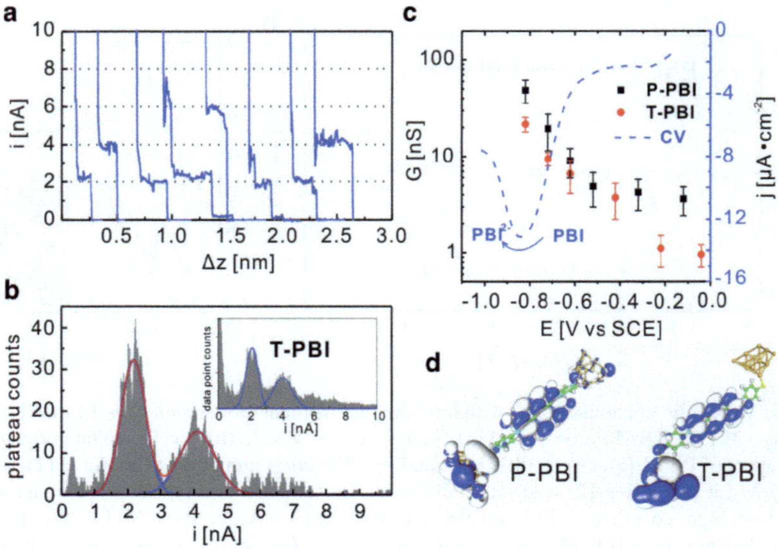

Fig. 28 (**a**) Typical current-distance (i-Δz) retraction curves for T-PBI molecules immobilized on Au(111)-(1 × 1) in 0.05 M $LiClO_4$, pH ~ 12, at $E_{bias} = 0.10$ V; $E_S = 0.82$ V (vs SCE). (**b**) Plateau count conductance histogram constructed from individual current–distance curves of T-PBI, 500 selected out of 2,500. The *inset* shows the corresponding all-data-point conductance histogram. (**c**) Single junction conductances of P-PBI and T-PBI in dependence on the substrate potential E_s. The *dashed line* corresponds to the single scan voltammogram recorded for a monolayer of immobilized T-PBI on Au(111)-(1 × 1) in 0.05 M $LiClO_4$, pH ~ 12, scan rate 0.1 V s^{-1}. (**d**) LUMO orbitals of P-PBI and T-PBI molecules attached to Au_{14} clusters (without the bulky *tert*-butyl-phenoxy groups in the bay area) [170]

The statistical analysis of these traces revealed all-data-point and plateau-counting histograms with a series of uniformly spaced current peaks (Fig. 28b). The first peak of each sequence could be identified as the single-junction conductance current. The data set of T-PBI plotted in Fig. 28b leads to 21.7 ± 3.8 nS at $E_S = -0.820$ V.

The potential dependence of the "Au–P-PBI(T-PBI)–Au" single-junction conductances was obtained at variable substrate (E_S) and tip (E_T) potentials in $-0.90 \text{ V} < E_S\ (E_T) < 0.00$ V, while keeping the bias voltage E_{bias} constant. Figure 28c shows semi-logarithmic plots of the single junctions conductances "Au–P-PBI(T-PBI)–Au" vs the substrate potential E_S. The conductances of P-PBI junctions are typically, i.e., within the entire potential range studied, a factor of 3 larger than those of T-PBI. We also observed that the conductances of both systems are rather independent of the electrode potential in the stability range of the neutral PBI species. Decreasing the substrate and tip potentials toward negative values, and approaching the potential regions of the reduction processes P-PBI|P-PBI$^{\bullet -}$, relative to P-PBI$^{\bullet -}$|P-PBI^{2-} (as for T-PBI), leads to a rapid increase of the junction conductance (up to two orders of magnitude). No plateaus or peaks were observed at $E < -0.90$ V. The recording of reliable data at $E < -0.90$ V is hampered due to the onset of adlayer desorption.

Transport calculations revealed a LUMO-controlled mechanism, with the current flowing through the π-orbitals of the connecting pyridyl/phenyl-units (Fig. 28d). The LUMO hybridization is controlled entirely by how this π-system is connected to the electrodes. With the pyridine coupling there is a direct overlap with the Au-surface states, while in the case of a thiol coupling a mediating S-atom has been introduced. Since the S-levels mix with the benzene π-orbitals only at energies more than 0.5 eV below the Fermi energy, the S-atom constitutes a barrier for the current flow through the LUMO, which expresses itself as a reduced coupling. The coupling of P-PBI to the adjacent electrodes is larger, as compared to T-PBI by ~ 2.6, which scales nicely with the experimentally observed threefold higher junction conductance of the pyridyl-terminated PBI derivative.

7.5 *Redox-Mediated Tunneling Current*

The above *transport* experiments in a *symmetric* "Au–T-PBI(P-PBI)–Au" junction are complemented by single-molecule *tunneling* experiments [170]. Here, a redox-molecule modified tunneling junction is created, with an *asymmetric* PBI derivative bound chemically either to the Au tip or to the Au(111) substrate surface. After establishing a stable tunneling junction, the STM feedback was switched off, and subsequently the tunneling current i_T was recorded at constant *xyz*- position as a function of the applied voltage, while keeping constant the voltage difference between tip and substrate (bias voltage). In other words, the Fermi levels of tip and substrate were shifted relative to the discrete molecular levels.

Figure 29a shows a typical i_T vs E_S tunneling curve for an "Au–T-PBI–Au" junction, constructed by averaging ten individual traces, $E_{bias} = 0.10$ V. The curve exhibits a pronounced maximum at $E_{max} = -0.740 \pm 0.075$ V, which is close to the formal potential of the surface-confined molecule, $E^0 = -0.790 \pm 0.075$ V, and does not scale with the scan rate v. The full-width-at-half-maximum (FWHM) is estimated as 0.27 ± 0.02 V. The current at the maximum is two orders of magnitude higher than the values at the two turning points $E_S = 0.00$ V and -1.200 V, which represent the stability regions of the neutral T-PBI species or its dianion T-PBI^{2-}, respectively. The large current of the maximum is also significantly higher than the electrochemical background signal (faradaic current, recorded upon retracting the tip in the *z*-direction out of the tunneling regime) and convincingly demonstrates an enhanced tunneling response modulated by the presence of the redox-active T-PBI molecule. The position of the current maximum depends linearly on the applied bias voltage, according to $E_{max} = -0.668\ \text{V} - 0.704 \times E_{bias}$ (Fig. 29b). The FWHM of the tunneling resonance broadens with increasing bias voltage. Clearly the enhanced tunneling current represents the "opening" or "gating" of an additional molecular tunneling channel, which resembles a transistor-like behavior. Due to the sharpness of the tip and the exponential distance dependence of the tunneling signal, the main contribution to the enhanced tunneling signal in

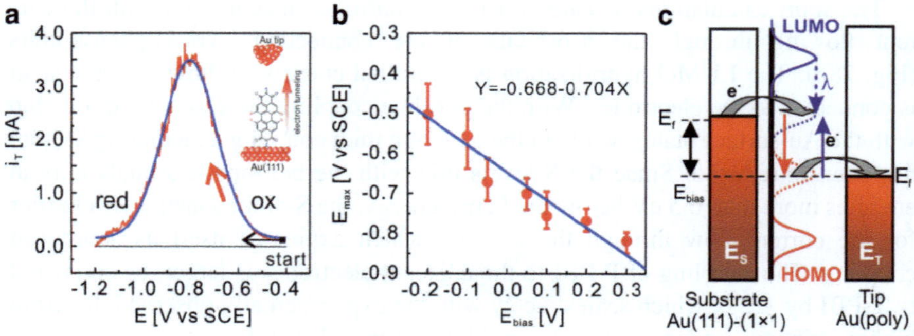

Fig. 29 (**a**) Tunneling current (i_T) vs substrate potential (E_S) trace (*red curve*) recorded in 0.05 M $LiClO_4$ at pH ~ 12 for a monolayer of T-PBI immobilized on Au (111)-(1 × 1). The bias voltage was fixed to 0.10 V. The scan started at $E_S = -0.42$ V, after switching off the tunneling feedback ($i_T = 0.1$ nA), sweep rate 1 V s^{-1}. The *solid blue trace* represents the fitted curve based on (4) with the following parameters: $\lambda_R = 0.22$ eV; $\eta = 0.55$ and $\xi = 0.38$. The *black curve* shows the tip voltammogram when the tip was retracted out of the tunneling regime. (**b**) Bias voltage, E_{bias}, dependence of the position of the maximum in the tunneling current, E_{max}, (**c**) Schematic energy level diagram of a sequential two-step electron transfer process mediated by a redox-active molecule [144, 145, 170]

the asymmetric junction is assumed to originate from a rather small number of redox-active molecules.

A maximum in the tunneling current is predicted by resonance tunneling, a mechanism based on a coherent two-step electron transfer, or a sequential two-(multi-) step electron transfer process (Fig. 29c) [144–147, 265, 266]. The former two regimes imply a shift of the maximum in the tunneling current by the reorganization Gibbs free energy λ_R with respect to the formal potential E^0. A maximum, located close to the formal potential, as observed in the present study, is predicted by the third mechanism. Additional support for this scenario is given by the linear dependence of E_{max} vs E_{bias} (Fig. 29b). Within the limits of low bias voltages E_{bias} and overvoltages $\eta = (E - E^0)$, as well as sufficiently strong molecule-electrode coupling (adiabatic limit), Ulstrup et al. proposed the following theoretical formalism [144–147]:

$$i_T = e\kappa\rho(eE_{\text{bias}})\frac{\omega_{\text{eff}}}{4\pi}\exp\left(-\frac{e(\lambda_R + E_{\text{bias}})}{4k_BT}\right)\left\{\cosh\left[\frac{e}{2k_BT}\left(\xi\eta + \left(\gamma - \frac{1}{2}\right)E_{\text{bias}}\right)\right]\right\}^{-1} \quad (4)$$

with

$$i_T(\eta) = \frac{i_{T\,\max}}{\cosh[a(\eta - \eta_{\max})]}, \quad (5)$$

$$a = \frac{e\xi}{2k_{\mathrm{B}}T} = 19.4\xi, \tag{6}$$

and

$$\eta_{\mathrm{max}} = \frac{1}{\xi}\left(\frac{1}{2} - \gamma\right)E_{\mathrm{bias}}. \tag{7}$$

The reorganization free energy λ_{R} represents the electronic–vibrational coupling, ξ and γ are fractions of the overpotential η and of the bias voltage E_{bias} at the site of the redox center, e is the elementary charge, k_{B} the Boltzmann constant, and ω_{eff} a characteristic nuclear vibration frequency. κ and ρ represent, respectively, the microscopic transmission coefficient and the density of electronic levels in the metal leads, which are assumed to be identical for both the reduction and the oxidation of the intermediate redox group. i_{Tmax} and η_{max} are the current and the overvoltage at the maximum.

The fits of (4)–(7) to the experimental curves plotted in Fig. 29a, b lead to $\lambda_{\mathrm{R}} = 0.22$ eV, $\xi = 0.38$, and $\gamma = 0.55$ [170]. The fits of more than 30 negative and positive going half-cycles yield average parameters $\lambda_{\mathrm{R}} = (0.17 \pm 0.08)$ eV, $\xi = (0.32 \pm 0.08)$, and $\gamma = (0.45 \pm 0.15)$, indicating a substantial potential drop at the position of the redox site. The value of the reorganization energy λ_{R} is reasonable, and strengthens the mechanism proposed. The good coincidence between the fitted and the measured trace provides additional support. A detailed analysis of the theoretical formalism expressed by (4) is given in [110].

The asymmetric tunneling configuration is distinctly different from the symmetric "Au–molecule–Au" configuration. The latter exhibits modifications of the junction conductance with the molecules bound to both leads, due to electronic structure changes of the redox-active unit ("inner-sphere reorganization energy"), while the former exhibits a maximum of the tunneling current due to the coexistence of the oxidized and the reduced forms, with dominant contributions of the dynamics of the solvent molecules in the gap ("outer-sphere reorganization energy"). The observed enhancement of the tunneling current in an electrochemical environment is not restricted to perylene-type junctions. It has also been observed for other redox systems, such as viologens [34, 110, 155], ferrocene [150], redox-proteins [109, 172, 174], and very recently for a hydroquinone derivative chemically bound to a gold electrode [175].

8 Quantized Charging of Nanometer-Sized Gold Clusters

Metal nanoparticles (NPs), which represent an important family of nanoscale objects with unique electronic, optical, catalytic, and magnetic properties, are distinctly different from both molecular and bulk materials. In the following we address recent studies on quantized charging of individual (local approach) and larger numbers of gold nanoclusters (macroscopic approach), in particular monolayer protected nanoclusters (MPC), in both ionic liquid and aqueous electrolyte.

8.1 Introduction

The electronic and electrochemical properties of metal nanoparticles (NP) can be tuned through the metal core and the ligand shell. Depending on the core size of the NPs [or monolayer-protected clusters (MPC), see Fig. 30a], one may distinguish between bulk continuum (core diameter$\gg$ 3–4 nm), quantized double-layer charging, and molecule-like charging (<1.5 nm) [336, 337]. In addition, the energetics of electron uptake to an MPC is determined by three different energy scales [338]. Quantum confinement is manifested in the spacing of electron levels ΔE, such as between HOMO and LUMO energy levels. Furthermore, in small crystals,

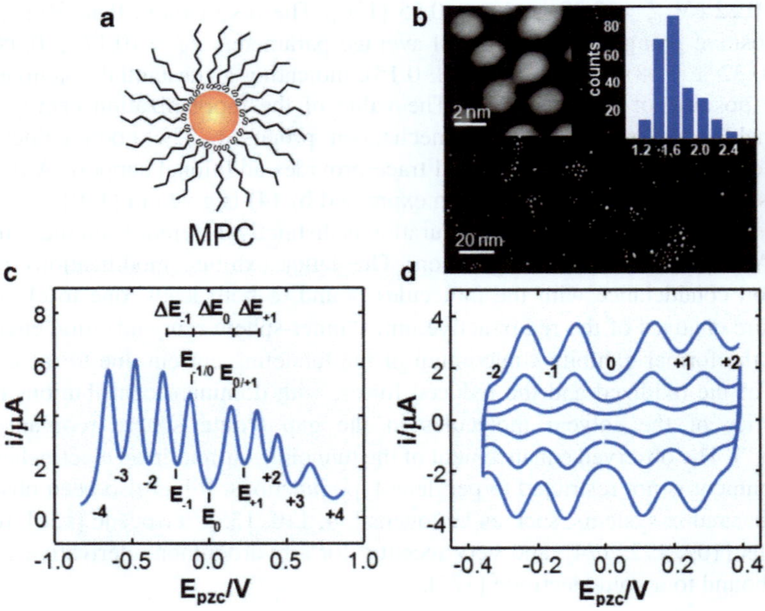

Fig. 30 (**a**) Schematic representation of an alkanedithiol monolayer protected cluster, MPC. (**b**) High resolution TEM image together with a core diameter distribution histogram for the $Au_{144}C6S$ MPC used. (**c**) Differential pulse voltammogram of a drop-cast $Au_{144}C6S$ film formed on a Pt electrode in ionic liquid (HMImFET). Pulse width 60 ms, pulse height 50 mV, period 200 ms, scan rate 20 mV s^{-1}, scan direction negative to positive. For the zero and plus/minus one charge states, relevant potentials, E, and stability regions, ΔE, have been indicated. For conciseness, all other regions are indentified by the assigned charge states z only. (**d**) Cyclic voltammograms of a drop-cast $Au_{145}C6S$ film formed on a Pt electrode in BMPyFSI, for charge states $|z| < 2$ as indicated. Scan rate 20, 50, 100 mV s^{-1} (inner to outer traces) [334, 335]

Coulomb repulsion between the added electrons results in a considerable energy cost E_C. The third relevant energy scale is k_BT. If ΔE and E_C are both smaller than k_BT, the behavior is bulk continuum-like. In metal nanocrystals with core-size diameters ranging between 1.5 nm and 3 ~ 4 nm, E_C is often considerably larger than either ΔE or k_BT. Their charging state response is then determined by the Coulomb blockade. In very small metal clusters and semiconductor NPs, both ΔE and E_C are smaller than k_BT. In consequence, discrete and molecule-like electronic energy levels are accessible upon charging. We notice that metal particles do not show quantum confinement effects comparable to semiconductor quantum dots, unless their core diameter is smaller than 1.5 nm. However, due to the sub-attofarad capacitance (aF $= 10^{-18}$ F) of MPCs (C_{MPC}), the electrostatic energy required to add an electron, E_C, substantially exceeds the thermal energy at room temperature (*RT*). This means that the charge of the core can be controlled, and discrete electron transfer is observed. In the presence of a sufficiently concentrated electrolyte, to neglect diffuse-layer contributions, the classical concentric spheres capacitor formula provides a good first approximation of the MPC capacitance [339, 340]:

$$C_{MPC} = 4\pi\varepsilon\varepsilon_0 \frac{r(r+d)}{d}, \tag{8}$$

where ε_0 is the permittivity in vacuum, ε the static dielectric constant of the ligand shell, d its thickness, and r the radius of the gold core.

For perfect monodisperse MPC, a succession of regularly spaced charging peaks is expected as the potential is varied:

$$\Delta E_z = \frac{e}{C_{MPC}}. \tag{9}$$

The subscript z refers to the charge state of the cluster, i.e., the sign and the number of one-electron charges stored on the MPC core; ΔE_z is then the potential range where the charge state z is stable.

In conclusion, for larger core diameters, electron addition/removal is a purely capacitive phenomenon, while for sub-nanometer core diameters, the energies are determined both by the spacing of the molecular energy levels and the capacitive charging. C_{MPC}, which is typically in the sub-attofarad range, determines the charging energy, and can be estimated from electrochemical measurements, such as differential pulse voltammetry (DPV), from the potential spacing between successive electron transfers ΔE_z. [341]

8.2 *Quantized Charging on Macroscopic Electrodes*

Murray et al. demonstrated in two seminal contributions that freely diffusing, monodisperse hexanethiol-capped 1.6-nm Au-NPs (~Au_{145}) exhibit sequential charging in electrochemical experiments at room temperature, employing cyclic voltammetry or DPV [341, 342]. Quinn et al. [343] and Mertens et al. [334, 335] improved the monodispersity in terms of size and composition. Both groups observed 15 separated charging peaks, each corresponding to single-electron transfers between diffusing MPC and a metal electrode. The peaks can be rationalized by quantized double-layer charging of the MPC in the electrolyte solution near the electrode surface, with the MPC viewed as a multivalent "redox species" [344, 345]. Murray et al. demonstrated that the charge state of the NPs can be tuned by varying the electrode potential, and that the charged NPs retained most of their charge upon removal from the solution and re-dissolution [346]. A number of studies have investigated the influence of capping ligand (layer thickness, chemical structure), electrolyte, and temperature [123, 347]. Hicks et al. demonstrated that the NPs capacitance can be modeled as the capacitance of two concentric conducting spheres separated by a dielectric [348]. The model was extended by Quinn et al. [337] and Su et al. [349], both taking medium effects into account.

Summarizing, the energetics of MPC (with a core size ranging between ~1 nm and 3–4 nm) charging in solution are determined by the capacitance. It is primarily a function of core size and the nature of the protecting monolayer. However, the capacitance can be significantly altered by medium effects, such as solvent and electrolyte ions.

Quantized double-layer charging was also observed with immobilized mono- [350] and multi-layer [351] assemblies of NPs on solid substrates, employing covalent or electrostatic linking procedures ([336, 347] and literature cited therein). Potential spacing ΔE_z and the NP capacitance in low-polarity electrolytes were found to be similar to the data from bulk-solution experiments. The charging response in high-polarity solvents, such as aqueous electrolytes, was distinctly different: ΔE_z was smaller and an ion-induced rectification was observed. This phenomenon was tentatively attributed to a coupled ion-transfer and electron-transfer process [337]. However, this hypothesis is still not yet proven. Redox- or otherwise active groups could be introduced, upon modification of the particles ligand shell, by kinetically controlled place-exchange reactions. [352].

We have recently synthesized monodisperse hexanethiolate-protected gold clusters, Au_{144}-C6S, and determined an average core diameter of 1.7 ± 0.3 nm, in support of the estimated molecular formula Au_{144}-$(C_6H_{11}S)_{52}$ (Fig. 30b) [334].

Solution-phase DPV of Au_{144}-C_6S dispersed in 10 mM [bis(triphenylphosphoranylidene)-ammoniumtetrakis-(pentafluorophenyl)-borate (BTPPATPFB)/toluene]:[acetonitrile] 2:1 revealed well-behaved, equally spaced and symmetric quantized double-layer charging peaks with $\Delta E \sim 0.270 \pm 0.010$ V. Applying the classical concentric spheres capacitor model (8) reveals an individual cluster capacitance of ~0.6 aF [334, 335].

Immobilization of these clusters on Pt(hkl) surfaces, employing bifunctional linkers with one pyridyl- and one thiol-based anchoring group, was achieved [334, 335]. The terminal pyridyl anchor ensures a strong chemical bond to the

Pt(hkl) substrate, while the thiol is attached to the NP via ligand exchange [353], which could be characterized in aqueous and organic electrolytes.

We have also demonstrated that well-behaved quantized charging of gold MPCs is possible in air- and water-stable room-temperature ionic liquids, such as 1-hexyl-3-methylimidazolium tris(penta-fluoroethyl)-trifluorophosphate (HMImFEP), Fig. 30c, d [334, 335]. As ionic liquids have very attractive features, including near-zero vapor pressure, considerable thermal stability, and an electrochemical stability window that often exceeds 4 V, this demonstration is particularly significant from a technological point of view.

8.3 Quantized Charging of Individual Nanoclusters

Single-electron Coulomb charging was observed for different types and sizes of individual metal NPs in vacuum or in air, employing specifically STM and STS [24, 39, 353–355]. Albrecht et al. showed in a proof-of-principle experiment the possibility of intrinsic multi-state switching of Au nanoclusters through electrochemical gating [123]. Motivated by this pioneering report, we created a ferrocenethiol-based self-assembled template monolayer on Au(111) electrodes with immobilized and uniformly-distributed gold clusters of 2.4 $\pm$ 0.5 nm diameter and a rather narrow size distribution (Fig. 31a) [150].

We carried out a series of systematic constant-bias-mode in situ STS experiments, with the gold tip positioned above a single gold cluster. We observed for the cluster-modified junctions up to seven clearly resolved narrow peaks in the tunneling current i_T plotted vs the substrate potential E_S (I_T versus E_S characteristics, Fig. 31b [150]). These data were analyzed by fitting a series of Gaussians to distinguish individual peaks, and to determine their positions, heights, and widths. An example is presented in Fig. 31b. The peak-to-peak spacing in the current–voltage curves, based on a statistical analysis of 73 individual traces, was found to be very regular (Fig. 31d) and equal to $\Delta E = 0.11 \pm 0.02$ V. The FWHM is also rather uniform, and amounts to 0.07 ± 0.03 V. On the other hand, we observed that the measured heights of the individual peaks vary by two orders of magnitude, and are distributed exponentially with a decay factor of ~0.5 nA. We noted that these regularly-spaced peaks were positioned on both sides of the formal potential of the ferrocene-driven redox process, e.g., at $E \approx E^0 = 0.51$ V. The heights of these peaks typically decrease with distance from E^0. In an attempt to generalize these observations, we constructed two-dimensional peak-position vs peak-height distributions based on all experimental data accessible. The plot in Fig. 31c clearly supports the trend formulated above. On the basis of the distribution of peak spacings and peak heights, we attribute our observations to a modulation of the tunneling current enhancement in the redox-active tunneling junction by a sequential capacitive charging of single gold nanoclusters. Following this hypothesis, we estimate [336, 339, 340, 344] the average single-particle capacitance as $C \approx e/\Delta E = 1.5 \pm 0.2$ aF.

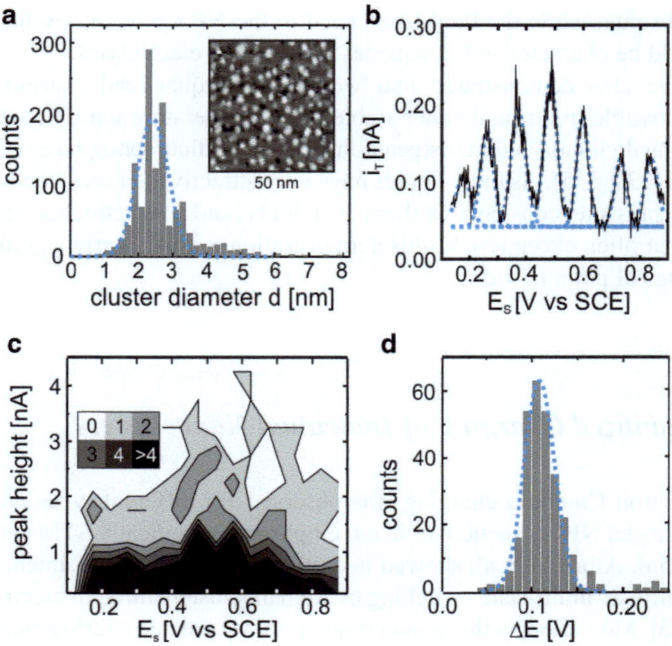

Fig. 31 (**a**) STM image ($i_T = 0.05$ nA, $E_S = 0.7$ V, $E_T = 0.1$ V; vertical scale 0.4 nm) and size distribution histogram of gold nanoclusters formed with a self-assembled template monolayer on Au (111) in 1.0 M $HClO_4$. The *dotted line* represents the Gaussian fit. The corresponding parameters are the average cluster diameter of 2.4 nm and the standard deviation $\sigma = 0.45$ nm. (**b**) Example of a constant bias current–voltage response of a single gold cluster as triggered by "electrolyte gating", $E_{bias} = 0.10$ V, $i_T = 0.05$ nA. The *dotted traces* represent its fit by a series of Gaussians. (**c**) Contour map of peak position vs peak height distribution for 361 peaks. The various shadings indicate areas with different amounts of counts per cell (0.05 V × 0.5 nA) inside. (**d**) Peak spacing histogram and Gaussian fit for a data set 361 individual traces, such as shown in (**c**). The corresponding parameter are the average spacing $\Delta E = 0.11$ V and $\sigma = 0.017$ V [150]

Our study and the work of Albrecht et al. [123] reported, as a unique result, peak-shaped modulations of the tunneling current. The peaks are equally spaced, as well as rather close and narrow. This offers new perspectives on multistate electronic switching in condensed media at room temperature. The observed feature of sequential capacitive charging is reminiscent of a successive charging of multivalent redox molecules through several oxidation/reduction states. However, the latter are usually much wider than the former, and limited to a rather small number of individual states. The interplay of this redox resonance with the sequential charging of Au NPs in an individual electrochemical tunneling junction has not yet been treated theoretically, although the Coulomb charging effect under electrochemical conditions in a multistate redox molecule was recently addressed. The results seem to be in agreement with our observations. However, we believe that an appropriate model will need to consider the successive charging of the particle Coulomb levels in combination with the solvent activation and relaxation, as modified by the redox-active host lattice [150, 334, 335].

9 Conclusions and Outlook

Employing an STM-BJ technique in various condensed phases, charge transport properties through nanoscale objects at an electrified solid–liquid interface were addressed. In particular, the following representative systems were discussed in more detail: gold atomic contacts, alkanedithiols, biphenyl derivatives, 4,4'-bipyridine, redox-active perylene bisimides, and surface-immobilized gold clusters. The experimental investigation of these examples, together with quantum chemistry and transport calculations, provide a detailed understanding of selected fundamental relationships between molecular structure and conductance properties of nanostructures upon polarization in a confined space. The uniqueness of the concept of "electrolyte gating" to modulate charge transport at the nanoscale in condensed media and at room temperature was demonstrated. In particular, the addressing of the intrinsic electronic properties of single redox-active molecules and gold clusters was presented. These conceptual studies open up fascinating perspectives toward novel applications, such as multistate electronic switching, current amplification, rectification, and other electronics functions at the nanoscale.

Further development, beyond the current state of the art, relies on a higher diversity and reliability of experimental techniques and functionalized systems, as well as on the identification of nanoscale signatures, which will continuously inspire many physicists, chemists, biologists, and engineers. Examples include the construction of functional 3D architectures based on 2D templates, the combination of local spectroscopy (IR, Raman) with transport problems, the creation of organic–inorganic hybrid structures by rational combination of organic synthesis with inorganic platforms, such as carbon- and/or silicon-based materials, and the implementation of soft electronics (molecular basis and condensed medium) into hard electronics (conventional semiconductor-based micro-fabrication). It is our belief that the field of nano- and molecular-based electronics will continue growing as a highly innovative interdisciplinary science, which has a significant influence on various other scientific and engineering areas.

Acknowledgments This work was supported by the Swiss National Science Foundation, the Volkswagen Foundation, FUNMOLS, and the DFG priority program 1243. The authors would particularly like to acknowledge the fruitful discussions with M. Mayor, S. F. L. Mertens, F. Würthner, and the theory groups of F. Evers, F. Pauly, C. Cuevas, and C. Lambert.

References

1. Aviram A, Ratner MA (1974) Chem Phys Lett 29:277
2. Cuevas JC, Scheer E (2010) Molecular electronics: an introduction to theory and experiment. World Scientific, Singapore
3. Tour JM (2003) Molecular electronics: commercial insights, chemistry, architecture and programming. World Scientific, Singapore
4. Joachim C, Gimzewski JK, Aviram A (2000) Nature 408:541

5. Nitzan A, Ratner MA (2003) Science 300:1384
6. Heath JR, Ratner MA (2003) Phys Today 56:43
7. Joachim C, Ratner MA (2005) Proc Natl Acad Sci USA 102:8801
8. Carroll RL, Gorman CB (2002) Angew Chem Int Ed 41:4379
9. Lindsay SM, Ratner MA (2007) Adv Mater 19:23
10. Salomon A, Cahen D, Lindsay S, Tomfohr J, Engelkes VB, Frisbie CD (2003) Adv Mater 15:1881
11. Tao NJ (2006) Nat Nanotechnol 1:173
12. Ulgut B, Abruna HD (2008) Chem Rev 108:2721
13. Metzger RM (2008) J Mater Chem 18:4364
14. Selzer Y, Allara DL (2006) Annu Rev Phys Chem 57:593
15. Heath JR (2009) Annu Rev Mater Res 39:1
16. van der Molen SJ, Liljeroth P (2010) J Phys Condens Matter 22:133001
17. James DK, Tour JM (2004) Chem Mater 16:4423
18. McCreery RL (2004) Chem Mater 16:4477
19. Chen F, Hihath J, Huang ZF, Li XL, Tao NJ (2007) Annu Rev Phys Chem 58:535
20. Akkerman HB, de Boer B (2008) J Phys Condens Matter 20:013001
21. Kroger J, Neel N, Limot L (2008) J Phys Condens Matter 20:22301
22. Haick H, Cahen D (2008) Prog Surf Sci 83:217
23. Nichols RJ, Haiss W, Higgins SJ, Leary E, Martin S, Bethell D (2010) Phys Chem Chem Phys 12:2801
24. Andres RP, Bein T, Dorogi M, Feng S, Henderson JI, Kubiak CP, Mahoney W, Osifchin RG, Reifenberger R (1996) Science 272:1323
25. Gimzewski JK, Joachim C (1999) Science 283:1683
26. Donhauser ZJ, Mantooth BA, Kelly KF, Bumm LA, Monnell JD, Stapleton JJ, Price DW, Rawlett AM, Allara DL, Tour JM, Weiss PS (2001) Science 292:2303
27. Repp A, Meyer G, Paavilainen S, Olsson FE, Persson M (2006) Science 312:1196
28. Cui XD, Primak A, Zarate X, Tomfohr J, Sankey OF, Moore AL, Moore TA, Gust D, Harris G, Lindsay SM (2001) Science 294:571
29. Wold DJ, Haag R, Rampi MA, Frisbie CD (2002) J Phys Chem B 106:2813
30. Fan FRF, Yang JP, Cai LT, Price DW, Dirk SM, Kosynkin DV, Yao YX, Rawlett AM, Tour JM, Bard AJ (2002) J Am Chem Soc 124:5550
31. Tao NJ (1996) Phys Rev Lett 76:4066
32. Xu BQ, Tao NJJ (2003) Science 301:1221
33. Haiss W, van Zalinge H, Higgins SJ, Bethell D, Hobenreich H, Schiffrin DJ, Nichols RJ (2003) J Am Chem Soc 125:15294
34. Li Z, Han B, Meszaros G, Pobelov I, Wandlowski T, Blaszczyk A, Mayor M (2006) Faraday Discuss 131:121
35. Dadosh T, Gordin Y, Krahne R, Khivrich I, Mahalu D, Frydman V, Sperling J, Yacoby A, Bar-Joseph I (2005) Nature 436:677
36. Liao JH, Bernard L, Langer M, Schonenberger C, Calame M (2006) Adv Mater 18:2803
37. Reed MA, Zhou C, Muller CJ, Burgin TP, Tour JM (1997) Science 278:252
38. Kergueris C, Bourgoin JP, Palacin S, Esteve D, Urbina C, Magoga M, Joachim C (1999) Phys Rev B 59:12505
39. Reichert J, Ochs R, Beckmann D, Weber HB, Mayor M, Lohneysen HV (2002) Phys Rev Lett 88:176804
40. Gonzalez MT, Wu SM, Huber R, van der Molen SJ, Schonenberger C, Calame M (2006) Nano Lett 6:2238
41. Lortscher E, Ciszek JW, Tour J, Riel H (2006) Small 2:973
42. Park J, Pasupathy AN, Goldsmith JI, Chang C, Yaish Y, Petta JR, Rinkoski M, Sethna JP, Abruna HD, McEuen PL, Ralph DC (2002) Nature 417:722
43. Osorio EA, Bjornholm T, Lehn JM, Ruben M, van der Zant HSJ (2008) J Phys Condens Matter 20:374121

44. Seferos DS, Trammell SA, Bazan GC, Kushmerick JG (2005) Proc Natl Acad Sci USA 102:8821
45. Chen J, Reed MA, Rawlett AM, Tour JM (1999) Science 286:1550
46. Tran E, Duati M, Whitesides GM, Rampi MA (2006) Faraday Discuss 131:197
47. Seminario JM, Zacarias AG, Tour JM (1999) J Am Chem Soc 121:411
48. Venkataraman L, Klare JE, Nuckolls C, Hybertsen MS, Steigerwald ML (2006) Nature 442:904
49. Venkataraman L, Klare JE, Tam IW, Nuckolls C, Hybertsen MS, Steigerwald ML (2006) Nano Lett 6:458
50. Wang CS, Batsanov AS, Bryce MR, Martin S, Nichols RJ, Higgins SJ, Garcia-Suarez VM, Lambert CJ (2009) J Am Chem Soc 131:15647
51. Kamenetska M, Quek SY, Whalley AC, Steigerwald ML, Choi HJ, Louie SG, Nuckolls C, Hybertsen MS, Neaton JB, Venkataraman L (2010) J Am Chem Soc 132:6817
52. Kim B, Beebe JM, Jun Y, Zhu XY, Frisbie CD (2006) J Am Chem Soc 128:4970
53. Kiguchi M, Miura S, Hara K, Sawamura M, Murakoshi K (2006) Appl Phys Lett:89
54. Mishchenko A, Zotti L, Vonlathen D, Bürkle M, Pauly F, Cuevas JC, Mayor M, Wandlowski Th (2010) J Am Chem Soc 133:184
55. Ko CH, Huang MJ, Fu MD, Chen CH (2010) J Am Chem Soc 132:756
56. Park YS, Widawsky JR, Kamenetska M, Steigerwald ML, Hybertsen MS, Nuckolls C, Venkataraman L (2009) J Am Chem Soc 131:10820
57. Taniguchi M, Tsutsui M, Shoji K, Fujiwara H, Kawai T (2009) J Am Chem Soc 131:14146
58. Yasuda S, Yoshida S, Sasaki J, Okutsu Y, Nakamura T, Taninaka A, Takeuchi O, Shigekawa H (2006) J Am Chem Soc 128:7746
59. Chen F, Li XL, Hihath J, Huang ZF, Tao NJ (2006) J Am Chem Soc 128:15874
60. Martin S, Haiss W, Higgins S, Cea P, Lopez MC, Nichols RJ (2008) J Phys Chem C 112:3941
61. Park YS, Whalley AC, Kamenetska M, Steigerwald ML, Hybertsen MS, Nuckolls C, Venkataraman L (2007) J Am Chem Soc 129:15768
62. Xing YJ, Park TH, Venkatramani R, Keinan S, Beratan DN, Therien MJ, Borguet E (2010) J Am Chem Soc 132:7946
63. Martin CA, Ding D, Sorensen JK, Bjornholm T, van Ruitenbeek JM, van der Zant HSJ (2008) J Am Chem Soc 130:13198
64. Li C, Pobelov I, Wandlowski T, Bagrets A, Arnold A, Evers F (2008) J Am Chem Soc 130:318
65. Haiss W, Wang CS, Grace I, Batsanov AS, Schiffrin DJ, Higgins SJ, Bryce MR, Lambert CJ, Nichols RJ (2006) Nat Mater 5:995
66. Kamenetska M, Koentopp M, Whalley AC, Park YS, Steigerwald ML, Nuckolls C, Hybertsen MS, Venkataraman L (2009) Phys Rev Lett 102:126803
67. Wu SM, Gonzalez MT, Huber R, Grunder S, Mayor M, Schonenberger C, Calame M (2008) Nat Nanotechnol 3:569
68. Chang S, He J, Kibel A, Lee M, Sankey O, Zhang P, Lindsay S (2009) Nat Nanotechnol 4:297
69. Huber R, Gonzalez MT, Wu S, Langer M, Grunder S, Horhoiu V, Mayor M, Bryce MR, Wang CS, Jitchati R, Schonenberger C, Calame M (2008) J Am Chem Soc 130:1080
70. Choi SH, Kim B, Frisbie CD (2008) Science 320:1482
71. Leary E, Higgins SJ, van Zalinge H, Haiss W, Nichols RJ (2007) Chem Commun 38:3939
72. Venkataraman L, Park YS, Whalley AC, Nuckolls C, Hybertsen MS, Steigerwald ML (2007) Nano Lett 7:502
73. Hybertsen MS, Venkataraman L, Klare JE, Whalley CA, Steigerwald ML, Nuckolls C (2008) J Phys Condens Matter 20:374115
74. Haiss W, Wang CS, Jitchati R, Grace I, Martin S, Batsanov AS, Higgins SJ, Bryce MR, Lambert CJ, Jensen PS, Nichols RJ (2008) J Phys Condens Matter 20:374119
75. Vonlanthen D, Mishchenko A, Elbing M, Neuburger M, Wandlowski T, Mayor M (2009) Angew Chem Int Ed 48:8886

76. Mishchenko A, Vonlanthen D, Meded V, Burkle M, Li C, Pobelov IV, Bagrets A, Viljas JK, Pauly F, Evers F, Mayor M, Wandlowski T (2010) Nano Lett 10:156
77. Diez-Perez I, Hihath J, Lee Y, Yu LP, Adamska L, Kozhushner MA, Oleynik II, Tao NJ (2009) Nat Chem 1:635
78. Ashwell GJ, Sujka M, Green A (2006) Faraday Discuss 131:23
79. Elbing M, Ochs R, Koentopp M, Fischer M, von Hanisch C, Weigend F, Evers F, Weber HB, Mayor M (2005) Proc Natl Acad Sci USA 102:8815
80. Salomon A, Arad-Yellin R, Shanzer A, Karton A, Cahen D (2004) J Am Chem Soc 126:11648
81. Guisinger NP, Greene ME, Basu R, Baluch AS, Hersam MC (2004) Nano Lett 4:55
82. Collier CP, Wong EW, Belohradsky M, Raymo FM, Stoddart JF, Kuekes PJ, Williams RS, Heath JR (1999) Science 285:391
83. Luo Y, Collier CP, Jeppesen JO, Nielsen KA, DeIonno E, Ho G, Perkins J, Tseng HR, Yamamoto T, Stoddart JF, Heath JR (2002) ChemPhysChem 3:519
84. Duan XF, Huang Y, Lieber CM (2002) Nano Lett 2:487
85. Blum AS, Kushmerick JG, Long DP, Patterson CH, Yang JC, Henderson JC, Yao YX, Tour JM, Shashidhar R, Ratna BR (2005) Nat Mater 4:167
86. Liao JH, Agustsson JS, Wu SM, Schonenberger C, Calame M, Leroux Y, Mayor M, Jeannin O, Ran YF, Liu SX, Decurtins S (2010) Nano Lett 10:759
87. Song H, Kim Y, Jang YH, Jeong H, Reed MA, Lee T (2009) Nature 462:1039
88. Reddy P, Jang SY, Segalman RA, Majumdar A (2007) Science 315:1568
89. Kubatkin S, Danilov A, Hjort M, Cornil J, Bredas JL, Stuhr-Hansen N, Hedegard P, Bjornholm T (2003) Nature 425:698
90. White HS, Kittlesen GP, Wrighton MS (1984) J Am Chem Soc 106:5375
91. Meulenkamp EA (1999) J Phys Chem B 103:7831
92. Kruger M, Buitelaar MR, Nussbaumer T, Schonenberger C, Forro L (2001) Appl Phys Lett 78:1291
93. Rosenblatt S, Yaish Y, Park J, Gore J, Sazonova V, McEuen PL (2002) Nano Lett 2:869
94. Repp J, Meyer G, Stojkovic SM, Gourdon A, Joachim C (2005) Phys Rev Lett 94:026803
95. Stipe BC, Rezaei MA, Ho W (1998) Science 280:1732
96. Heinrich AJ, Lutz CP, Gupta JA, Eigler DM (2002) Science 298:1381
97. Liljeroth P, Repp J, Meyer G (2007) Science 317:1203
98. Bumm LA, Arnold JJ, Cygan MT, Dunbar TD, Burgin TP, Jones L, Allara DL, Tour JM, Weiss PS (1996) Science 271:1705
99. Mantooth BA, Weiss PS (2003) Proc IEEE 91:1785
100. Weiss PS (2008) Acc Chem Res 41:1772
101. Love JC, Estroff LA, Kriebel JK, Nuzzo RG, Whitesides GM (2005) Chem Rev 105:1103
102. Jackel F, Watson MD, Mullen K, Rabe JP (2004) Phys Rev Lett 92:188303
103. Puigmarti-Luis J, Minoia A, Uji-i H, Rovira C, Cornil J, De Feyter S, Lazzaroni R, Amabilino DB (2006) J Am Chem Soc 128:12602
104. Stabel A, Herwig P, Mullen K, Rabe JP (1995) Angew Chem Int Ed 34:1609
105. Gesquiere A, De Feyter S, De Schryver FC, Schoonbeek F, van Esch J, Kellogg RM, Feringa BL (2001) Nano Lett 1:20
106. Poirier GE (1997) Chem Rev 97:1117
107. Fabre B (2010) Acc Chem Res 43:1509
108. Della Pia A, Chi Q, Jones DD, Macdonald JE, Ulstrup J, Eliott M (2010) Nano Lett. doi: 10.1021/nl10333aq
109. Chi QJ, Farver O, Ulstrup J (2005) Proc Natl Acad Sci USA 102:16203
110. Pobelov IV, Li ZH, Wandlowski T (2008) J Am Chem Soc 130:16045
111. Ye T, Kumar AS, Saha S, Takami T, Huang TJ, Stoddart JF, Weiss PS (2010) ACS Nano 4:3697
112. Moore AM, Dameron AA, Mantooth BA, Smith RK, Fuchs DJ, Ciszek JW, Maya F, Yao YX, Tour JM, Weiss PS (2006) J Am Chem Soc 128:1959

113. Chen F, He J, Nuckolls C, Roberts T, Klare JE, Lindsay S (2005) Nano Lett 5:503
114. Li ZH, Pobelov I, Han B, Wandlowski T, Blaszczyk A, Mayor M (2007) Nanotechnology 18:044018
115. Fujihira M, Suzuki M, Fujii S, Nishikawa A (2006) Phys Chem Chem Phys 8:3876
116. Ulrich J, Esrail D, Pontius W, Venkataraman L, Millar D, Doerrer LH (2006) J Phys Chem B 110:2462
117. Jang SY, Reddy P, Majumdar A, Segalman RA (2006) Nano Lett 6:2362
118. Wierzbinski E, Slowinski K (2006) Langmuir 22:5205
119. Sek S, Misicka A, Swiatek K, Maicka E (2006) J Phys Chem B 110:19671
120. Wold DJ, Frisbie CD (2001) J Am Chem Soc 123:5549
121. Morita T, Lindsay S (2007) J Am Chem Soc 129:7262
122. Cervera J, Mafe S (2010) ChemPhysChem 11:1654
123. Albrecht T, Mertens SFL, Ulstrup J (2007) J Am Chem Soc 129:9162
124. Meszaros G, Kronholz S, Karthauser S, Mayer D, Wandlowski T (2007) Appl Phys A 87:569
125. Li CZ, Bogozi A, Huang W, Tao NJ (1999) Nanotechnology 10:221
126. Holliday BJ, Mirkin CA (2001) Angew Chem Int Ed 40:2022
127. Park SJ, Taton TA, Mirkin CA (2002) Science 295:1503
128. Yi ZW, Trellenkamp S, Offenhausser A, Mayer D (2010) Chem Commun 46:8014
129. Guo XF, Small JP, Klare JE, Wang YL, Purewal MS, Tam IW, Hong BH, Caldwell R, Huang LM, O'Brien S, Yan JM, Breslow R, Wind SJ, Hone J, Kim P, Nuckolls C (2006) Science 311:356
130. Feldman AK, Steigerwald ML, Guo XF, Nuckolls C (2008) Acc Chem Res 41:1731
131. Amlani I, Rawlett AM, Nagahara LA, Tsui RK (2002) Appl Phys Lett 80:2761
132. van der Molen SJ, Liao JH, Kudernac T, Agustsson JS, Bernard L, Calame M, van Wees BJ, Feringa BL, Schonenberger C (2009) Nano Lett 9:76
133. He J, Forzani ES, Nagahara LA, Tao NJ, Lindsay S (2008) J Phys Condens Matter 20:374120
134. Agrait N, Yeyati AL, van Ruitenbeek JM (2003) Phys Rep 377:81
135. van Ruitenbeek JM, Alvarez A, Pineyro I, Grahmann C, Joyez P, Devoret MH, Esteve D, Urbina C (1996) Rev Sci Instrum 67:108
136. Moreland J, Ekin JW (1985) J Appl Phys 58:3888
137. Muller CJ, Vanruitenbeek JM, Dejongh LJ (1992) Phys Rev Lett 69:140
138. Gruter L, Gonzalez MT, Huber R, Calame M, Schonenberger C (2005) Small 1:1067
139. Grunder S, Huber R, Wu SM, Schonenberger C, Calame M, Mayor M (2010) Chimia 64:140
140. Gimzewski JK, Moller R (1987) Phys Rev B 36:1284
141. Xu BQ, Xiao XY, Yang XM, Zang L, Tao NJ (2005) J Am Chem Soc 127:2386
142. Mishchenko A, Li C, Hong W (2011) Small (in preparation)
143. Halbritter A, Makk P, Mackowiak S, Csonka S, Wawrzynizk M, Martinek J (2010) arXiv/1006.1811:1
144. Zhang J, Chi Q, Kuznetsov AM, Hansen AG, Wackerbarth H, Christensen HEM, Andersen JET, Ulstrup J (2002) J Phys Chem B 106:1131
145. Zhang JD, Kuznetsov AM, Medvedev IG, Chi QJ, Albrecht T, Jensen PS, Ulstrup J (2008) Chem Rev 108:2737
146. He HX, Zhu JS, Tao NJ, Nagahara LA, Amlani I, Tsui R (2001) J Am Chem Soc 123:7730
147. Kuznetsov AM, Medvedev IG, Ulstrup J (2007) J Chem Phys 127:104708
148. Hulea IN, Brom HB, Houtepen AJ, Vanmaekelbergh D, Kelly JJ, Meulenkamp EA (2004) Phys Rev Lett 93:166601
149. Tian JH, Yang Y, Zhou XS, Schollhorn B, Maisonhaute E, Chen ZB, Yang FZ, Chen Y, Amatore C, Mao BW, Tian ZQ (2010) ChemPhysChem 11:2745
150. Li ZH, Liu YQ, Mertens SFL, Pobelov IV, Wandlowski T (2010) J Am Chem Soc 132:8187
151. Di Ventra M, Pantelides ST, Lang ND (2000) Appl Phys Lett 76:3448
152. Di Ventra A, Lang ND, Pantelides ST (2002) Chem Phys 281:189
153. Han WH, Durantini EN, Moore TA, Moore AL, Gust D, Rez P, Leatherman G, Seely GR, Tao NJ, Lindsay SM (1997) J Phys Chem B 101:10719

154. Haiss W, Albrecht T, van Zalinge H, Higgins SJ, Bethell D, Hobenreich H, Schiffrin DJ, Nichols RJ, Kuznetsov AM, Zhang J, Chi Q, Ulstrup J (2007) J Phys Chem B 111:6703
155. Leary E, Higgins SJ, van Zalinge H, Haiss W, Nichols RJ, Nygaard S, Jeppesen JO, Ulstrup J (2008) J Am Chem Soc 130:12204
156. He J, Chen F, Lindsay S, Nuckolls C (2007) Appl Phys Lett 90:072112
157. Xu BQQ, Li XLL, Xiao XYY, Sakaguchi H, Tao NJJ (2005) Nano Lett 5:1491
158. Albrecht T, Guckian A, Ulstrup J, Vos JG (2005) Nano Lett 5:1451
159. Albrecht T, Moth-Poulsen K, Christensen JB, Guckian A, Bjornholm T, Vos JG, Ulstrup J (2005) Faraday Discuss:265
160. Albrecht T, Moth-Poulsen K, Christensen JB, Hjelm J, Bjornholm T, Ulstrup J (2006) J Am Chem Soc 128:6574
161. Albrecht T, Guckian A, Kuznetsov AM, Vos JG, Ulstrup J (2006) J Am Chem Soc 128:17132
162. Seo K, Konchenko AV, Lee J, Bang GS, Lee H (2008) J Am Chem Soc 130:2553
163. Ricci AM, Calvo EJ, Martin S, Nichols RJ (2010) J Am Chem Soc 132:2494
164. Visoly-Fisher I, Daie K, Terazono Y, Herrero C, Fungo F, Otero L, Durantini E, Silber JJ, Sereno L, Gust D, Moore TA, Moore AL, Lindsay SM (2006) Proc Natl Acad Sci USA 103:8686
165. Xiao XY, Nagahara LA, Rawlett AM, Tao NJ (2005) J Am Chem Soc 127:9235
166. He J, Fu Q, Lindsay S, Ciszek JW, Tour JM (2006) J Am Chem Soc 128:14828
167. Wassel RA, Credo GM, Fuierer RR, Feldheim DL, Gorman CB (2004) J Am Chem Soc 126:295
168. Xiao XY, Brune D, He J, Lindsay S, Gorman CB, Tao NJ (2006) Chem Phys 326:138
169. Li XL, Hihath J, Chen F, Masuda T, Zang L, Tao NJ (2007) J Am Chem Soc 129:11535
170. Li C, Mishchenko A, Li Z, Pobelov I, Wandlowski T, Li XQ, Wurthner F, Bagrets A, Evers F (2008) J Phys Condens Matter 20:374122
171. Morita T, Lindsay S (2008) J Phys Chem B 112:10563
172. Alessandrini A, Salerno M, Frabboni S, Facci P (2005) Appl Phys Lett 86:133902
173. Alessandrini A, Corni S, Facci P (2006) Phys Chem Chem Phys 8:4383
174. Chi QJ, Zhang JD, Jensen PS, Christensen HEM, Ulstrup J (2006) Faraday Discuss 131:181
175. Petrangolini P, Alessandrini A, Berti L, Facci P (2010) J Am Chem Soc 132:7445
176. Landauer R (1957) IBM J Res Dev 1:223
177. Imry Y (1986) Physics of mesoscopic systems. In: Grinstein G, Mazenko G (ed) Directions in condensed matter physics. World Scientific, Singapore
178. Scheer E, Joyez P, Esteve D, Urbina C, Devoret MH (1997) Phys Rev Lett 78:3535
179. Cuevas JC, Yeyati AL, Martin-Rodero A (1998) Phys Rev Lett 80:1066
180. Scheer E, Agrait N, Cuevas JC, Yeyati AL, Ludoph B, Martin-Rodero A, Bollinger GR, van Ruitenbeek JM, Urbina C (1998) Nature 394:154
181. Krans JM, Muller CJ, Yanson IK, Govaert TCM, Hesper R, Vanruitenbeek JM (1993) Phys Rev B 48:14721
182. Li CZ, Tao NJ (1998) Appl Phys Lett 72:894
183. Krans JM, Vanruitenbeek JM, Fisun VV, Yanson IK, Dejongh LJ (1995) Nature 375:767
184. Yanson AI, vanRuitenbeek JM (1997) Phys Rev Lett 79:2157
185. Sirvent C, Rodrigo JG, Vieira S, Jurczyszyn L, Mingo N, Flores F (1996) Phys Rev B 53:16086
186. Smit RHM, Untiedt C, Yanson AI, van Ruitenbeek JM (2001) Phys Rev Lett 87:266102
187. Calvo MR, Fernandez-Rossier J, Palacios JJ, Jacob D, Natelson D, Untiedt C (2009) Nature 458:1150
188. Rubio G, Agrait N, Vieira S (1996) Phys Rev Lett 76:2302
189. Rubio-Bollinger G, Bahn SR, Agrait N, Jacobsen KW, Vieira S (2001) Phys Rev Lett 87:026101
190. Torres JA, Saenz JJ (1996) Phys Rev Lett 77:2245
191. Sanchez-Portal D, Artacho E, Junquera J, Ordejon P, Garcia A, Soler JM (1999) Phys Rev Lett 83:3884

192. Huisman EH, Trouwborst ML, Bakker FL, de Boer B, van Wees BJ, van der Molen SJ (2008) Nano Lett 8:3381
193. Smit RHM, Noat Y, Untiedt C, Lang ND, van Hemert MC, van Ruitenbeek JM (2002) Nature 419:906
194. Csonka S, Halbritter A, Mihaly G (2006) Phys Rev B 73:075405
195. Csonka S, Halbritter A, Mihaly G, Shklyarevskii OI, Speller S, van Kempen H (2004) Phys Rev Lett 93:016802
196. Kiguchi M, Stadler R, Kristensen IS, Djukic D, van Ruitenbeek JM (2007) Phys Rev Lett 98:146802
197. Novaes FD, da Silva AJR, da Silva EZ, Fazzio A (2006) Phys Rev Lett 96:016104
198. Thijssen WHA, Marjenburgh D, Bremmer RH, van Ruitenbeek JM (2006) Phys Rev Lett 96:026806
199. Kiguchi M, Djukic D, van Ruitenbeek JM (2007) Nanotechnology 18:035205
200. Tal O, Krieger M, Leerink B, van Ruitenbeek JM (2008) Phys Rev Lett 100:196804
201. Li CZ, Sha H, Tao NJ (1998) Phys Rev B 58:6775
202. He HX, Tao NJ (2002) Adv Mater 14:161
203. Xu BQ, He HX, Tao NJ (2002) J Am Chem Soc 124:13568
204. Jelinek P, Perez R, Ortega J, Flores F (2006) Phys Rev Lett 96:046803
205. Shu C, Li CZ, He HX, Bogozi A, Bunch JS, Tao NJ (2000) Phys Rev Lett 84:5196
206. Xu BQ, He HX, Boussaad S, Tao NJ (2003) Electrochim Acta 48:3085
207. Kiguchi M, Konishi T, Miura S, Murakoshi K (2007) Nanotechnology 18:424011
208. Zhou XS, Wei YM, Liu L, Chen ZB, Tang J, Mao BW (2008) J Am Chem Soc 130:13228
209. Li C, Wandlowswki Th (2011), Electrochim. Acta, in preparation
210. Li C, Wandlowski Th (2011) Small, in preparation
211. Dakkouri A, Kolb DM (1999) Reconstruction of gold surface. In: Wieckowski A (ed) Interfacial electrochemistry: theory, experiment and application. Marcel Dekker, New York
212. Kolb DM, Schneider J (1986) Electrochim Acta 31:929
213. Dretschkow T, Wandlowski T (1997) Ber Der Bunsenges Phys Chem Chem Phys 101:749
214. Wandlowski T, Ataka K, Pronkin S, Diesing D (2004) Electrochimica Acta 49:1233
215. He HX, Shu C, Li CZ, Tao NJ (2002) J Electroanal Chem 522:26
216. Abellan J, Chicon R, Arenas A (1998) Surf Sci 418:493
217. Barnett RN, Hakkinen H, Scherbakov AG, Landman U (2004) Nano Lett 4:1845
218. Okamoto M, Takayanagi K (1999) Phys Rev B 60:7808
219. De Maria L, Springborg M (2000) Chem Phys Lett 323:293
220. Hakkinen H, Barnett RN, Landman U (1999) J Phys Chem B 103:8814
221. Lang G, Heusler KE (1995) J Electroanal Chem 391:169
222. Clausen-Schaumann H, Seitz M, Krautbauer R, Gaub HE (2000) Curr Opinion Chem Bio 4:524
223. Janshoff A, Neitzert M, Oberdorfer Y, Fuchs H (2000) Angew Chem Int Ed 39:3213
224. Hugel T, Seitz M (2001) Macromol Rapid Comm 22:989
225. Rief M, Grubmuller H (2002) ChemPhysChem 3:255
226. Engel A, Gaub HE (2008) Ann Rev Biochem 77:127
227. Huang ZF, Chen F, Bennett PA, Tao NJ (2007) J Am Chem Soc 129:13225
228. Huang ZF, Xu BQ, Chen YC, Di Ventra M, Tao NJ (2006) Nano Lett 6:1240
229. Evans E (1998) Faraday Discuss 111:1
230. Evans E (2001) Annu Rev Biophys Biomol Struct 30:105
231. Tsutsui M, Shoji K, Taniguchi M, Kawai T (2008) Nano Lett 8:345
232. Velez P, Dassie SA, Leiva EPM (2008) Chem Phys Lett 460:261
233. Zhao JW, Murakoshi K, Yin X, Kiguchi M, Guo Y, Wang N, Liang S, Liu H (2008) J Phys Chem C 112:20088
234. Lipkowski J, Shi ZC, Chen AC, Pettinger B, Bilger C (1998) Electrochim Acta 43:2875
235. Magnussen OM (2002) Chem Rev 102:679
236. Pascual JI, Mendez J, Gomezherrero J, Baro AM, Garcia N, Landman U, Luedtke WD, Bogachek EN, Cheng HP (1995) Science 267:1793

237. Untiedt C, Bollinger GR, Vieira S, Agrait N (2000) Phys Rev B 62:9962
238. Garcia-Mochales P, Serena PA, Garcia N, CostaKramer JL (1996) Phys Rev B 53:10268
239. Garcia-Mochales P, Serena PA (1997) Phys Rev Lett 79:2316
240. Cuevas JC, Yeyati AL, Martin-Rodero A, Bollinger GR, Untiedt C, Agrait N (1998) Phys Rev Lett 81:2990
241. Mann B, Kuhn H (1971) J Appl Phys 42:4398
242. Li XL, He J, Hihath J, Xu BQ, Lindsay SM, Tao NJ (2006) J Am Chem Soc 128:2135
243. Haiss W, Nichols RJ, van Zalinge H, Higgins SJ, Bethell D, Schiffrin DJ (2004) Phys Chem Chem Phys 6:4330
244. Haiss W, van Zalinge H, Bethell D, Ulstrup J, Schiffrin DJ, Nichols RJ (2006) Faraday Discuss 131:253
245. Haiss W, Martin S, Leary E, van Zalinge H, Higgins SJ, Bouffier L, Nichols RJ (2009) J Phys Chem C 113:5823
246. Cui XD, Primak A, Zarate X, Tomfohr J, Sankey OF, Moore AL, Moore TA, Gust D, Nagahara LA, Lindsay SM (2002) J Phys Chem B 106:8609
247. Cui XD, Zarate X, Tomfohr J, Sankey OF, Primak A, Moore AL, Moore TA, Gust D, Harris G, Lindsay SM (2002) Nanotechnology 13:5
248. Beebe JM, Engelkes VB, Miller LL, Frisbie CD (2002) J Am Chem Soc 124:11268
249. Engelkes VB, Beebe JM, Frisbie CD (2004) J Am Chem Soc 126:14287
250. Zhou JF, Chen F, Xu BQ (2009) J Am Chem Soc 131:10439
251. Scaini D, Castronovo M, Casalis L, Scoles G (2008) ACS Nano 2:507
252. Gonzalez MT, Brunner J, Huber R, Wu SM, Schonenberger C, Calame M (2008) New J Phys 10:065018
253. Martin CA, Ding D, van der Zant HSJ, van Ruitenbeek JM (2008) New J Phys 10:065008
254. York RL, Nguyen PT, Slowinski K (2003) J Am Chem Soc 125:5948
255. Weiss EA, Chiechi RC, Kaufman GK, Kriebel JK, Li ZF, Duati M, Rampi MA, Whitesides GM (2007) J Am Chem Soc 129:4336
256. Lee T, Wang WY, Klemic JF, Zhang JJ, Su J, Reed MA (2004) J Phys Chem B 108:8742
257. Kushmerick JG, Holt DB, Pollack SK, Ratner MA, Yang JC, Schull TL, Naciri J, Moore MH, Shashidhar R (2002) J Am Chem Soc 124:10654
258. Beebe JM, Kim B, Frisbie CD, Kushmerick JG (2008) ACS Nano 2:827
259. Huisman EH, Guedon CM, van Wees BJ, van der Molen SJ (2009) Nano Lett 9:3909
260. Xia JL, Diez-Perez I, Tao NJ (2008) Nano Lett 8:1960
261. Hihath J, Arroyo CR, Rubio-Bollinger G, Tao NJ, Agrait N (2008) Nano Lett 8:1673
262. Wang WY, Lee T, Reed MA (2003) Phys Rev B 68:035416
263. Holmlin RE, Haag R, Chabinyc ML, Ismagilov RF, Cohen AE, Terfort A, Rampi MA, Whitesides GM (2001) J Am Chem Soc 123:5075
264. Slowinski K, Chamberlain RV, Miller CJ, Majda M (1997) J Am Chem Soc 119:11910
265. Chidsey CED (1991) Science 251:919
266. Smalley JF, Feldberg SW, Chidsey CED, Linford MR, Newton MD, Liu YP (1995) J Phys Chem 99:13141
267. Arnold A, Weigend F, Evers F (2007) J Chem Phys 126:174101
268. Im HS, Bernstein ER (1988) J Chem Phys 88:7337
269. Ando S, Hironaka T, Kurosu H, Ando I (2000) Magn Reson Chem 38:241
270. Wang J, Cooper G, Tulumello D, Hitchcock AP (2005) J Phys Chem A 109:10886
271. Pauly F, Viljas JK, Cuevas JC, Schon G (2008) Phys Rev B 77:155312
272. Kondo H, Nara J, Kino H, Ohno T (2008) J Chem Phys 128:064701
273. Benniston AC, Harriman A, Li P, Patel PV, Sams CA (2008) Chem Eur J 14:1710
274. Smalley JF, Sachs SB, Chidsey CED, Dudek SP, Sikes HD, Creager SE, Yu CJ, Feldberg SW, Newton MD (2004) J Am Chem Soc 126:14620
275. Lee MH, Speyer G, Sankey OF (2007) J Phys Condens Matter 19:215204
276. Tomfohr J, Sankey OF (2004) J Chem Phys 120:1542
277. Finch CM, Sirichantaropass S, Bailey SW, Grace IM, Garcia-Suarez VM, Lambert CJ (2008) J Phys Condens Matter 20:022203

278. Zotti LA, Kirchner T, Cuevas JC, Pauly F, Huhn T, Scheer E, Erbe A (2010) Small 6:1529
279. Vonlanthen D, Rotzler J, Neuburger M, Mayor M (2010) Eur J Org Chem 1:120
280. Rotzler J, Vonlanthen D, Barsella A, Boeglin A, Fort A, Mayor M (2010) Eur J Org Chem 6:1096
281. Pauly F, Viljas JK, Cuevas JC (2008) Phys Rev B 78:035315
282. Perdew JP (1986) Phys Rev B 33:8822
283. Ahlrichs R, Bar M, Haser M, Horn H, Kolmel C (1989) Chem Phys Lett 162:165
284. Eichkorn K, Treutler O, Ohm H, Haser M, Ahlrichs R (1995) Chem Phys Lett 242:652
285. Pauly F, Viljas JK, Huniar U, Hafner M, Wohlthat S, Burkle M, Cuevas JC, Schon G (2008) New J Phys 10:125019
286. Schmitteckert P, Evers F (2008) Phys Rev Lett 100:086401
287. Vonlathen D, Rudnev A, Mishchenko A, Käslin A, Rotzler J, Neuburger M, Wandlowski Th, Mayor M (2011) Europ J Chem. 17:7236–7250 doi:10.1002/chem.201003763
288. Xue YQ, Ratner MA (2004) Phys Rev B 69:085403
289. Baheti K, Malen JA, Doak P, Reddy P, Jang SY, Tilley TD, Majumdar A, Segalman RA (2008) Nano Lett 8:715
290. Li C, Manrique DZ, Lambert CJ, Wandlowski Th (2011) ACS Nano, in preparation 2011
291. Steel PJ (1990) Coordin Chem Rev 106:227
292. Stang PJ, Olenyub B (2000) In: Nalwa HS (ed) Handbook of nanostructured materials, vol 15. New York, Academic Press
293. Boag NM, Coward KM, Jones AC, Pemble ME, Thompson JR (1999) Acta Cryst Sec C 55:672
294. Almennigen A, Bastiansen O (1958) K Nor Vidensk Selesk Skr 4:1
295. Spotswood T, Tanzer CI (1967) Austr J Chem 20:1227
296. Manutova YS, Maltseva LS, Kamaaev FG, Leontév B, Mikhamedkhanova S, Otroshchenko OS, Sadykov AS (1973) Izv Akad Nauk SSSR Ser Khim 7:1510
297. Ould-Moussa L, Poizat O, Castella-Ventura M, Buntinx G, Kassab E (1996) J Phys Chem 100:2072
298. Bagrets A, Arnold A, Evers F (2008) J Am Chem Soc 130:9013
299. Quek SY, Kamenetska M, Steigerwald ML, Choi HJ, Louie SG, Hybertsen MS, Neaton JB, Venkataraman L (2009) Nat Nanotechnol 4:230
300. Li XL, Xu BQ, Xiao XY, Yang XM, Zang L, Tao NJ (2006) Faraday Discuss 131:111
301. Zhou XS, Chen ZB, Liu SH, Jin S, Liu L, Zhang HM, Xie ZX, Jiang YB, Mao BW (2008) J Phys Chem C 112:3935
302. Leary E, Hobenreich H, Higgins SJ, van Zalinge H, Haiss W, Nichols RJ, Finch CM, Grace I, Lambert CJ, McGrath R, Smerdon J (2009) Phys Rev Lett 102:086801
303. Wandlowski T, Ataka K, Mayer D (2002) Langmuir 18:4331
304. Mayer D, Dretschkow T, Ataka K, Wandlowski T (2002) J Electroanal Chem 524:20
305. Wang X, Liu Z, Zhuang MD, Zhang HM, Xie ZX, Wu DY, Ren B, Tian ZQ (2007) Appl Phys Lett 91:101105
306. Liu Z, Wang X, Dai K, Jin S, Zeng ZC, Zhuang MD, Yang ZL, Wu DY, Ren B, Tian ZQ (2009) J Raman Spectrosc 40:1400
307. Yanson AI, Bollinger GR, van den Brom HE, Agrait N, van Ruitenbeek JM (1998) Nature 395:783
308. Trouwborst ML, Huisman EH, Bakker FL, van der Molen SJ, van Wees BJ (2008) Phys Rev Lett 100:175502
309. Stadler R, Thygesen KS, Jacobsen KW (2005) Phys Rev B 72:241401
310. Futamata M (2001) J Phys Chem B 105:6933
311. Corni S (2007) IEEE Trans Nanotechnol 6:561
312. Wurthner F (2004) Chem Commun 14:1564
313. Grimsdale AC, Mullen K (2005) Angew Chem Int Ed 44:5592
314. Ranke P, Bleyl I, Simmerer J, Haarer D, Bacher A, Schmidt HW (1997) Appl Phys Lett 71:1332

315. Ego C, Marsitzky D, Becker S, Zhang JY, Grimsdale AC, Mullen K, MacKenzie JD, Silva C, Friend RH (2003) J Am Chem Soc 125:437
316. Horowitz G, Kouki F, Spearman P, Fichou D, Nogues C, Pan X, Garnier F (1996) Adv Mater 8:242
317. Briseno AL, Mannsfeld SCB, Reese C, Hancock JM, Xiong Y, Jenekhe SA, Bao Z, Xia Y (2007) Nano Lett 7:2847
318. Tamizhmani G, Dodelet JP, Cote R, Gravel D (1991) Chem Mater 3:1046
319. Schmidt-Mende L, Fechtenkotter A, Mullen K, Moons E, Friend RH, MacKenzie JD (2001) Science 293:1119
320. Goltner C, Pressner D, Mullen K, Spiess HW (1993) Angew Chem Int Ed 32:1660
321. Sauer M (2003) Angew Chem Int Ed 42:1790
322. Wilson TM, Tauber MJ, Wasielewski MR (2009) J Am Chem Soc 131:8952
323. Osswald P, Wurthner F (2007) J Am Chem Soc 129:14319
324. Salbeck J, Kunkely H, Langhals H, Saalfrank RW, Daub J (1989) Chimia 43:6
325. Su W, Jiang J, Lu W, Luo Y (2006) Nano Lett 6:2091
326. Cao H, Jiang J, Ma J, Luo Y (2008) J Am Chem Soc 130:6674
327. Wurthner F, Sautter A, Schmid D, Weber PJA (2001) Chem Eur J 7:894
328. Wurthner F, Stepanenko V, Chen ZJ, Saha-Moller CR, Kocher N, Stalke D (2004) J Org Chem 69:7933
329. Baggerman J, Jagesar DC, Vallee RAL, Hofkens J, De Schryver FC, Schelhase F, Vogtle F, Brouwer AM (2007) Chem Eur J 13:1291
330. Chen ZJ, Baumeister U, Tschierske C, Wurthner F (2007) Chem Eur J 13:450
331. Taylor J, Brandbyge M, Stokbro K (2003) Phys Rev B 68:121101
332. Mowbray DJ, Jones G, Thygesen KS (2008) J Chem Phys 128:111103
333. Palma JL, Cao C, Zhang XG, Krstic PS, Krause JL, Cheng HP (2010) J Phys Chem C 114:1655
334. Mertens SFL, Blech K, Sologubenko AS, Mayer J, Simon U, Wandlowski T (2009) Electrochim Acta 54:5006
335. Mertens SFL, Meszaros G, Wandlowski T (2010) Phys Chem Chem Phys 12:5417
336. Murray RW (2008) Chem Rev 108:2688
337. Laaksonen T, Ruiz V, Liljeroth P, Quinn BM (2008) Chem Soc Rev 37:1836
338. Grabert H, Devoret MH (1972) Single electron tunneling. Plenum, New York
339. Chen SW, Murray RW, Feldberg SW (1998) J Phys Chem B 102:9898
340. Chen SW, Murray RW (1999) J Phys Chem B 103:9996
341. Chen SW, Ingram RS, Hostetler MJ, Pietron JJ, Murray RW, Schaaff TG, Khoury JT, Alvarez MM, Whetten RL (1998) Science 280:2098
342. Ingram RS, Hostetler MJ, Murray RW, Schaaff TG, Khoury JT, Whetten RL, Bigioni TP, Guthrie DK, First PN (1997) J Am Chem Soc 119:9279
343. Quinn BM, Liljeroth P, Ruiz V, Laaksonen T, Kontturi K (2003) J Am Chem Soc 125:6644
344. Templeton AC, Wuelfing MP, Murray RW (2000) Acc Chem Res 33:27
345. Hicks JF, Miles DT, Murray RW (2002) J Am Chem Soc 124:13322
346. Pietron JJ, Hicks JF, Murray RW (1999) J Am Chem Soc 121:5565
347. Zabet-Khosousi A, Dhirani AA (2008) Chem Rev 108:4072
348. Hicks JF, Templeton AC, Chen SW, Sheran KM, Jasti R, Murray RW, Debord J, Schaaf TG, Whetten RL (1999) Anal Chem 71:3703
349. Su B, Zhang MQ, Shao YH, Girault HH (2006) J Phys Chem B 110:21460
350. Chen SW (2000) J Phys Chem B 104:663
351. Deng FJ, Chen SW (2005) Phys Chem Chem Phys 7:3375
352. Aguila A, Murray RW (2000) Langmuir 16:5949
353. Dubois JGA, Gerritsen JW, Schmid G, van Kempen H (1996) Physica B 218:262
354. Nijhuis CA, Oncel N, Huskens J, Zandvliet HJW, Ravoo BJ, Poelsema B, Reinhoudt DN (2006) Small 2:1422
355. Homberger M, Simon U (2010) Philos Trans R Soc A 368:1405

Top Curr Chem (2012) 313: 189–216
DOI: 10.1007/128_2011_175

Published online: 28 June 2011

Tunneling Spectroscopy of Organic Monolayers and Single Molecules

K.W. Hipps

Abstract Basic concepts in tunneling spectroscopy applied to molecular systems are presented. Junctions of the form M-A-M, M-I-A-M, and M-I-A-I′-M, where A is an active molecular layer, are considered. Inelastic electron tunneling spectroscopy (IETS) is found to be readily applied to all the above device types. It can provide both vibrational and electron spectroscopic data about the molecules comprising the A layer. In IETS there are no strong selection rules (although there are preferences) so that transitions that are normally IR, Raman, or even photon-forbidden can be observed. In the electronic transition domain, spin and Laporte forbidden transitions may be observed. Both vibrational and electronic IETS can be acquired from single molecules. The negative aspect of this seemingly ideal spectroscopic method is the thermal line width of about 5 k_BT. This limits the useful measurement of vibrational IETS to temperatures below about 10 K. In the case of most electronic transitions where the intrinsic linewidth is much broader, useful experiments above 100 K are possible. One further limitation of electronic IETS is that it is generally limited to transitions with energy less than about 20,000 cm^{-1}. IETS can be identified by peaks in d^2I/dV^2 vs bias voltage plots that occur at the same position (but not necessarily same intensity) in either bias polarity.

Elastic tunneling spectroscopy is discussed in the context of processes involving molecular ionization and electron affinity states, a technique we call orbital mediated tunneling spectroscopy, or OMTS. OMTS can be applied readily to M-I-A-M and M-I-A-I′-M systems, but application to M-A-M junctions is problematic. Spectra can be obtained from single molecules. Ionization state results correlate well with UPS spectra obtained from the same systems in the same environment. Both ionization and affinity levels measured by OMTS can usually be correlated with one electron oxidation and reduction potentials for the molecular species in solution. OMTS can be identified by peaks in dI/dV vs bias voltage plots that *do not* occur at the same position in either bias polarity. Because of the intrinsic

K.W. Hipps
Department of Chemistry and Materials Science and Engineering Program, Washington State University, Pullman 99164-4630, WA, USA
e-mail: hipps@wsu.edu

width of the ionization and affinity transitions, OMTS can be applied at temperatures above 500 K.

This is not a comprehensive review of more than 20 years of research and there are many excellent papers that are not cited here. An absence of a citation is not a reflection on the quality of the work.

Keywords Inelastic · Elastic · Tunneling · Spectroscopy · Orbital mediated · Vibrational · Electronic · Ionization levels · Affinity levels

Contents

1 Introduction

As attested to by the breadth of topics in this volume, the subject of electron transport in molecular devices is a broad and active area. Within this general area, one often finds common fundamental problems relating to the nature of the entity producing the observed electronic response. Some of these issues include:

- Attachment sites for a molecule within a single molecule junction
- Orientation of individual molecules within a self-assembled monolayer (SAM)
- Conformation of molecule(s) associated with a given electronic response
- Binding of molecules(s) to electrode surfaces
- Chemical transformations in molecules associated with device formation
- Location of molecular electronic energy levels in relation to device properties

All of these issues can be addressed through spectroscopy, but the structure of the devices in question severely limits the applicable techniques. By nature, electronic devices with single molecules, or single monolayers, as their active element are buried interface problems. The generally metallic electrodes inhibit access to the molecular materials by electromagnetic radiation. While there are examples of resonance Raman and photoemission studies of buried interfaces [1–8], these generally require large area devices that are quite difficult to make uniformly, that contain a monolayer SAM, and are impossible (because of signal-to-noise issues) as a single-molecule device. There are also a number of studies on what one would call half-buried interfaces (those in which the molecular layer is exposed) [9–13]. These latter studies are not appropriate for characterization of functioning devices. On the other hand, tunneling spectroscopy *requires* a complete functioning electronic circuit.

Moreover, tunneling spectra from single molecules can be obtained (see below). Thus, tunneling spectroscopy is an excellent tool to address fundamental questions about the molecular layer or single molecule in a molecular electronic device.

We will find that tunneling spectroscopy can provide vibrational and electronic spectra. The vibrational transitions are not subject to the same selection rules in as photon spectroscopy – IR active, Raman active, and even inactive vibrations are sometimes observed. Thus we will be able to address questions about molecular bonding, conformation, and (to a limited extent) molecular orientation. In the electronic realm, both electronic transitions and ionization (and affinity) processes can be observed. The constraint on non-ionizing electronic transitions is that transitions requiring more than 2 eV are often difficult to observe. Thus, tunneling spectroscopy will be most useful for the study of molecular excitations normally associated with the near-infrared region of the spectrum. First ionization processes falling into a window of roughly 4–8 eV below the vacuum level are readily studied, as are affinity levels falling between 2 and 5 eV below the vacuum level. Thus, organic systems with oxidation or reduction potentials falling in the window from roughly −2.5 to +3.0 V (vs SCE) are likely to yield useful insights when probed by tunneling spectroscopy.

In what follows we will explore the fundamentals of tunneling spectroscopy as applied to molecular systems. Our primary emphasis will be on devices, primarily metal-active molecular layer-metal (M-A-M), and metal-insulator-active molecular layer-metal (M-I-A-M), and M-I-A-I′-M devices. The "active molecular layer" may be a single organic molecule, a SAM, or a monolayer of organic material with imperfect aggregation. By "device" we will generally mean stand-alone electronic structures or scanning tunneling microscope (STM)-based measurements, since the STM is a special case of an M-I-A-M structure.

2 Electron Tunneling

The nanoscale world is exciting because it is governed by rules differing from those in the macroscopic, or even microscopic, realm. It is a world where quantum mechanics dominates the scene, and events on the single-molecule scale are critical. What we know about the behavior of material on our scale is no longer true on the nanometer scale, and our formularies must be re-written. In order to study this quantum world, a quantum-mechanical probe is essential. Electron tunneling provides that quantum-mechanical tool.

In the Newtonian world, a particle can never be in a region where its potential energy is greater than its total energy. To do so would require a negative kinetic energy – a clear impossibility since $mv^2/2 \geq 0$. As the scale shrinks to molecular dimensions, of the order of 1 nm, classical concepts fail and quantum mechanics is required. Thus, it is possible for a particle to move between two classically allowed regions by way of a region where its potential energy is greater than its total energy – this is the phenomenon of tunneling. While it can occur for relatively heavy particles

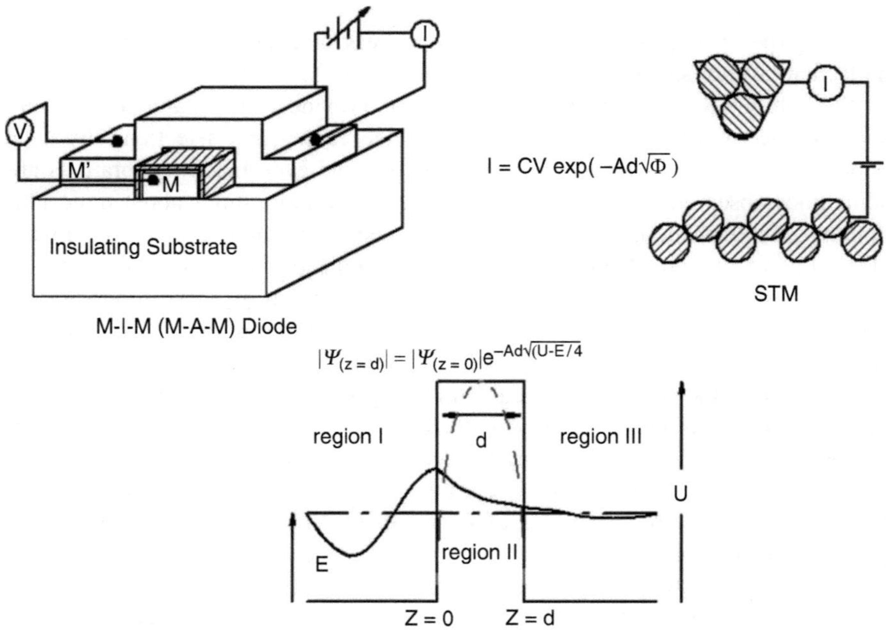

Fig. 1 Schematic drawings of a tunnel diode, an STM, and the electronic energy diagram appropriate for both. *U* is the height of the potential barrier, *E* is the energy of the incident electron, *d* is the thickness of the barrier, *A* is approximately 1.02 Å/(eV)$^{1/2}$ if *U* and *E* are in electron volts and *d* is in angstroms, ψ_0 is the wavefunction of the incident electron, and ψ_d is the wavefunction after transmission through the barrier. *I* is the measured tunneling current, *V* is the applied bias, and *M* and *M′* are the electrode metals

such as protons, it is far more probable for light particles such as electrons. Electron tunneling is a particularly useful probe, because it is easy to control the rate of flow and the energy of electrons, and to set up precisely controlled regions through which the electron must tunnel. An early example of an electron-tunneling device was the metal-insulator-metal (M-I-M) tunnel diode, Fig. 1 ([14] – these authors are the fathers of all tunneling spectroscopy [15]; [16–20]). Since most of the active molecular species of interest are insulators at low bias, this is also a diagram for an M-A-M diode. Also shown in Fig. 1 are the corresponding features of an STM [21–28]. Both devices rely on exactly the same physics. Within the conductors (metal electrodes in the M-I-M' case, substrate and atomically sharp tip in the STM case) the electrons are in classically allowed regions I and III. Within these regions, their total energy *E* is greater than their potential energy. However, the potential is greater than *E* in the gap between conductors (the insulator in the M-I-M case, the vacuum or solvent gap in the STM case), region II in Fig. 1. This region is classically forbidden, but quantum-mechanically allowed. A simple quantum-mechanical calculation quickly demonstrates that the probability of transmission through the barrier decreases exponentially with the thickness of the barrier and the square root of the potential (barrier height) relative to the electron energy. If distance, *d*, is measured in Angstroms (0.1 nm) and energies (E and U) are measured in electron volts, then the constant A in Fig. 1 is approximately 1.02 [19, 21].

In reality, as the barrier becomes narrower, it deviates from the square shape. One often used model is the parabolic barrier (dashed line in Fig. 1). When the barrier is composed of molecules, not only is the barrier shape difficult to predict, but the effective mass of the electron can deviate significantly from the free-electron mass. In order to take these differences into account, a more sophisticated treatment of the tunneling problem, based on the WKB method, can be used [21, 29–31]. Even if the metals are the same, differences in deposition methods, surface crystallographic orientation, and interaction with the active layer generally result in slightly different work functions on either side of the barrier.

How this quantum-mechanical model relates to M-I-M devices can be seen by considering Fig. 2, and by applying a particle-in-a-box model to the metal electrodes. At $T = 0$ K, valence electrons fill two at a time, according to the Pauli principle, until charge neutrality is reached. The energy of the last electron to be added, that with the highest energy, is said to be located at the Fermi energy, E_F. The energy required to remove that least bound electron to the vacuum level is the work function, Φ. The Fermi energy is the electrochemical potential of an electron in a metal. Thus, whenever two metals are brought into equilibrium, their Fermi energies equalize. Since at thermodynamic equilibrium, all levels are equally filled, and almost all are doubly filled, no net transfer of charge across the barrier can occur. If, however, a bias voltage of V is applied between the electrodes, a small band of states (of width equalling the applied voltagein width) in one electrode (left in Fig. 2) is

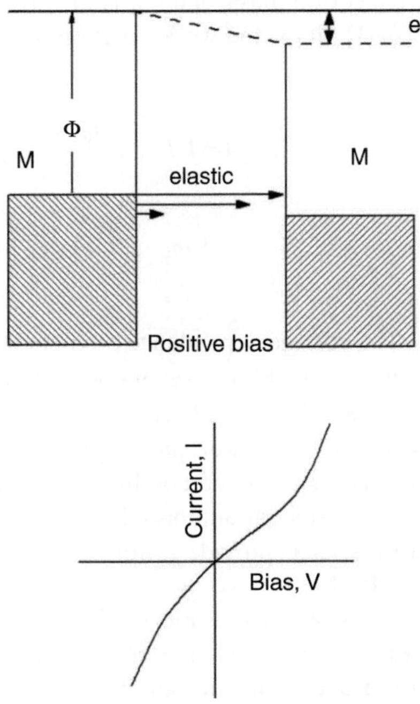

Fig. 2 Elastic tunneling through a structureless barrier

at an appropriate energy for electrons to tunnel from those states into vacant states of the other electrode (right in Fig. 2) without loss in energy. This is elastic tunneling.

If one assumes that the bias voltage is small compared to the barrier height (thus U–E is approximately equal to the work function, Φ), that the barrier is a simple square barrier, and that the density of states near E_F varies slowly, then the tunneling current is roughly given by

$$I \cong cV \exp(-\mathrm{d}\sqrt{\Phi}). \quad (1)$$

The exponential dependence of tunneling current on electrode separation is the essential element of the STM, a device that can produce exquisitely well-resolved images of molecules and atoms. Modern scanning tunneling microscopes are capable of resolving single atoms at temperatures ranging from near 0 K to above 600 K. STM images have been acquired in ultra-high vacuum (UHV), in air, even in electrochemical cells. The STM has allowed us to visualize the nanoscale world in a way that is essential for understanding processes on that scale. Because STM imaging has been extensively reviewed, we will simply refer the reader to the literature for a deeper understanding of the method [21, 22, 25, 28, 32–39].

Equation (1) suggests that tunnel junctions should be ohmic. This is true only for very small bias. A much better description of the tunneling current results when the effects of barrier shape, changes in barrier with applied potential, and effective mass of the electron are all included. An example of such an improved relationship is given by (2), where J is the current density, α is a unitless parameter used to account empirically for non-rectangular barrier shape and deviations in the effective electron mass, and Φ_B is the mean barrier height given by $\Phi_B = (\Phi_L + \Phi_R)/2$ [30, 31]. Φ_L is the work function of the left-hand metal:

$$J = \frac{e}{4\pi^2\hbar d^2}\left[\left(\Phi_B - \frac{eV}{2}\right)\exp\left(-\frac{(8m)^{1/2}\alpha d\left(\Phi_B - \frac{eV}{2}\right)}{\hbar}\right) - \left(\Phi_B + \frac{eV}{2}\right)\exp\left(-\frac{(8m)^{1/2}\alpha d\left(\Phi_B + \frac{eV}{2}\right)}{\hbar}\right)\right]. \quad (2)$$

The voltage dependence predicted by (2) leads to a highly nonlinear $I(V)$ curve if plotted over a bias of several volts. If the barrier is sufficiently asymmetrical at zero bias, the $I(V)$ curve becomes asymmetrical as well as nonlinear. Such a curve is shown in the lower part of Fig. 2. Equations for more complicated zero-bias barriers, such as the combination of a trapezoid and a square barrier, have been given by several authors [40, 41]. Equations like (2) and those for more complex barriers can provide information about barrier height and barrier thickness [30, 40–45].

However, there is more information in the tunneling current than just the surface geometry (e.g., STM), barrier height and shape, or barrier thickness. If a structured barrier is considered, one in which there are discrete electronic and/or vibrational states, the tunneling current can also reflect this energetic variation. Changes in current with applied bias voltage at constant electrode separation provide spectroscopic information about the surface of the electrodes and any material located in the barrier region. Moreover, there are several different types of interactions that can lead to distinctly different spectroscopic methods. These electron tunneling spectroscopies can be generally classed as based upon either inelastic or elastic electron tunneling processes. Our primary interest in this chapter is on the spectroscopic information that can be obtained about molecular barriers through electron tunneling. This encompasses both inelastic and elastic processes.

3 Inelastic Electron Tunneling Spectroscopy

Let us return to the simple model depicted in Fig. 2, but include the possibility of a manifold of vibrational states and both ground and excited electronic states. We will find that inelastic electron tunneling spectroscopy (IETS) gives the energy level differences familiar to the conventional photon spectroscopist [19, 20, 25]. IETS is based upon inelastic scattering, as is shown in Fig. 3. The moving electronic charge interacts with the time-varying molecular charges (electronic or vibrational) to induce excitation of the molecule in the barrier with concomitant loss of energy by the electron. This interaction has both long- and short-range interaction components; thus, selection rules are more relaxed than in IR or electron energy loss spectroscopy (EELS) [46–48]. As we will show later, there are very few (if any) hard selection rules in IETS. Instead, we often speak of selection preferences. In IETS we see some bands that appear in IR or Raman spectra, and others that are totally optically forbidden. Electronic IETS transitions with $\Delta S \neq 0$ are often observed. Thus, IETS is an excellent complementary vibrational spectroscopy for high-symmetry molecules, where many of the modes are optically forbidden.

An example of how Raman, IR, and IETS complement each other is given in Fig. 4, which is drawn using data originally published in [49]. While the resolution afforded by IETS (even at 4 K) is not as good as that provided by IR spectroscopy, it is not much worse than seen in Raman and is sufficiently good to identify individual normal modes. Of most interest is the fact that many more fundamentals are seen in IETS than in the IR or Raman spectra of this D_{3h} symmetry ion.

The vibrational analysis of the TCM ion gives $\Gamma_{vib} = 2A'_1 + A'_2 + 4E' + 2A''_2 + E''$. Of these, the A'_1 and E'' modes are only Raman-active, the E' modes are both IR- and Raman-active, the A''_2 modes are IR-active only, and the A'_2 mode is inactive in both IR and Raman [49]. Since the inactive mode is a bending motion, it is expected to occur in the region below 800 cm^{-1}. This region of the IR, Raman, and tunneling spectrum is shown in Fig. 4. Note that the low-frequency A''_2 mode is observed in the tunneling spectrum. The assignments in Fig. 4 were made based upon data acquired by all three spectroscopic methods. Moreover, the peak positions from three different isotopomers (the natural abundance, ^{13}C-enriched, and ^{15}N-enriched species) were measured, and an 11-parameter empirical force-field

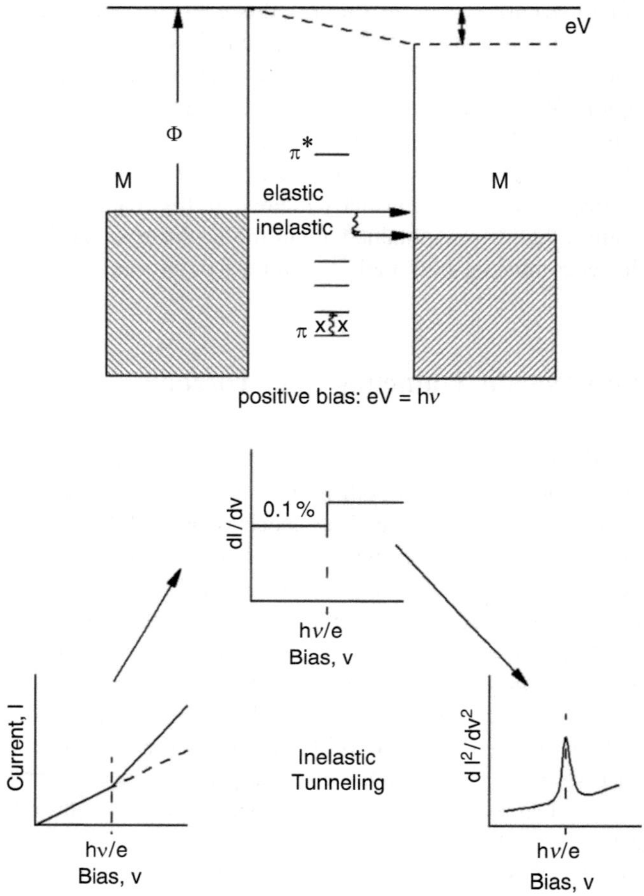

Fig. 3 Energy diagram for an M-A-M diode showing elastic and inelastic tunneling processes (*top*). The HOMO (π) and LUMO (π*) orbital energies and a few vibrational levels are indicated. Applied bias energy (eV) is just sufficient to allow inelastic tunneling with excitation of the first vibrational level, eV = hν. Also shown (*bottom*) are the *I*(*V*) curve, conductance-*V* curve, and the IETS spectrum that would result from both elastic processes and the first inelastic channel. (Reproduced by permission of the American Chemical Society from [19])

was fitted to the data. This procedure provided a $\pm 2\ \text{cm}^{-1}$ standard deviation fit to all the normal modes of the ion [49]. Note that the inactive A'_2 mode appears as a medium-intensity band in the tunneling spectrum, making it easy to identify and to evaluate isotopic shifts.

Another interesting facet of the vibrational IETS is the weakness of overtone and combination bands. There are sound theoretical reasons to expect that overtone bands should be very weak in IETS [46, 47]. To our knowledge, there has been no theoretical investigation of the intensities of combination bands in tunneling spectra. To be sure, there are experimental papers that contain tunneling band assignments that include assignments as combination and overtone bands. Most

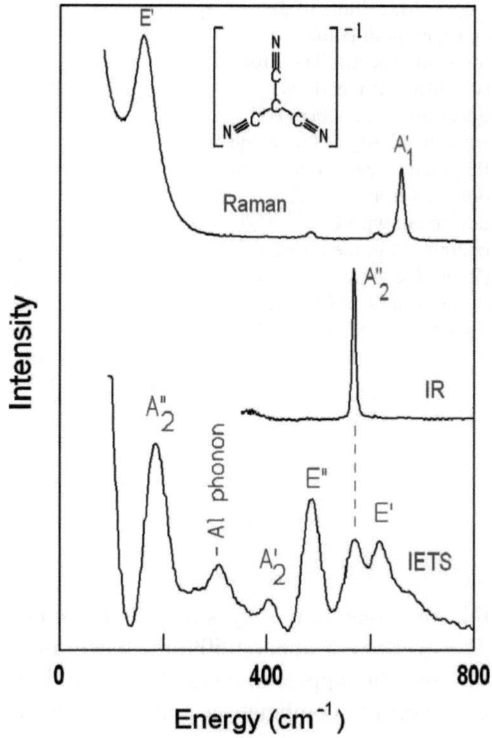

Fig. 4 Comparison of Raman, IR, and IETS spectra of the tricyanomethanide (TCM) ion. IETS taken from an Al-AlOx-CsTCM-Pb device. Figure drawn using data from [49]

of these are unreliable in that they were either made by simply choosing the energetically nearest equivalent IR or Raman band assignment, or were used when the real issue was chemical composition. Consider two examples. The best known overtone band in all of IETS was the overtone of the Al–O motion that occurs near 1,900 cm^{-1}. In the mid-1980s, however, Adler [50] and Gauthier [51] showed that this band was actually the Al–H fundamental stretching motion! Another example comes from the otherwise excellent work of Yang and White [52]. In their study of thiourea, they suggest that unspecified combination bands may be responsible for the "extra bands" seen in the NH stretching region of the tunneling spectrum. In the NH stretching region no individual bands are resolved, and it is difficult to determine how many transitions occur near 3,300 cm^{-1}. A more likely explanation for the observed lumpy band is that the NH region is complex because of inhomogeneity in adsorption sites. The observed texture might also be due to coupling between lattice and internal motions (Fig. 5).

Based on empirical observation, a general statement about overtones and combination bands might be: "Overtones do occur, but they are very weak. Combination bands are seldom observed." Kirtley, for example, says that overtones are about a factor of 200 weaker than fundamentals in the case of the benzoate ion [47, 53]. Ramsier, Henriksen, and Gent identify a single clear overtone in the tunneling spectrum of the phosphite ion (HPO_3^{-2}) [54]. The fundamental associated with

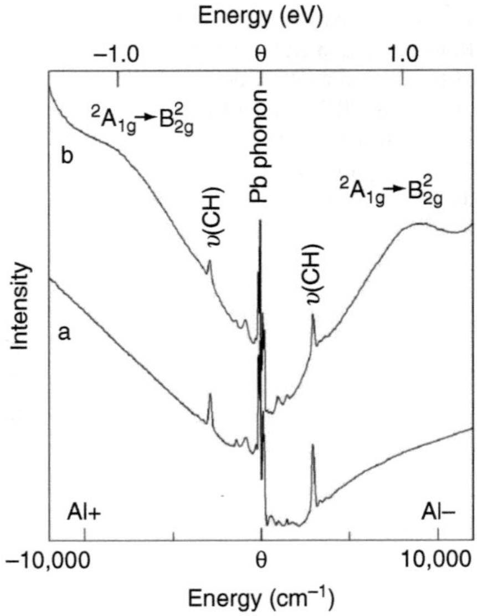

Fig. 5 Magnitude of the constant modulation tunneling spectra (NTI) for two different junctions. Spectrum a was obtained from an Al-Al_2O_3-(en)silane-Pb junction. Spectrum b was obtained from a similar device but with Co^{2+} complexed to the (en)silane. (Reproduced by permission of the American Chemical Society from [19])

this overtone is a very strong P–H bending band, and occurs at 1,034 cm^{-1}. The overtone is about 1/50th the height of the fundamental. All the other bands in the spectrum appear to be fundamentals. In our own work, we have seen little clear evidence for combination and overtone bands. An exception to the rule about overtones is afforded by the intense CH stretch in the tetraethylammonium ion (2,971 cm^{-1}) [55]. This band has a very weak overtone at 5,940 cm^{-1}. The integrated intensity of the overtone is about 1/60 of the fundamental band.

In the early days of IETS there were several attempts to calculate the IETS intensities for real molecules (see for example [46–48, 56]). These calculations required significant computational power (in their day), and produced results that were qualitatively correct but quantitatively wrong. Perhaps the most useful result of these studies was the observation of an orientational preference. For dipoles close to one metal surface, maximum intensity was obtained when the dipole was normal to the metal surface. For dipoles near the center of the barrier, the reverse was true. Today, the ability to compute vibrational IETS has advanced significantly, especially in the case of single-molecule tunneling [57–62]. An example of this progress is provided by the work of Paulsson [60], Fig. 6, where the computed IETS of a single alkanedithiol stretched between two gold contacts of varying separation is displayed. Tao and coworkers provide one of many examples of IETS obtained experimentally from a single molecular junction [63].

One can develop an understanding of IETS without performing difficult calculations. Let us consider the IETS spectrum on an energetic basis, while viewing Fig. 3. If the applied voltage is less than $h\nu/e$, the inelastic channel is closed because the energetically accessible final states are already filled (Pauli principle). At $V = h\nu/e$ ($h\nu$ = vibrational energy) the inelastic channel opens, because now it is just energetically possible for an electron with the Fermi energy of the left electrode

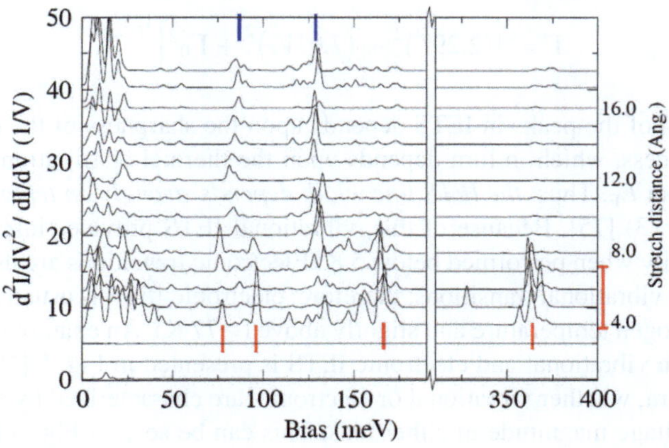

Fig. 6 Calculated inelastic electron tunneling spectra with the different stretching distances offset by 2.5 V^{-1}. The geometries with gauche defects are marked in the right margin for stretching 4.0–7.0 Å. Highlighted frequency bands are (1) *top*, ~82 (degenerate C–S) and 130 (C–C) meV, and (2) *bottom*, 75 (nondegenerate C–S), 95 (rock), 165 (wag), and 365 (C–H) meV. (Reprinted with permission from [60])

to scatter inelastically off the ground state molecule, leaving behind a vibrationally excited molecule. The final state of that scattered electron has just enough energy to find a vacant state in the right-hand metal. Further increases in V result in additional occupied metal states, which can scatter inelastically and find empty states of appropriate energy. This leads to an associated increase in current due to this channel. As is depicted in the lower half of Fig. 3, there is a break in the $I(V)$ curve at $V = h\nu/e$. If one measures the conductance, dI/dV, the opening of the inelastic channel is signaled by a step. Plotting d^2I/dV^2 vs V produces a peak at $V = h\nu/e$. Both vibrational and electronic transitions may be observed as peaks in the d^2I/dV^2 vs V plots. One can not only probe states of the barrier region but also of the metal electrodes. Thus, phonons and magnons of metal and semiconductor films have been studied by IETS [64–68].

In its simplest form, an IET spectrum is a plot of d^2I/dV^2 vs V [18–20, 25, 69–74]. It turns out that using $d^2I/dV^2/(dI/dV)$ as the y axis provides spectra having flatter baselines, and is most appropriate for high-bias work [71–74]. These are called normalized tunneling intensities (NTI) or constant-modulation spectra. Simple tunneling spectra are measured by applying both a variable bias, V, and a small modulation component, V_f, at frequency, f. A lock-in amplifier is used to detect the $2f$ signal which is proportional to d^2I/dV^2. The instrumentation required for obtaining normalized intensities, NTI, is a bit more complex [71–73]. In general, the bias voltage may be converted to the more conventional wavenumbers through the factor of 8,066 cm^{-1}/V. Vf is the rms amplitude in modulation always influences the observed signal strength and resolution. The signal increases as V_f^2, but the experimental line width is proportional to V_f [75, 76]. If T is measured in Kelvin, V_f in millivolts, and Γ_0 is the natural line width in cm^{-1}, then the half-width at $1/e$ height for an IETS band observed with normal metal contacts is given in cm^{-1} by

$$\Gamma = \left[(2.29T)^2 + (7.07V_f)^2 + {\Gamma_0}^2\right]^{1/2}. \tag{3}$$

The width of the peaks in IETS depends upon the sharpness of the onset of the inelastic process, which in turn depends upon the thermal distribution of electron energies about E_F. Thus, *the IETS line width depends strongly on temperature and as shown by* (3) [75]. Because of this, vibrational IETS provides infrared-quality resolution only when performed below 5 K. Electronic transitions are usually much broader than vibrational transitions; therefore, electronic IETS is usually performed at liquid nitrogen temperature and slightly above ($\geq$77 K). An example of a system showing both vibrational and electronic IETS is presented in Fig. 5 [19].

IET spectra, whether vibrational or electronic, are characterized by appearing at the same voltage magnitude in either bias. This can be seen in Fig. 5 for both the labeled C–H stretch and the spin-allowed but Laporte-forbidden electronic transition. The relative intensities as a function of bias, however, can differ. The data presented in Fig. 5 was taken from an M-I-A-M′ junction, and the electrons traveling first through the insulator before scattering have a greater transmission probability than those inelastically scattering prior to traversing the insulator. This is, of course, due to the exponential dependence of the current on the barrier height (1 or 2). For an M-A-M junction, a more symmetrical distribution of intensity with bias is expected, but the peak positions should be unchanged.

The invariance of IETS in an M-A-M junction vs an M-I-A-M device is exceptionally well demonstrated by the work of Reed [30]. Figure 7 shows the Au–alkanedithiol–Au structure he used to create a single barrier tunnel diode. The IET spectra obtained from this device were stable and repeatable upon successive bias sweeps. The spectrum at 4.2 K is characterized by three pronounced peaks in the 0–200 mV region at 33, 133, and 158 mV. From comparison with previously reported IR, Raman, and high-resolution electron energy-loss (HREEL) spectra of

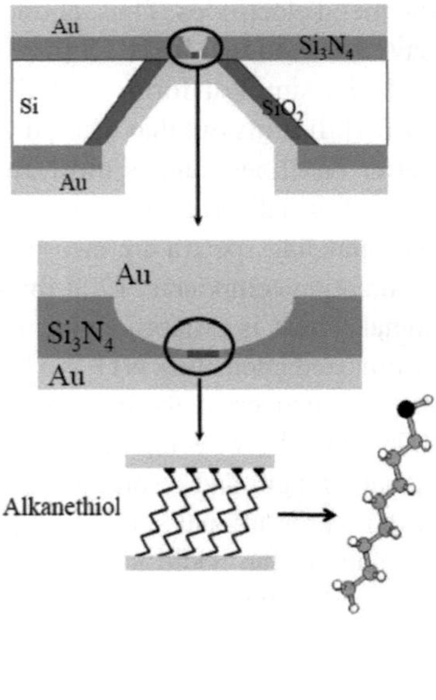

Fig. 7 Schematics of a nanometer scale M-A-M diode (not drawn to scale in relative thickness). *Top schematic* is the cross section of a silicon wafer with a nanometer scale pore etched through a suspended silicon nitride membrane. *Middle and bottom schematics* show a Au/SAM/Au junction formed in the pore area. (Reprinted with permission from [30])

SAM-covered gold surfaces, these three peaks were assigned to Au–S, C–C, and C–H modes of surface-bound alkanethiolates [30]. The absence of a strong S–H signal at 329 mV suggests that most of the thiol groups have reacted with the gold bottom and top contacts. Peaks are also reproducibly observed at 80, 107, and 186 mV. We note that all alkanethiolate peaks, without exception or omission, occur in the spectra.

Another example comes from the work of Kushmerick [77]. He used the crossed gold wire configuration, shown in Fig. 7, to obtain IETS spectra of C11 alkanethiol, OPE, and OPV2. The peak positions observed agreed well with those expected for the individual compounds. He also found that the strong transitions were those with a large transverse dipole or polarizability component perpendicular to the Au contacts. This is consistent both with early theoretical treatments and also with the more recent work of Troisi and Ratner [58]. It is also interesting to note that contact with metallic nanoparticles, even if they are not the electrode, can enhance the IETS intensities and change line-shapes for some modes [78, 79]. One also observes dramatic changes in IETS intensity in the STM environment, with many bands being unobservable [80] (Fig. 8).

As a final note on the general character of electronic IET spectra, we point out that vibronic structuring of electronic IETS has been known in M-I-A-M structures for many years [55, 74]. In recent times, it has also been demonstrated in the STM environment [81].

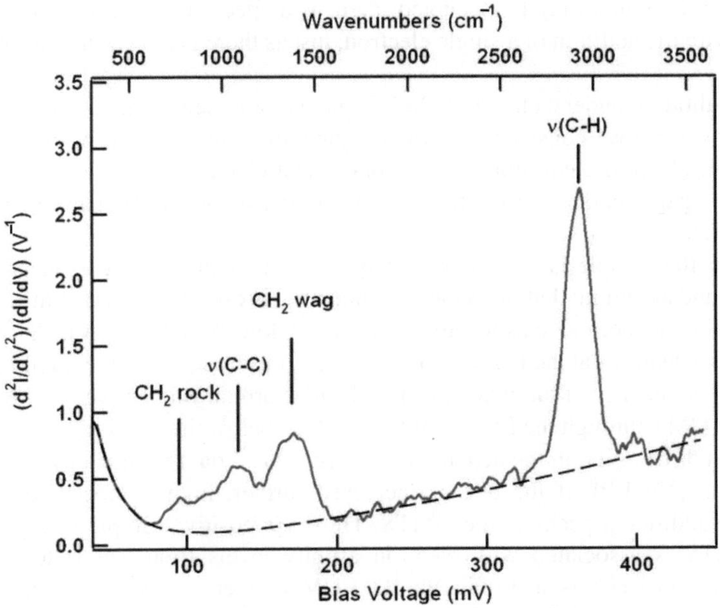

Fig. 8 IETS obtained from an Au-C11thiol-Au crossed wire junction. (Reprinted with permission from [77], pp 20–25)

4 Elastic Electron Tunneling Spectroscopy

Until about 1988, essentially all of tunneling spectroscopy in tunnel diodes was IETS. In 1989 Mazur and Hipps began observing strange vibrational line shapes and huge new signals that were equal or greater in intensity than electronic IETS, but that could not be explained by a molecular excitation process. These new transitions produce peaks in d*I*/d*V* (rather than d^2I/dV^2), and are due to direct tunneling via unoccupied or occupied molecular orbitals. The exact mechanism of interaction can vary from case to case. It might be true resonance tunneling where the effective residence time of the tunneling electron on the molecule is negligible compared to nuclear motion. It might be a real oxidation or reduction of the molecule followed by thermally induced return to the original charge state (electron hopping), or it might be a redox that occurs too rapidly for thermal relaxation, such as occurs in ultra-violet photoelectron spectroscopy (UPS) or inverse photoemission spectroscopy (IPS). Because there are a number of different physical processes that can give rise to these bands, we chose to call the spectroscopy associated with measuring these transitions orbital mediated tunneling spectroscopy (OMTS) [82–84]. Because the time-scales are usually unknown, we sometimes refer to transient redox processes. The technique might equally well be called ionization and affinity-level spectroscopy. Ionization spectroscopy is the measurement of the energy required to remove electrons from a filled (or partially filled) orbital. Affinity-level spectroscopy measures the energy released when an electron is captured by an atom or molecule. Since there are generally several vacant orbitals that may be occupied, there is a spectrum of affinity levels associated with the addition of a single electron, just as there is a spectrum of ionization levels.

A qualitative understanding of OMTS may be obtained with reference to Fig. 9. In this example we consider the simplest junction configuration, wherein a molecule is in electronic equilibrium with one metal electrode, and a relatively large insulating gap separates it from the other electrode. We will consider more complex cases later.

When the sample is biased positively ($V_b > 0$) with respect to the tip, as in Fig. 9c, and assuming that the molecular potential is essentially that of the substrate [85], only the normal elastic current flows at low bias ($\leq$1.5 V). As the bias increases, electrons at the Fermi surface of the tip approach, and eventually surpass, the absolute energy of an unoccupied molecular orbital (the LUMO at +1.78 V in Fig. 9c). OMT through the LUMO at $\Phi - 1.78$ V below the vacuum level produces a peak in d*I*/d*V*, seen in the actual STM based OMTS data for nickel(II) octaethylporphyrin (NiOEP). If the bias is increased further, higher unoccupied orbitals produce additional peaks in the OMTS. Thus, the positive sample bias portion of the OMTS is associated with electron affinity levels (transient reductions). In reverse (opposite) bias, as in Fig. 9b, the LUMO never comes into resonance with the Fermi energy, and no peak due to unoccupied orbitals is seen. However, occupied orbitals are probed in reverse bias. In the NiOEP case, the HOMO at

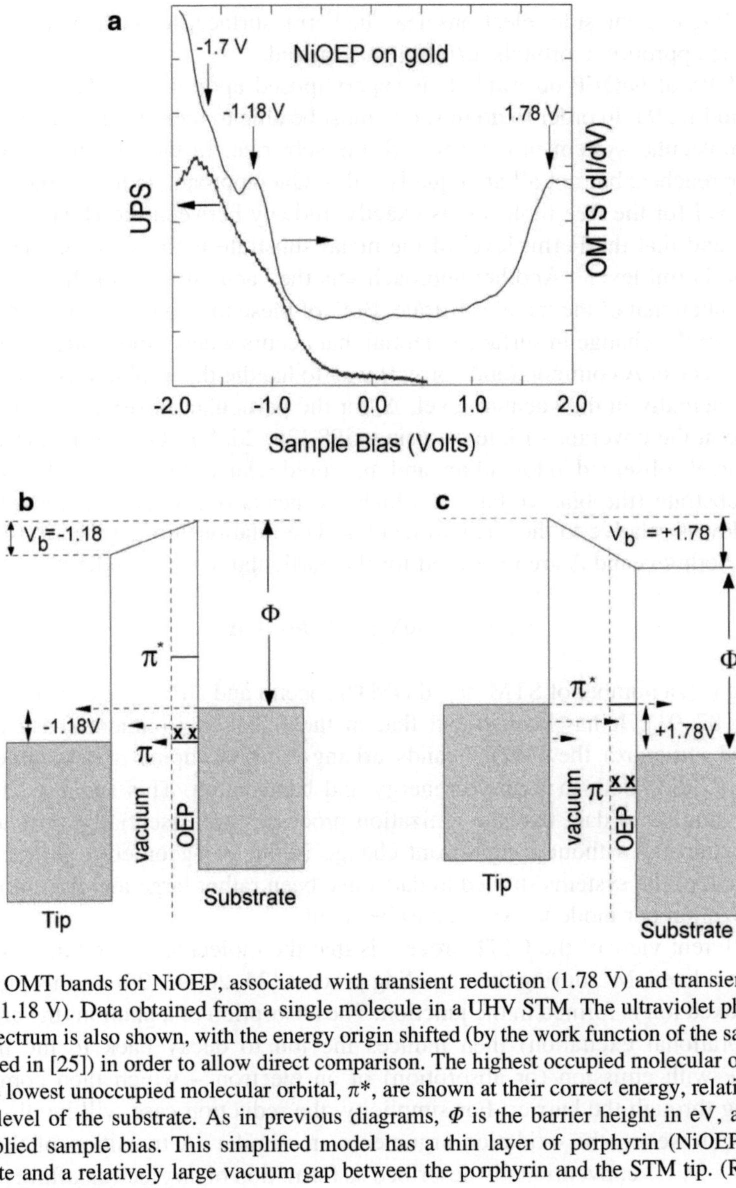

Fig. 9 OMT bands for NiOEP, associated with transient reduction (1.78 V) and transient oxidation (−1.18 V). Data obtained from a single molecule in a UHV STM. The ultraviolet photoelectron spectrum is also shown, with the energy origin shifted (by the work function of the sample, as discussed in [25]) in order to allow direct comparison. The highest occupied molecular orbital, π, and the lowest unoccupied molecular orbital, π^*, are shown at their correct energy, relative to the Fermi level of the substrate. As in previous diagrams, Φ is the barrier height in eV, and V_b is the applied sample bias. This simplified model has a thin layer of porphyrin (NiOEP) on the substrate and a relatively large vacuum gap between the porphyrin and the STM tip. (Reprinted with permission from [26])

Φ + 1.18 V below the vacuum level produces a peak in dI/dV at −1.18 V sample bias. It is also clear from Fig. 9 that there are other occupied MOs, with one near Φ + 1.70 V giving a well-defined shoulder. Note that peaks are observed in dI/dV (and not I). This is because, once current starts to flow through orbital-mediated channels, increasing the bias doesn't turn it off. On the other hand, the probability of tunneling is greatest for electrons near the Fermi surface; so, as the Fermi surface passes the appropriate orbitals, dI/dV is maximized.

The UPS of NiOEP on Au(111) is superimposed upon the OMTS of the same system in Fig. 9a. In order to do this, one must be able to identify the vacuum level of the molecular system in contact with the substrate. In the literature one finds three approaches, but not all are equally valid. One approach is to assume that the Fermi level for the free molecule is exactly midway between the HOMO and the LUMO, and that the Fermi level of the metal substrate is the same as that "free-molecule Fermi level." Another approach sets the vacuum level of the free molecule to match that of the metal substrate. Both of these methods are flawed because they ignore the change in surface potential that occurs when a molecule adsorbs on a metal surface. A common (and correct) way to handle this problem is to measure the shift actually in the vacuum level, Δ, for the particular adsorbate on the given substrate at the coverage of interest using UPS [25, 26, 86]. Once Δ is known, the energy levels observed in tunneling, and measured relative to the Fermi level of the metal substrate (the bias voltage at which the peaks occur) can be converted to energy levels relative to the vacuum level, ε. The relationship is given in (4) [25]. Ideally, both Φ_M and Δ are measured for the particular system studied:

$$\varepsilon_{\text{OMTS}} = -eV_{\text{peak}} + \Phi_M + \Delta. \tag{4}$$

Using (4), a number of STM-based OMTS spectra and UPS have been compared [25, 26, 87–91]. It has been found that in the STM environment (basically an M-I-A-M structure) the OMTS bands arising from occupied states agree well with the UPS – both in terms of energy and band-width. This suggests, for the systems studied to date that the ionization processes are essentially vertical processes occurring without a significant change in nuclear geometry. On the other hand, most of the systems studied to date have been rather large and the amount of reorganization per mode is expected to be small.

A different view of the OMT process is that the molecule, M, is fully reduced, M^-, or oxidized, M^+, during the tunneling process [25, 26, 92–95]. In this picture a fully relaxed ion is formed in the junction. The absorption of a phonon (the creation of a vibrational excitation) then induces the ion to decay back to the neutral molecule with emission (or absorption) of an electron – which then completes tunneling through the barrier. For simplicity, the reduction case will be discussed in detail; however, the oxidation arguments are similar. A transition of the type $M + e^- \rightarrow M^-$ is conventionally described as formation of an electron affinity level. The most commonly used measure of condensed-phase electron affinity is the half-wave reduction potential measured in non-aqueous solvents, $E_{1/2}$. Often these values are tabulated relative to the saturated calomel electrode (SCE). In order to correlate OMTS data with electrochemical potentials, we need them referenced to an electron in the vacuum state. That is, we need the potential for the half reaction:

$$\text{M(solution)} + e^-(\text{vac}) \rightarrow \text{M}^-(\text{solution}).$$

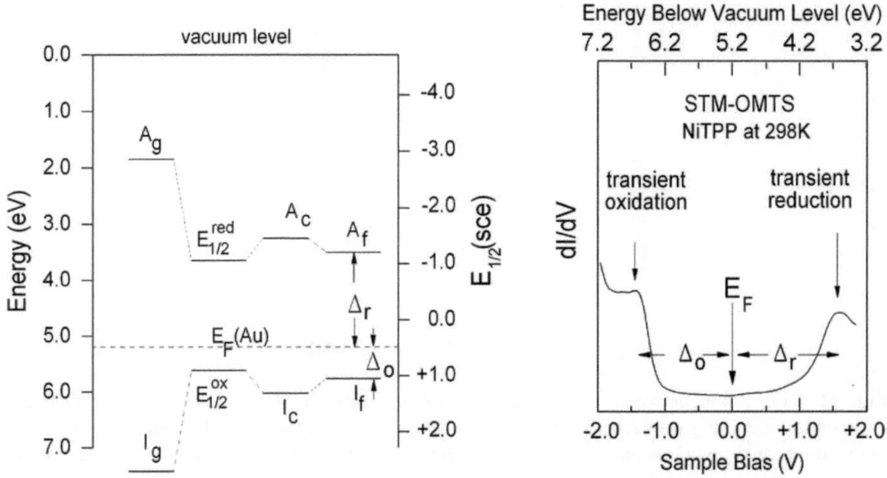

Fig. 10 Electrochemical energy level model for orbital mediated tunneling. A_g and A_c are the gas- and crystalline-phase electron affinities, $E_{1/2}$(SCE) is the electrochemical potential referenced to the saturated calomel electrode, and provides the solution-phase electron affinity. E_F is the Fermi level of the substrate (Au here). The corresponding positions in the OMT spectrum are shown by Δ_r and Δ_o and correspond to the electron affinity and ionization potential of the adsorbate film modified by interaction with the supporting metal, A_f. The spectrum is that of nickel(II) tetraphenyl-porphyrin on Au (111). (Reprinted with permission from [26])

These values can be closely approximated, from those referenced to SCE, by adding 4.71 V to $E_{1/2}$(SCE) [25, 26, 93–97]. That is, $E_{1/2}(\text{vacuum}) = E_{1/2}(\text{SCE}) + 4.71$ V. This connection between solution-phase electrochemical potentials and vacuum-level-based spectroscopies, such as OMTS and UPS, is extremely useful, but the derivation is rather complex, and the results are only approximate. For example, the difference in electron affinity in the gas phase and in solution is primarily due to solvation stabilizing the reduced form. The reader wishing to understand better its origins is encouraged to consult [96, 97]. The various energy conventions are depicted in Fig. 10, where the connection between OMTS bands and electron affinities is made within the context of the model. This diagram is based on one presented by Loutfy et al. [97] and expanded to OMTS by Mazur and Hipps [25, 82–84, 88, 91, 98]. By using the measured values of E_F and Δ (from UPS), the OMTS bands can be located, both relative to the vacuum level and also to electrochemical potentials.

Redox potentials in the solid state are expected to differ from those in solution [97]. Moreover, there will be shifts in the potentials of a thin film, relative to that of a solid, due to interactions with the metal support and counter electrode, including image-charge effects. There may be an opposite signed shift due to the absence of a covering layer of adsorbed molecules [99]. Another complication is the fact that electrochemical potentials are equilibrium values, and therefore reflect the energy associated with the formation of an ion in its equilibrium geometry. OMTS transitions, as discussed above, may occur so rapidly that the ion is formed in an excited

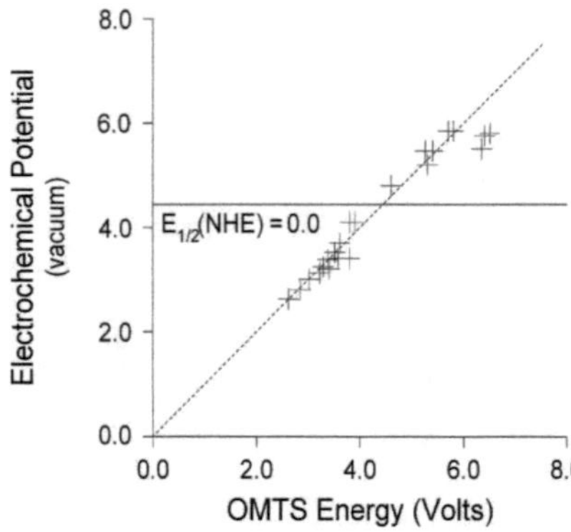

Fig. 11 Correlation between electrochemical potentials and OMTS bands for more than ten compounds including polyacenes, phthalo-cyanines, and porphyrins. OMTS data were acquired both from tunnel junctions and STM measurements. The standard potential relative to the normal hydrogen electrode associated with the half reaction M(solution) + e^{-}(vac) → M^{-}(solution) is the *y* axis. The three outliers are assigned to the ring oxidation of porphyrins. (Reprinted with permission from [26])

state – a vertical transition in the Franck–Condon sense. For a wide range of materials and film thicknesses (sub-monolayer to about 1 nm) studied to date, a fortuitous cancellation of polarization terms and differences between vertical and equilibrium affinities has resulted in many OMTS bands laying close to the positions predicted from electrochemistry (see Fig. 11). This correlation is especially good for *unoccupied* orbitals. Unfortunately, OMTS bands associated with *occupied* orbitals generally lay *deeper* than predicted by solution-phase electrochemical oxidation potentials, and closer to the gas-phase ionization potentials.

The near transferability of electrochemical values to thin film band positions for affinity levels, but not ionization levels, indicates that the polarization energy terms differ for these processes. This is a failure in the simple model used to generate Fig. 10, where it was assumed that only the sign of the polarization energy changed. This failure is particularly large for porphyrins. Given the trends in stabilization of ion energies by the surrounding molecules and image charges induced in the metal substrate, we would expect the ionization potential of thin-film NiOEP to be about 0.5–1.0 eV less than for the gas phase. Instead, the ionization energies measured from a thin film are nearly identical to those reported from the gas phase [91, 100]. These discrepancies also occur with UPS observations (which agree with the OMTS), suggesting that the problem is in the model, not the technique.

While the first electrochemical reduction potential provides an estimate for A_c (assuming it is a reversible process), the second and higher reduction potentials do not provide the spectrum of single electron affinity levels. Rather, they provide information about two-electron, three-electron, and higher electron reduction processes, and, therefore, depend on electron pairing energy. Thus, the utility of solution-phase reduction potentials for estimating solid-state affinity levels is limited to the lowest affinity level. The same argument applies to oxidation potentials beyond the first. OMTS, on the other hand, probes the single-electron reduction energies for the spectrum of states of the negative ion, and the single-electron ionization energies for the spectrum of states of the positive ion. Thus, OMTS can be used to determine ionization spectra and affinity levels beyond the first transitions of each type [20, 25, 88, 98].

4.1 Elastic Tunneling in M-I-A-I′-M Junctions

While almost all elastic tunneling spectroscopy has been performed either in the M-I-A-M configuration (either in tunnel diodes or in the STM), these are not the only possible geometries. One geometry sometimes encountered in STM based measurements is the M-I-A-I′-M configuration. The analysis of the resonances in elastic tunneling in this case becomes more complex because of two factors: (1) the potential at the active molecule is NOT the potential of either electrode, and is often impossible to independently measure, and (2) the *rates* of tunneling into and out of the active molecule can become comparable, and the possibility of true charging of the molecule occurs. This situation is most often seen in the case of quantum dot and C_{60} tunneling spectroscopy [22, 101–104], but has been observed in the case where the STM substrate was oxidized prior to adsorption of the active molecule [81]. A discussion of the difficulties associated with interpreting spectra obtained in these cases is beyond the scope of this chapter, and the reader is referred to the review authored by Selzer and Allara [101]. In general, however, one expects to see occupied and unoccupied states appearing in both polarities, and one may also observe ionization and affinity levels of ionization products formed by the tunneling process.

As the size of adsorbed molecules increases, especially in cases like proteins, which have an electroactive portion buried inside a relatively electronically inert sheath, the molecule might become indistinguishable from a small quantum dot. In this case one would expect that the potential at which oxidation or reduction occurred would begin to depend on tip-molecule separation, and not be simply related to spectroscopic or electrochemical values. Lindsey and coworkers attempted to measure experimentally the ratio of the local potential (at a molecule experiencing OMTS) to the applied potential (between tip and substrate) in the case of porphyrins, but were unable to demonstrate a significant dependence [95]. Deng and Hipps set out to test the tip-molecule distance dependence of the OMTS in the case of nickel(II) tetraphenylporphyrin (NiTPP) adsorbed on Au(111) under UHV conditions [85]. Adjusting the set-point prior to spectral measurement allowed for the control of the tip-sample distance. A sequence of dI/dV(V), I(V), and I(z) curves were acquired over a wide range of setpoint currents and bias voltages. The I(z) data was measured to provide a means of converting set-point values to relative tip displacements. Determining the peak positions and peak shapes from the dI/dV curves was difficult, because of the strong variation in both the resonant and elastic (background) intensities with tip-sample separation. Stroscio and Feenstra [105–107] considered this problem several years ago, and determined that this difficulty could often be eliminated by using the logarithmic derivative, dln*I*/dln*V* (*V*), as the spectral intensity function. Ukrainstev discussed some problems with this method [108], but they were not relevant in the NiTPP case. Deng and Hipps found that changes in tip-sample distance over several Angstroms and a factor of 20 in set-point current produce no measurable changes in orbital energy splitting [85]. It was suggested that it may be possible to measure reliably STM-OMTS on molecules larger than tetraphenylporphyrin without concern about the tip-molecule separation distance, provided that the effective gap impedance (between electro-active moiety in the molecule and the substrate) does not drop below about 500 MΩ. Recent OMTS studies from the Heietschold group show that the spectrum of F_{16}CoPc on Ag(111) is also independent of tip-molecule separation [109]. On the other hand, Gopakumar et al. [110] found in the STM-OMTS of d^8 (Ni, Pt, Pd) phthalocyanines adsorbed on HOPG (0001) a systematic shift of the HOMO toward the Fermi level when the tip-sample distance decreases, while the unfilled levels near E_F remain pinned. Hietschold [109] suggests that the difference between the effects might be explained based on a more capacitive organic/HOPG interface, compared with the more ohmic organic/metal interface. A strong organic-metal interaction involves a tunneling current immediately transferred into the substrate, as a result of the molecular orbital overlapping with the metal surface states, while a weakly interacting substrate such as HOPG, that is dominated by π electron density, leads to the formation of charging effects at the interface. In any case, STM-based OMTS studies should always be accompanied by tip-surface distance dependent studies.

Two final concerns must be addressed: surface oxidation state and temperature dependence. Whenever one deposits a redox-active species on a metal surface, the oxidation state of the adsorbate (and therefore the OMTS bands) may change. One example is the adsorption of a biaxially substituted dicyano cobalt phthalocyanine salt, $MCoPc(CN)_2$ (where M = K or Cs), on gold to form the reduced species CoPc [111]. A second example is provided by the adsorption of TCNE on gold, silver, and copper. In that order, the charge state of TCNE on the surface ranges from 0 to −3, and the OMTS reflects these changes.

The second issue of interest is the temperature dependence of elastic tunneling spectroscopy. Because the bands are intrinsically wide, spectra measure at 5 K are similar in line shape to those measured near 300 K. In any case, the integrated normalized intensities,

$$\int_{\text{band}} \left[\left(\frac{\mathrm{d}I}{\mathrm{d}V} \right) \Big/ \left(\frac{I}{V} \right) \right] \mathrm{d}V,$$

are expected to be temperature-invariant as long as the ionization process is electronically (as opposed to vibronically) allowed. This was recently tested for the first time, and found to hold for the case of CoPc on Au(111) between 100 K and 300 K [112].

4.2 *Elastic Tunneling in Single-Molecule Junctions*

The M-A-M structure is equivalent to the M-I-M, or even the M-I-A-M structures, when analyzing IETS, but NOT for OMTS. OMTS bands depend upon the absolute energy of the molecular states with respect to the Fermi energies of the metals, while IETS bands only depend upon molecular energy differences. Thus, knowledge of the actual potential in which a molecule resides is essential for OMTS, but nearly irrelevant for IETS [113]. Unfortunately, there is very little known experimentally, other than that the local fields in junctions formed by breaking or by electromigration are very much larger than expected for planar electrodes [113]. Theoretical calculations have given us much deeper insight into the properties of these junctions [61, 114–122].

If the M-A-M device is truly symmetrical, then one would expect to see both ionization and reduction processes in both bias directions. If one assumes a simple electrostatic model for a molecule between two metal plates, then one would predict that the ionization and affinity states would appear at voltages such that $V = 2E/e$. However, for single-molecule junctions the electric field is NOT uniform across the molecule. One finds that shortening the distance between two metallic electrodes results in stronger coupling and lowers the potential between them [120]. Moreover, the applied electric field causes a reorganization of charge within the molecule. This internal reorganization produces dipole moments at the molecule–metal interface that can lead to a large part of the potential drop occurring at the interfaces [120, 121], but for other molecules the drop can occur in the molecular core [121]. If the molecular species is not inversion-symmetric, and the attachment to the electrodes identical on each end, different voltage drops will occur at each electrode. Another complicating issue is the choice of Fermi level, which is sometimes arbitrarily chosen as midway between HOMO and LUMO of the molecules [120] and sometimes determined by charge neutrality of the contact-molecule complex [122]. Of equal importance is the fact that these single-molecule junctions are generally formed through covalent bonding between the ends of the molecule and the electrode. Thus, one may no longer use the states of the isolated molecule, but must consider those of the new species, that includes a significant admixture of the density of states of each metal. As pointed out by Heurich and coworkers [122], many molecular orbitals participate in a single conduction channel, and the current may not be dominated by contributions from the closest MOs.

The take-home message here is that conductivity measurements in single-molecule junctions are difficult to analyze without the support of quantum mechanical calculations that include the metal electrodes. This is very much the domain of specialists, and the simple rules discussed for analyzing elastic tunneling spectra in other junction types generally do not apply for metal–single-molecule–metal junctions.

References

1. Hipps KW, Keder JK (1983) A direct comparison of surface Raman spectroscopy and inelastic electron tunneling spectroscopy. J Phys Chem 87:3187–3189
2. Hipps KW, Dowdy J, Hoagland JJ (1991) Surface Raman spectra of a biased and buried ultra-thin CuPc layer. Langmuir 7:5–7
3. Tsang JC, Yokota Y, Matz R, Rubloff G (1984) Raman spectroscopy of PtSi formation at the Pt/Si(100) interface. Appl Phys Lett 44:430–432
4. Heske C, Eich D, Fink R, Umbach E, van Buuren T, Bostedt C, Terminello LJ, Kakar S, Grush MM, Callcott TA, Himpsel FJ, Ederer DL, Perera RCC, Riedl W, Karg F (1999) Observation of intermixing at the buried Cds/Cu(In, Ga)Se2 thin film solar cell heterojunction. Appl Phys Lett 74:1451–1453
5. Jarrige I, Jonnard P, Frantz-Rodriguez N, Danaie K, Bosseboeuf A (2002) Study of the NiTi/SiO_2 interface: analysis of the electronic distributions. Surf Interface Anal 34:694–697
6. Huang LJ, Rajesh K, Lau WM, Ingrey S, Landheer D, Noel JP, Lu Z (1995) Interfacial properties of metal–insulator–semiconductor capacitors on GaAs(110). J Vac Sci Technol A 13:792–796
7. Yang G-R, Zhao Y-P, Wang B, Barnat E, McDonald J, Lu T-M (1998) Chemical interactions at Ta/fluorinated polymer buried interfaces. Appl Phys Lett 72:1846–1847
8. Brillson LJ (2001) Nanoscale luminescence spectroscopy of defects at buried interfaces and ultrathin films. J Vac Sci Technol B 19:1762–1769
9. Grozea D, Turak A, Feng XD, Lu ZH, Johnson D, Wood R (2002) Chemical structure of Al/LiF/Alq interfaces in organic light-emitting diodes. Appl Phys Lett 81:3173–3175
10. Jun Y, Zhu X-Y (2004) FTIR spectroscopy of buried interfaces in molecular junctions. J Am Chem Soc 126:13224–13225
11. Chen C, Wang J, Even MA, Chen Z (2002) Sum frequency generation vibrational spectroscopy studies on “Buried” polymer/polymer interfaces. Macromolecules 35:8093–8097
12. Li Q, Hua R, Chou KC (2008) Electronic and conformational properties of the conjugated polymer MEH-PPV at a buried film/solid interface investigated by two-dimensional IR-visible sum frequency generation. J Phys Chem B 112:2315–2318
13. Blackstock JJ, Donley CL, Stickle WF, Ohlberg DA, Yang JJ, Stewart DR, Williams RS (2008) Oxide and carbide formation at titanium/organic monolayer interfaces. J Am Chem Soc 130:4041–4047
14. Hurst HG, Ruppel W (1964) Tunnel currents through Al_2O_3 films with contacts of various work functions. Z Naturforsch 19a:573–579
15. Lambe J, Jaklevic RC (1968) Molecular vibration spectra by inelastic electron tunneling. Phys Rev 165:821–832
16. Thomas DE, Klein JM (1963) Tunneling current structure resolution by differentiation. Rev Sci Instrum 34:920–924
17. Jaklevic RC, Lambe J (1996) Molecular vibration spectra by electron tunneling. Phys Rev Lett 17:1139–1140
18. Hansma PK (ed) (1982) Tunneling spectroscopy. Plenum, New York
19. Hipps KW, Mazur U (1993) Inelastic electron tunneling: an alternative molecular spectroscopy. J Phys Chem 97:7803–7814
20. Mazur U, Hipps KW (2001) Inelastic electron tunneling spectroscopy. In: Chalmers J, Griffiths P (eds) Handbook of vibrational spectroscopy, vol 1. Wiley, New York, pp 812–829
21. Hamers RJ (1996) Scanned probe microscopies in chemistry. J Phys Chem 100: 13103–13120
22. Moore AA, Weiss P (2008) Functional and spectroscopic measurements with scanning tunneling microscopy. Annu Rev Anal Chem 1:857–882
23. Wahl P, Diekhöner L, Schneider MA, Kern K (2008) Background removal in scanning tunneling spectroscopy of single atoms and molecules on metal surfaces. Rev Sci Instrum 79:043104-4

24. Giancarlo LC, Fang H, Avila L, Fine LW, Flynn GW (2000) Molecular photography in the undergraduate laboratory: identification of functional groups using scanning tunneling microscopy. J Chem Educ 77:66–71
25. Hipps KW (2006) Scanning tunneling spectroscopy. In: Vij DR (ed) Handbook of applied solid state spectroscopy. Springer, New York, ISBN: 0-387-32497-6
26. Hipps KW, Scudiero L (2005) Electron tunneling, a quantum probe for the quantum world of nanotechnology. J Chem Ed 82:704–711
27. Hamers RJ (1989) Atomic-resolution surface spectroscopy with the scanning tunneling microscope. Ann Rev Phys Chem 40:531–559
28. Lehmpuhl DW (2003) Incorporating scanning probe microscopy into the undergraduate chemistry curriculum. J Chem Educ 80:478–479
29. Vilan A (2007) Analyzing molecular current-voltage characteristics with the Simmons tunneling model: scaling and linearization. J Phys Chem C 111:4431–4444
30. Wang W, Lee T, Reed MA (2005) Electron tunneling in self-assembled monolayers. Rep Prog Phys 68:523–544
31. Simmons JG (1963) Generalized formula for electric tunneling effect between similar electrodes separated by a thin film. J Appl Phys 34:1793–1803
32. Güntherodt H-J, Wiesendanger R (eds) (1992) Scanning tunneling microscopy I. Springer, Berlin
33. Güntherodt H-J, Wiesendanger R (eds) (1992) Scanning tunneling microscopy II. Springer, Berlin
34. Magonov SN, Whangbo M-H (1996) Surface analysis with STM and AFM. VCH, New York
35. Stroscio JA, Kaiser WJ (eds) (1993) Scanning tunneling microscopy. Academic, New York
36. Meyer E, Jarvis SP, Spencer ND (2004) Scanning probe microscopy in materials science. MRS Bull 29(7):443–448
37. Bhushan B, Fuchs H, Hosaka S (eds) (2004) Applied scanning probe methods. Springer, Berlin
38. Bonnel D (ed) (2000) Scanning probe microscopy and spectroscopy. Wiley, New York
39. Bard AJ, Mirkin MV (eds) (2001) Scanning electrochemical microscopy. Dekker, New York
40. Hipps KW, Mazur U (1980) An inelastic electron tunneling spectroscopy study of some iron cyanide complexes. J Phys Chem 84:3162–3172
41. Korman CS, Lau JC, Johnson AM, Coleman RV (1979) Studies of aromatic-ring compounds adsorbed on alumina and magnesia using inelastic electron tunneling. Phys Rev B 19: 994–1002
42. Holmlin RE, Haag R, Chabinyc ML, Ismagilov RF, Cohen AE, Terfort A, Rampi M, Whitesides GM (2001) Electron transport through thin organic films in metal-insulator-metal junctions based on self-assembled monolayers. J Am Chem Soc 123:5075–5085
43. Lee T, Wang W, Klemic JF, Zhang J, Su J, Reed MA (2004) Comparison of electronic transport characterization methods for alkanethiol self-assembled monolayers. J Phys Chem B 108:8742–8787
44. Wang W, Lee T, Reed MA (2003) Mechanism of electron conduction in self-assembled alkanethiol monolayer devices. Phys Rev B 68:035416–035423
45. DelRio FW, Steffens KL, Jaye C, Fischer DA, Cook RF (2010) Elastic, adhesive, and charge transport properties of a metal-molecule-metal junction: the role of molecular orientation, order, and coverage. Langmuir 26:1688–1699
46. Kirtley J (1980) Inelastic electron tunneling spectroscopy. ACS Symp Ser 137:217–245
47. Kirtley J, Soven P (1979) Multiple-scattering theory of intensities in inelastic-electron-tunneling spectroscopy. Phys Rev B 19:1812–1817
48. Knockenmuss R, Hipps KW (1982) Some proposed modifications in the theory of inelastic electron tunneling spectroscopy and the source of parameters utilized. J Phys Chem 86: 4477–4480
49. Hipps KW, Aplin AT (1985) The tricyanomethanide ion: an infrared, Raman, and tunneling spectroscopy study including isotopic substitution. J Phys Chem 89:5459–5464

50. Igalson J, Adler JG (1983) Effect of hydrogen-isotope implantation on aluminum oxide tunnel-junction barriers. Phys Rev B 28:4970–4975
51. Gauthier S, De Cheveigne S, Klein J, Belin M (1984) Evidence for the formation of aluminum-hydrogen bonds in tunneling junctions. Phys Rev B 29:1748–1754
52. Yang WJ, White HW (1982) Comparison of experimental and theoretical intensities of inelastic electron tunneling spectra for thiourea. Surf Sci 118:303–320
53. Weinberg WH (1982) Inelastic electron tunneling spectroscopy of supported homogeneous cluster compounds. Vib Spectra Struct 11:1–50
54. Ramsier RD, Henriksen PN, Gent AN (1988) Adsorption of phosphorus acids on alumina. Surf Sci 203:72–88
55. Hipps KW, Mazur U (1987) $^4A_2 \rightarrow {}^4T_2$ and $^4A_2 \rightarrow {}^4T_1$ electronic transitions in cobalt(II) tetrachloride: an FT-IR and inelastic electron tunneling spectroscopy study. J Am Chem Soc 109:3861–3865
56. Godwin LM, White HW, Ellialtioglu R (1981) Comparison of experimental and theoretical inelastic electron tunneling spectra for formic acid. Phys Rev B 23:5688–5694
57. Arroyo CR, Frederiksen T, Rubio-Bollinger G, Vélez M, Arnau A, Sánchez-Portal D, Agraït N (2010) Characterization of single-molecule pentanedithiol junctions by inelastic electron tunneling spectroscopy and first-principles calculations. Phys Rev B 81:075405/1-075405/5
58. Troisi A, Ratner MA (2005) Modeling the inelastic electron tunneling spectra of molecular wire junctions. Phys Rev B 72:033408/1-033408/4
59. Mathias K, Jun J, Yi L (2008) A molecular view on electron transport in molecular electronic devices. J Comput Theor Nanosci 5:401–421
60. Paulsson M, Krag C, Frederiksen T, Brandbyge M (2009) Conductance of alkanedithiol single-molecule junctions: a molecular dynamics study. Nano Lett 9:117–121
61. Paulsson M, Frederiksen T, Brandbyge M (2010) Molecular electronics: insight from first-principles transport simulations. Chimia 64:350–355
62. Okabayashi N, Paulsson M, Ueba H, Konda Y, Komeda T (2010) Site selective inelastic electron tunneling spectroscopy probed by isotope labeling. Nano Lett 10:2950–2955
63. Hihath J, Bruot C, Tao N (2010) Electron-phonon interactions in single octanedithiol molecular junctions. ACS Nano 4:3823–3830
64. Yamaguchi K (1992) Inelastic electron tunneling through antiferromagnetic semiconductors MPS3 (M = Fe, Mn). In: Ping J, Hou-Zhi Z (eds) International conference on the physics of semiconductors, 21st, vol 1. World Scientific, Singapore, pp 661–664
65. Ando Y, Murai J, Miyazaki T (1999) Analysis of the interface in ferromagnet/insulator junctions by inelastic-electron-tunneling-spectroscopy. J Magn Magn Mater 198–199: 161–163
66. van de Veerdonk RJ, Moodera JS, de Jonge WJ (1999) Characterization of magnetic tunnel junctions using IETS. J Magn Magn Mater 198–199:152–154
67. Yamaguchi K (1993) Phonon structure of single-crystal gallium thiophosphide (Ga2/3PS3) from inelastic electron tunneling spectroscopy. Phys Status Solidi B 179:K11–K15
68. Yamaguchi K (1995) Phonons of indium selenide single crystals studied by inelastic electron tunneling spectroscopy. Phys Status Solidi B 190:409–413
69. Dizhur EM, Fedorov AV (2004) Digital processing in tunneling spectroscopy. arXiv:cond-mat/0408206v2
70. Honciuc A, Metzger RM, Gong A, Spangler CW (2007) Elastic and inelastic electron tunneling spectroscopy of a new rectifying monolayer. J Am Chem Soc 129:8310–8319
71. Hipps KW, Mazur U (1988) Constant-resolution tunneling spectroscopy. Rev Sci Instrum 59:1903–1905
72. Hipps KW, Mazur U (1987) All-digital inelastic electron tunneling spectro-meter utilizing the IEEE-488 instrument bus and an IBM PC-XT controller. Rev Sci Instrum 58:265–268
73. Seman TR, Mallik RR (1999) Electronic signal regulator for constant resolution inelastic electron tunneling spectroscopy. Rev Sci Instrum 70:2808–2814

74. Hipps KW, Mazur U (1987) Vibrational and low-lying electronic transitions in tetraalkylammonium salts of tetrabromocobaltate(2-), tetrachlorocobaltate(2-) and tetrathiocyanatocobaltate(2-) as observed by Raman, IR and tunneling spectroscopies. J Phys Chem 91:5218–5224
75. Reynolds S, Oxley DP (1996) Measurement and modeling of vibrational-mode lineshape and linewidth in inelastic electron tunneling spectroscopy. Surf Sci 368:324–329
76. Hipps KW, Peter S (1989) Line shape in normal metal tunneling spectroscopy. J Phys Chem 93:5717–5721
77. Kushmerick JG, Blumb S, Long DP (2006) Metrology for molecular electronics. Anal Chim Acta 568:20–27
78. Yu L, Zangmeister CD, Kushmerick JG (2007) Origin of discrepancies in inelastic electron tunneling spectra of molecular junctions. Phys Rev Lett 98:206803–206804
79. Rahimi M, Hegg M (2009) Probing charge transport in single-molecule break junctions using inelastic tunneling. Phys Rev B 79:081404(R)
80. Franke KJ, Schulze G, Pascual JI (2010) Excitation of Jahn-Teller active modes during electron transport through single C_{60} molecules on metal surfaces. J Phys Chem Lett 1: 500–504
81. Wu SW, Nazin GV, Chen X, Qiu XH, Ho W (2004) Control of relative tunneling rates in single molecule bipolar electron transport. Phys Rev Lett 93: 236802/1-236802/4
82. Mazur U, Hipps KW (1994) Unoccupied orbital mediated tunneling: resonance-like structures in the tunneling spectra of polyacenes. J Phys Chem 98:5824–5829
83. Mazur U, Hipps KW (1995) Resonant tunneling bands and electrochemical reduction potentials. J Phys Chem 99:6684–6688
84. Mazur U, Hipps KW (1999) Orbital-mediated tunneling, inelastic electron tunneling, and electrochemical potentials for metal phthalocyanine thin films. J Phys Chem B 103:9721–9727
85. Deng W, Hipps KW (2003) Tip-sample distance dependence in the STM-based orbital-mediated tunneling spectrum of nickel(II) tetraphenylporphyrin deposited on Au(111). J Phys Chem B 107:10736–10740
86. Ishii H, Seki K (1997) Energy level alignment at organic/metal interfaces studied by UV photoemission: breakdown of traditional assumption of a common vacuum level at the interface. Trans Electron Devices 44:1295–1301
87. Tsiper EV, Soos Z, Gao W, Kahn A (2002) Electronic polarization at surfaces and thin films of organic molecular crystals: PTCDA. Chem Phys Lett 360:47–52
88. Scudiero L, Barlow DE, Mazur U, Hipps KW (2001) Scanning tunneling microscopy, orbital mediated tunneling spectroscopy, and ultraviolet photoelectron spectroscopy of metal(II) tetraphenylporphyrins deposited from vapor. J Am Chem Soc 123:4073–4080
89. Proehl H, Toerker M, Sellam F, Fritz T, Leo K (2001) Comparison of ultraviolet photoelectron spectroscopy and scanning tunneling spectroscopy measurements on highly ordered ultrathin films of hexa-*peri*-hexabenzocoronene on Au(111). Phys Rev B 63:205409/1-205409/6
90. Barlow DE, Scudiero L, Hipps KW (2004) An STM study of the structure and orbital mediated tunneling spectra of CoPc and CoTPP on Au(111): mixed composition films. Langmuir 20:4413–4421
91. Scudiero L, Barlow DE, Hipps KW (2002) Scanning tunneling microscopy, orbital mediated tunneling spectroscopy, and ultraviolet photoelectron spectroscopy of nickel(II) octaethylporphyrin deposited from vapor. J Phys Chem B 106:996–1003
92. Sumi H (1998) V-I characteristics of STM processes as a probe detecting vibronic interactions at a redox state in large molecular adsorbates such as metalloproteins. J Phys Chem B 102:1833–1844
93. Kuznetsov AM, Ulstrup J (2000) Mechanisms of in situ scanning tunnelling microscopy of organized redox molecular assemblies. J Phys Chem A 104:11531–11540

94. Schmickler W, Tao N (1997) Measuring the inverted region of an electron transfer reaction with a scanning tunneling microscope. Electrochim Acta 42:2809–2815
95. Han W, Durantini EN, Moore TA, Moore AL, Gust D, Rez P, Letherman G, Seely G, Tao N, Lindsay SM (1997) STM contrast, electron-transfer chemistry, and conduction in molecules. J Phys Chem B 100:10719–10725
96. Richardson DE (1990) Thermochemical interpretation of electrode potentials for transition metal complexes. Inorg Chem 29:3213–3223
97. Loutfy RO, Hsiao CK, Ong BS, Keoshkerian B (1984) Thermochemical interpretation of electrode potentials for transition metal complexes. Can J Chem 62:1877–1886
98. Mazur U, Hipps KW (1994) Resonant tunneling in metal phthalocyanines. J Phys Chem 98:8169–8172
99. Hill IG, Kahn A, Soos ZG, Pascal RA Jr (2000) Charge-separation energy in films of π-conjugated organic molecules. Chem Phys Lett 327:181–188
100. Westcott BL, Gruhn N, Michelsen L, Lichtenberger D (2000) Experimental observation of non-aufbau behavior: photoelectron spectra of vanadyloctaethylporphyrinate and vanadylphthalocyanine. J Am Chem Soc 122:8083–8089
101. Selzer Y, Allara DA (2006) Single-molecule electrical junctions. Annu Rev Phys Chem 57:593–623
102. Li B, Zeng C, Zhao J, Yang J, Hou JD, Zhu Q (2006) Single-electron tunneling spectroscopy of single C_{60} in double-barrier tunnel junction. J Chem Phys 124:064709–064720
103. Bakkers EP, Hens Z, Kouwenhoven LP, Gurevich L, Vanmaekelbergh D (2002) A tunneling spectroscopy study on the single-particle energy levels and electron-electron interactions in CdSe quantum dots. Nanotechnology 13:258–262
104. Katz D, Millo O, Kan S, Banin U (2001) Control of charging in resonant tunneling through InAs nanocrystal quantum dots. Appl Phys Lett 79:117–119
105. Stroscio JA, Feenstra RM, Fein AP (1986) Electronic structure of the silicon(111) 2×1 surface by scanning-tunneling microscopy. Phys Rev Lett 57:2579–2582
106. Feenstra RM (1984) Tunneling spectroscopy of the (110) surface of direct-gap III-V semiconductors. Phys Rev B 50:4561–4570
107. Stroscio JA, Feenstra RM (1993) In: Stroscio JA, Kaiser WJ (eds) Scanning tunneling microscopy. Methods of experimental physics, vol 27. Academic, New York
108. Ukraintsev VA (1996) Data evaluation technique for electron-tunneling spectroscopy. Phys Rev B 53:11176–11185
109. Toader M, Gopakumar TG, Shukrynau P, Hietschold M (2010) Exploring the F16CoPc/Ag (110) interface using scanning tunneling microscopy and spectroscopy. Part 2: adsorption-induced charge-transfer effect. J Phys Chem C 114:21548–21554
110. Gopakumar TG, Meiss J, Pouladsaz D, Hietschold M (2008) HOMO-LUMO gap shrinking reveals tip-induced polarization of molecules in ultrathin layers: tip-sample distance-dependent scanning tunneling spectroscopy on d8 (Ni, Pd, and Pt) phthalocyanines. J Phys Chem C 112:2529–2537
111. Mazur U, Leonetti M, English WA, Hipps KW (2004) Spontaneous solution phase redox deposition of a dense cobalt(II) phthalocyanine monolayer on gold. J Phys Chem B 108: 17003–17006
112. Gyarfas B, Wiggins B, Hipps KW (2010) Temperature independence of orbital mediated tunneling in cobalt(II) phthalocyanine. J Phys Chem C 114:13349–13353
113. Rahimi M, Troisi A (2009) Probing local electric field and conformational switching in single-molecule break junctions. Phys Rev B 79:113413–113414
114. Wang C, Luo Y (2003) Current–voltage characteristics of single molecular junctions: dimensionality of metal contacts. J Chem Phys 119:4923–4928
115. Kula M, Jiang J, Luo Y (2008) A molecular view of electron transport in molecular electronic devices. J Comput Theor Nanosci 5:401–421
116. Joachim C, Ratner MA (2005) Molecular electronics: some views on transport junctions and beyond. Proc Natl Acad Sci 102:8801–8808

117. Petrov EG (2005) Tunneling through localized states of a single molecule. Mol Cryst Liq Cryst 426:49–58
118. Lin L, Leng J, Song X, Li Z, Luo Y, Wan C (2009) Effect of aromatic coupling on electronic transport in bimolecular junctions. J Phys Chem C 113:14474
119. Jiang J, Kula M, Luo Y (2006) A generalized quantum chemical approach for elastic and inelastic electron transports in molecular electronics devices. J Chem Phys 124: 034708–034710
120. Li Z, Zou B, Wang C (2006) Electronic transport properties of molecular bipyridine junctions: effects of isomer and contact structures. Phys Rev B 73:075326–075327
121. Xue Y, Ratner MA (2003) Microscopic study of electrical transport through individual molecules with metallic contacts. I. Band lineup, voltage drop, and high-field transport. Phys Rev B 68:115406–115418
122. Heurich J, Cuevas JC, Wenzel W, Schön G (2002) Electrical transport through single-molecule junctions: from molecular orbitals to conduction channels. Phys Rev Lett 88: 256803–256804

Top Curr Chem (2012) 313: 217–268
DOI: 10.1007/128_2011_222

Published online: 9 August 2011

Single Molecule Logical Devices

Nicolas Renaud, Mohamed Hliwa, and Christian Joachim

Abstract After almost 40 years of development, molecular electronics has given birth to many exciting ideas that range from molecular wires to molecular qubit-based quantum computers. This chapter reviews our efforts to answer a simple question: how smart can a single molecule be? In our case a molecule able to perform a simple Boolean function is a child prodigy. Following the Aviram and Ratner approach, these molecules are inserted between several conducting electrodes. The electronic conduction of the resulting molecular junction is extremely sensitive to the chemical nature of the molecule. Therefore designing this latter correctly allows the implementation of a given function inside the molecular junction. Throughout the chapter different approaches are reviewed, from hybrid devices to quantum molecular logic gates. We particularly stress that one can implement an entire logic circuit in a single molecule, using either classical-like intramolecular connections, or a deformation of the molecular orbitals induced by a conformational change of the molecule. These approaches are radically different from the hybrid-device approach, where several molecules are connected together to build the circuit.

N. Renaud (✉)
Department of Chemistry, Northwestern University, 2145 Sheridan Road, Evanston, IL 60208-3113 USA
e-mail: n-renaud@northwestern.edu

M. Hliwa
Nanoscience Group and MANA Satellite, CEMES/CNRS, 29 rue J. Marvig, BP 94347, 31055 Toulouse Cedex, France

Faculty of Sciences, Ben M'Sik University Hassan II-Mohammedia, Casablanca, Morocco
e-mail: Hliwa@cemes.fr

C. Joachim
Nanoscience Group and MANA Satellite, CEMES/CNRS, 29 rue J. Marvig, BP 94347, 31055 Toulouse Cedex, France
e-mail: joachim@cemes.fr

Keywords Intramolecular kirchhoff laws · Molecular electronics · Molecular logic gates · Single molecule electronic circuits · Quantum hamiltonian computing

Contents

1 Single Molecule Electronics

Microelectronics relies on the massive utilization of logic gates [1]. These key subunits perform logical operations on a set of binary logical inputs and calculate a single binary logical output [2]. These logical gates can be associated with each other to create more complex circuits that can, for example, add digital numbers. These circuits can then be associated with each other to get to the next level of complexity, and so on. This stacked architecture is only possible because of two ingredients: a very reliable elementary block: the transistor [3] and a set of very simple rules to connect these blocs together: the Boolean symbolic analysis of electronic circuits [4]. Printed integrated circuits [5] finally provide a cheap means of mass production that has allowed logic gates to sneak into our daily life. The main reason behind the popularity of the transistor is its extraordinary reliability. Due to its bistable characteristic, noise-induced deviations of a control gate voltage are naturally corrected and do not change its conductive state [6]. This property, called restoration, allows stacking millions of transistors together without worrying about the noise propagation in the circuit. The design of this circuit is itself assured by the symbolic analysis developed in the mid-1930s that allows finding the simplest circuit possible to perform a given operation [4]. The starting point of this analysis is extremely simple: if two switches are controlled each one by a logical input, then their serial and parallel association performs respectively the AND and the OR logic functions. Despite their simplicity, these two rules are the basis of modern micro-electronics, and the circuits drawn 60 years ago are not very different than the three millions that have been produced worldwide since you have started to read this chapter [7]. This impressive mass production is due to the lithography techniques used to print integrated circuits that allow the production of millions of transistors in a few hours. Beside, thanks to the major improvements made in this area since Nicola Tesla's first AND gate, printing a logic circuit is now barely more expensive than printing a single word on a newspaper [7]. However, the death of this approach was announced decades ago. The continuous miniaturization of transistors, that have given birth to the now well-known Moore's law [8], is threatened by both the diffraction limit of the lithography techniques and quantum effects arising when the dimensions of transistors are shrunk to a few dozen atoms [9]. Other solutions must be studied in order to keep increasing computing power by surface area in the future. In this framework, many alternative routes to solid-state electronics have been proposed in the last few decades [10]. Microwave-based computers were developed and used for approximately a decade in the 1960s [11, 12]. Based on the phase control of an LC circuit, these computers could not compete with the switching speed of transistors, and have been abandoned. Despite their numerous advantages, all-optical computers [13, 14], where the information is not carried by an electrical current but by a light beam, have also been abandoned due to their high production cost. The solution presented in this chapter was presented by Ari Aviram and Mark Ratner in 1974. Studying the conduction of organic films, they decided to investigate theoretically the conduction of a single molecule instead of an entire film: single-molecule electronics was born. "Molecular Rectifiers" [15], their seminal article, presents a molecular diode that uses the donor–acceptor structure of a molecule to let the electronic current flow in one direction, i.e., from the donor to the acceptor but not the other way. This conceptual breakthrough is the starting point of an entirely new approach that tries to embody a given electronic function in a single molecule, circumventing the two issues that threaten solid-state circuits. First, the synthesis of this device, achieved in 1990 [16, 17], is not limited by the diffraction criterion. Second, quantum effects are used here to advantage, and are not seen as an annoying source of disagreement. Measuring the conduction of a single molecule was far from being possible in the mid-1970s, and no one could tell if the molecular rectifier proposed by Aviram and Ratner could actually work. The invention of the scanning tunneling microscope in 1981 [18, 19] provided the experimental tool required to turn single-molecule electronics from a theoretical idea to an intense field of research. In the mid-1980s the use of the STM spreads from the IBM Zürich laboratory, its place of birth, to other IBM research labs and a few universities, and reached Aviram's laboratory in 1986. Feeling the potential of this new microscopy, Aviram decided to test his brand-new synthesized molecular switches under an STM tip. The principle of these switches was rather simple. Under an electric field, two hydrogen atoms moved reversibly on the molecule, modifying its electronic conduction [20]. Depositing these molecules on a gold surface, Aviram and coworkers were the first to measure the characteristic of a switch made out of a single molecule. However, the very abrupt I–V transition observed while positioning the STM tip apex on a few molecules and increasing the bias voltage was not due to an intramolecular proton transfer effect (as awaited) but was caused by the migration of a few Au atoms coming from the surface through the molecular monolayer. The paper was withdrawn [21]. The Au atoms are stabilized by the monolayer and create a nice tunneling path, leading almost to one quantum of conductance for the "tip apex–molecular monolayer–Au surface" tunnel junction. This migration phenomenon misleads a lot of consequent experiments even in 2010. It was transformed in a bonus with the observation of memristor effects [22]. Many other components [23], like molecular wires, amplifiers, transistors, or ammeters, were proposed in the spirit of Aviram and Ratner [24]. Aviram proposed to connect several rectifiers together to create logic circuits and memories, initiating the so-called hybrid-molecular approach [25]. Metallic nano-wires connect molecular devices that can be diodes, transistors, etc., to construct an electronic circuit following the classical architecture rules used in the solid-state approach: many such devices have been designed theoretically. The corresponding experimental set-up would require either to use several STM tips or to position the molecules exactly on top of nano-wires. This nano-manipulation is still a challenge. To solve this problem, Forest Carter proposed to incorporate all the active components and their interconnections in a single molecule [26]. This approach, called classical monomolecular electronics, not only solved the interconnection problem but also gave birth to the idea that a single (if complex) molecule could perform a complex logic function by itself. This idea also introduces a conceptual shift: in the monomolecular approach, logic gates become the smallest units of the electronic architecture, and simpler components like transistors are not needed. This leads directly to the question: what is the most complex logic function that can be implemented in a molecule? To explore this limit, gigantic molecules have been imagined [27]. When connected to several electrodes, these molecules can, for example, add or subtract digital numbers, but, due to their size, only a small current can go from one electrode to another. The quantum monomolecular approach tries to find a solution to this issue. Using the characteristic of the conduction at the molecular scale correctly, this approach succeeded in implementing complex functions in a small molecule [28].

In this chapter, the fundamentals of molecular conduction are presented in Sect. 2. This section is not a review of this intense field of research, but only presents the main concepts that are used in the following sections to create molecular devices. Section 3 presents three hybrid-molecular devices: the mechanism of the Aviram and Ratner rectifier is explained, as well as a molecular switch and a C_{60}-based amplifier. Section 4 explores the monomolecular approach and the intramolecular rules it leads to. Following Carter's seminal idea, classical monomolecular logic gates, incorporating several donor–acceptor diodes in the same molecule, are presented: using quantum properties of electronic transmission, we will see that these molecular rectifiers are actually not required separately. Section 5 presents the quantum monomolecular approach, where the logic functions are implemented in a small molecule controlling the energy and the spatial delocalization of its molecular orbitals.

2 Multiwire Multichannel Conduction Through a Single Molecule

To implement a logical function using the electronic transmission of a single molecule, the mechanism underlying the conduction at the molecular scale must be understood. Very soon after the invention of the STM, theoretical studies investigated the electronic transmission through a potential barrier, leading to the well-known exponential decay of the current intensity with the width of the barrier [29]. Introducing a molecule between the two electrodes, one introduces a complex object with many molecular orbitals. The first step toward molecular electronics is to describe accurately the impact of the molecular orbitals in the transmission properties of the junction. In their seminal work [30], Landauer and later Büttiker introduced the theory of quantum conduction through a mesoscopic system. According to this approach, the low-bias current intensity passing through a ballistic quantum wire is given by

$$I = \frac{2e}{\hbar}\int_{-\infty}^{\infty} dE\,(f_1(E - eV/2) - f_2(E + eV/2))T(E) \qquad (1)$$

where $f_1(E,V)$ and $f_2(E,V)$ are the Fermi function of the two electrodes at the bias voltage V and $T(E)$ is the transmission probability through the junction at the scattering energy, E. Many methods have been proposed to compute the $T(E)$ generalizing (1) to molecular junctions. Following the Lippman–Schwinger argument [31], most of them are based on the solution of the time-independent Schrödinger equation (TISE). The Green function is one of the most popular ways to manipulate the TISE and is the basis of many methods used to compute the $T(E)$ [32–42]. Other methods, based for example on the electrode's self-energies or on graph–theoretical arguments [43, 44], have also been developed and lead to similar results. The physical interpretation is made afterward, seeking a relationship between the $T(E)$ and other physical quantities, like the electron transfer rate [45–48] or local currents [49]. Several time-dependent approaches have also been proposed [50–58] but without trying to find a new physical insight in these line shapes and without comparing their results with other time-independent methods. The aim of this section is not to give an exhaustive review of all these techniques, but only to give the reader sufficient background to grasp the main features of the electronic conduction through a molecular junction. In this framework we will briefly review two different methods to compute the $T(E)$. The first, electron scattering quantum chemistry (ESQC) [59–64], is based on the TISE. This transfer-matrix based method uses spatial propagators to treat the electrodes mapping the molecular orbitals at their end. The second method is based on the time-dependent Schrödinger equation (TDSE) using the Fano model to account for the electrodes [65–69]. The TISE-based methods are numerically more efficient than the TDSE-approach, which requires the propagation of the wave function.

However, TDSE permits an analysis of the time-dependent population of the molecular orbitals during the tunneling, which is unavailable with the TISE.

2.1 The Single Impurity Problem

A model system to study electronic conduction through a molecule is represented in Fig. 1 [70, 71]. A simple molecule, composed of a single orbital $|\phi\rangle$ is connected to two linear atomic chains composed of N orbitals each. We suppose that $|\phi\rangle$ interacts only with the last state of each electrode via a coupling denoted by v_i. We suppose in the following that the electrodes are defined by $h = 2$ eV and $e_0 = 0$ eV. Depending on the ratio between h and v_i, different regimes can be explored: the tunneling regime for $|v_i| << |h|$ and the pseudoballistic regime for $|v_i| \cong |h|$.

2.1.1 The Time-Independent Approach

The ESQC method allows a fast calculation of the electronic transmission from an exact solution of the TISE. The basis set here is supposed orthogonal, but the method can be generalized to nonorthogonal ones. Following [31], solving this equation is equivalent to the propagation of an initial wave packet located on the

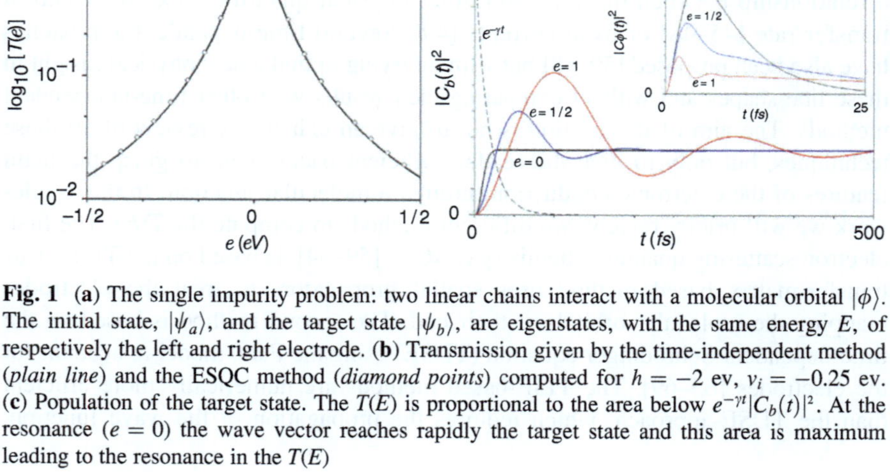

Fig. 1 (**a**) The single impurity problem: two linear chains interact with a molecular orbital $|\phi\rangle$. The initial state, $|\psi_a\rangle$, and the target state $|\psi_b\rangle$, are eigenstates, with the same energy E, of respectively the left and right electrode. (**b**) Transmission given by the time-independent method (*plain line*) and the ESQC method (*diamond points*) computed for $h = -2$ ev, $v_i = -0.25$ ev. (**c**) Population of the target state. The $T(E)$ is proportional to the area below $e^{-\gamma t}|C_b(t)|^2$. At the resonance ($e = 0$) the wave vector reaches rapidly the target state and this area is maximum leading to the resonance in the $T(E)$

left electrode at $t = -\infty$ that scatters through the impurity and reaches the right electrode for $t = \infty$. The propagation in the periodical electrodes is performed using spatial propagators, $P_{(n+1)(n-1)}$ from the state $(n + 1)$ to the state $(n - 1)$ of the electrode. The propagation through the impurity gives rise to an effective spatial propagator, $P_{2,1}$ whose elements are defined using the projected impurity Green's function on the last states of each electrode. The total transfer matrix, from the far left side to the far right side is then defined by

$$t(E) = P_{N;(N-2)}P_{(N-2);(N-4)} \dots P_{2;-1} \dots P_{(-N+4);(-N+2)}P_{(-N+2);-N}. \quad (2)$$

To ease the computation of this product, the spatial propagators on the electrode are diagonalized following $U_{n;(n-1)}P_{n;(n-2)}U_{(n-1);(n-2)} = 1$. This exact digitalization allows expressing the transfer matrix as $t(E) = U_{-1;2}P_{2;-1}U_{0;-1}$.The transmission coefficient is then simply given by $T(E) = |t11(E)|^{-2}$ where $t_{11}(E)$ is the first diagonal element of the 2×2 $T(E)$ matrix. For the single impurity problem all these different matrices are exactly known, and $T(E)$ can be calculated analytically. Interested readers are referred to [59] for more details. The transmission coefficient of the single impurity obtained following the ESQC method is represented in Fig. 1b for $v_i = 0.25$ eV. In the same figure is plotted the transmission coefficient obtained with the time-dependent method presented below. The comparison between these two curves and the related discussion can be found at the end of the next section.

2.1.2 The Time-Dependent Approach

Let us now consider the solution of the TDSE for this system (with $\hbar = 1$): $|\psi(t)\rangle = e^{-iHt}|\psi_a\rangle$. We suppose that the initial state of the evolution, $|\psi_a\rangle$, is an eigenstate of energy E of the left electrode. Starting from this initial state, $|\psi(t)\rangle$ evolves in time and space, and eventually reaches an eigenstate, $|\psi_b\rangle$, of the right electrode. Since we are dealing with elastic scattering, this eigenstate is associated with energy E. The impact of the continuum of states (introduced by the semi-infinite electrodes) on this temporal evolution is taken into account by a Fano model that shifts the energy of $|\phi\rangle$ by an imaginary constant [68, 69]. Due to their delocalization over the electrodes, $|\psi_a\rangle$ and $|\psi_b\rangle$ interact very weakly with the impurity. Due to these weak couplings, one oscillation frequency dominates the evolution of $|\psi(t)\rangle$ from $|\psi_a\rangle$ to $|\psi_b\rangle$ [72, 73]. The population of $|\psi_b\rangle$, $|C_b(t)|^2$, can be approximated accurately by the damped sinusoidal $|C_b(t)|^2 \cong ae^{-\Delta t}\sin^2(\Omega t)$ [74]. The corresponding population is represented in Fig. 2b and is discussed below. The transmission coefficient, $T(E)$ is obtained by averaging this population during the lifetime of $|\psi(t)\rangle$. Supposing that this lifetime follows an exponential law, $e^{-\gamma t}$, then the expression of the $T(E)$ reads

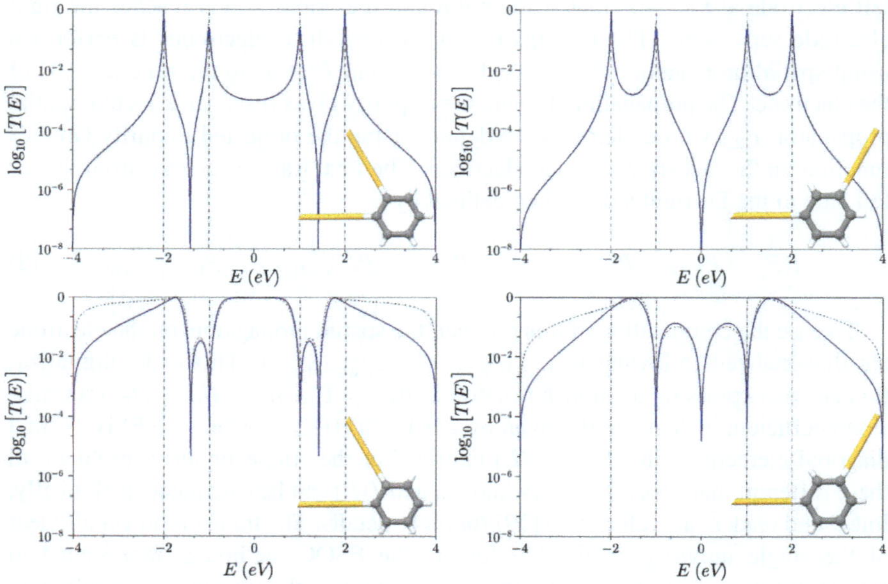

Fig. 2 Electronic conduction of a benzene ring between two conducting electrodes. These calculations are performed by the time-dependent method presented here (*solid line*) and by the ESQC method (*dashed line*). The electrodes are connected either in ortho (*left column*) or meta (*right column*) position. Two regimes are investigated: tunneling with $v_i = -0.25$ ev (*upper row*), pseudoballistic with $v_i = -2$ ev (*lower row*). The *vertical dashed lines* represent the energy of the benzene's molecular orbitals

$$T(E) = \left| \frac{\int_0^\infty C_b(t)\mathrm{e}^{-\gamma t}\mathrm{d}t}{\int_0^\infty \mathrm{e}^{-\gamma t}\mathrm{d}t} \right|^2 \approx a^2 \frac{\Omega^2 + \Delta^2}{\gamma^2}. \tag{3}$$

An expression for γ can be found by physical arguments by imposing the resonance condition, i.e., $T(E = e) = 1$ [74]. The probability $T(E)$ given by (3) is plotted in Fig. 1b for $v_i = 0.25$ eV. It agrees perfectly with the time-dependent method presented above. These two curves present a resonance, where the junction is completely transparent for the incident electrons. When their incident energies match the site-energy of the impurity, the electrons can flow freely through the barrier; otherwise they are almost completely reflected. By comparing the line shape of the $T(E)$ and the target state population, the time-dependent approach gives an interesting physical insight into this phenomenon. At resonance, the evolution of $|C_b(t)|^2$ is similar to a critically damped regime, where it converges to its asymptotic value faster than for any other value of E. In contrast, the off-resonance cases correspond to an under-damped regime, where $|\psi(t)\rangle$ oscillates between the two electrodes before reaching its asymptotic state. The resonance then corresponds to the fastest evolution from one side of the junction to the other. The population of the impurity is also represented in the inset of Fig. 1c. This population is larger at resonance than off-resonance, which agrees with the super-exchange picture of the tunneling process. Inelastic processes are then more likely to happen at resonance, where the electron spends enough time on the molecule to excite its vibrational modes, rather than off-resonance, where it barely resides in the molecular orbital. These inelastic effects are briefly discussed in Sect. 2.4.

2.2 Molecular Junction: The Benzene Ring

The two methods presented above can be generalized to larger systems, where the molecule is composed of many molecular orbitals. After briefly presenting their generalizations, these two methods are applied to the well-known case of the π network of a benzene ring between two conducting leads, in a tight-binding model. The electrodes are supposed to be connected either in the *ortho* or *meta* configuration, in the tunneling or the pseudoballistic regime. In the time-independent picture the only modification to go from the single impurity case to a more complex molecule is the expression of the effective propagator through the impurity, $P_{2;-1}$. Its expression has to account for all the molecular eigenstates of this "large impurity" and their interaction with the last state of each electrode.

The generalization of the time-dependent approach is more complex, since the TDSE cannot be solved analytically. However, thanks to the weak coupling between $|\psi_a\rangle$, $|\psi_b\rangle$ and the molecular orbitals, the Bloch effective Hamiltonian [72] can be used to obtain an accurate expression of the target state population as $C_b(t) \cong a_{\text{eff}}(\mathrm{e}^{-\mathrm{i}\lambda_+ t} - \mathrm{e}^{-\mathrm{i}\lambda_- t})$. Here, the $\lambda_\pm$ are the dominant eigenvalues and a_{eff} is the effective amplitude that can be determined using a Bloch criterion [73, 75, 76]. The transmission probability is then calculated following the transformation introduced in (3) with Ω and Δ given here by $\Omega = \mathrm{Re}(\lambda_- - \lambda_+)$ and $\Delta = |\mathrm{Im}(\lambda_- - \lambda_+)|$. The expression for γ is the same here as for the single impurity case, and the generalized Fano dissipative matrix can be found in [68, 69]. When applied to the network of the benzene ring, these two methods give the $T(E)$ spectrum presented in Fig. 2 for $v_i = 0.25$ eV (upper row) and $v_i = 2$ eV (lower row). Very good agreement is seen between the two methods. As for the single impurity case, resonances, characterized by $T(E) = 1$, are found when the energy of the incident electrons matches the energy of one eigenstate of the molecule. If this state connects the two electrodes, thanks to its delocalization over the molecule, it creates a resonant tunneling channel, where the electrons with the appropriate energy can flow freely. In the time-dependent approach a resonance corresponds to a maximum of the oscillation frequency of the wave vector between $|\psi_a\rangle$ and $|\psi_b\rangle$. This equivalence between a resonance and the maximum oscillation frequency will be used in Sect. 5 to design resonant molecular logic gates that control the energy and the localization of the molecular orbitals. Aside from these resonances, this molecular junction also presents deep interferences, where the junction is completely opaque for the incident electrons, i.e., $T(E) = 0$. The time-dependent approach leads to a very simple explanation of these well-known interferences [48, 77, 78]. When reaching the molecule from the left electrode, several energetic pathways, defined by the molecular orbitals of the molecule, are available for the electron to cross the junction and reach the right electrode. This situation is represented in Fig. 3. Instead of choosing one single pathway, the electrons take all these pathways simultaneously to cross the junction.

When relocalizing on the target state, the different contributions of the delocalized electron can interfere destructively. This is the mechanism underlying the interference patterns located at $E = \pm\sqrt{2}$ eV in the ortho configuration and at $E = 0$ in the meta configuration, energies for which the interference condition is met. In the time-dependent picture these interferences correspond to a zero of the oscillation frequency of $|\psi(t)\rangle$ from $|\psi_a\rangle$ to $|\psi_b\rangle$. Another type of interference is observed at $E = \pm 1$ in the pseudoballistic regime and for both configurations. These interferences are not due to the presence of several interfering pathways, but to the presence of a single molecular orbital that is connected to only one of the two electrodes. This dangling eigenstate creates a "no-way-out" for an incident electron whose energy is at resonance with this molecular orbital. In the time-dependent picture, $|\psi(t)\rangle$ then oscillates between $|\psi_a\rangle$ and this dangling state, and the maximum population of $|\psi_b\rangle$ remains extremely weak. In the specific case of the benzene ring, these dangling eigenstates, located at $E = \pm 1$, are degenerate with a resonant tunneling pathway. A sufficient splitting of this degeneracy,

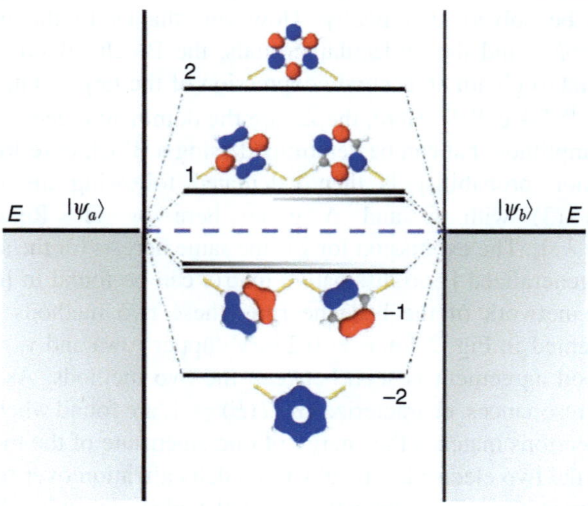

Fig. 3 Graphical representation of the scattering through the benzene ring in the meta configuration. The six molecular orbitals create six pathways for the electron to cross the junction. If their energy is resonant with one of these pathway, they flow freely through the barrier leading to $T(E) = 1$. The frequency-related interference observed in the meta-configuration on Fig. 2 at $E = 0$ comes from a destructive relocalization of the wave packet on $|\psi_b\rangle$. The amplitude-related interference located at $E = \pm 1$ comes from the dangling molecular orbitals at these energies that are connected to only one of the two electrodes. They create a "no-way-out" pathway preventing the electron from crossing the junction

induced by a relatively strong coupling between the molecule and the electrodes, is required for this interference patterns to appear. That is why they are only observed in the pseudoballistic regime and not in the tunneling regime.

2.3 Connecting a Molecule to N Electrodes

To implement a complex logic function with several inputs and outputs in a single molecule, this latter has to be connected to more than two electrodes. Beyond the experimental challenge it proposes [79], this multielectrode scattering process leads to an interesting theoretical problem. The generalization of ESQC to this multielectrode case has been the first method to tackle this problem. The so-called NESQC technique [60] is still based on the transfer-matrix formalism, and follows the same general mechanism presented above. The transfer matrix in this case is an N $\times$ N matrix, and the ratio between its elements defines the scattering amplitude between two of the N electrodes. The time-dependent method presented in the previous section can easily be generalized to account for the presence of the multiple electrodes generalizing the Fano dissipative matrix to the N-continuum case. The electronic conduction between two electrodes is then computed following exactly the same method as previously. These two methods are compared in Fig. 4 for the simple case of four electrodes connected to a naphthalene network, where each electrode interacts with one orbital of one carbon atom through a coupling $v_i = 0.5$ eV. As for the tunneling case of the benzene ring, the two methods agree perfectly,

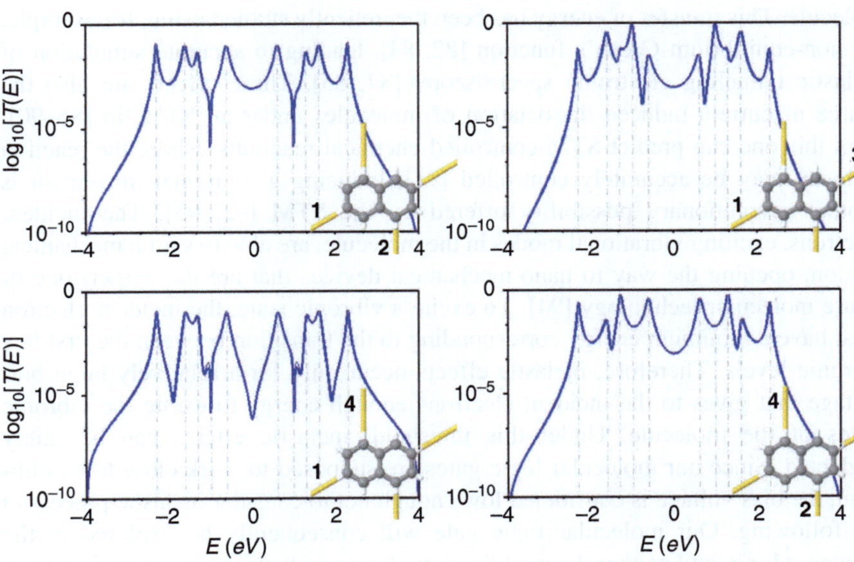

Fig. 4 Electronic conduction between four identical electrodes via a naphthalene calculated with the time-dependent method

and the two curves are superposed. One can notice the strong differences between the electronic conductions of different pairs of electrodes. For example, a resonance between two electrodes can become an interference between two others. This is the case when the corresponding molecular orbital is connected to two electrodes but not to a third, leading to an no way out interference. The Interference due to a zero frequency are also located at different energetic values for different pairs of electrode. Due to their respective interactions with the molecular orbital, the interference condition is different for each pair of electrodes. Very simple rules, based on the Dewar theory of alternating hydrocarbons, can be established to predict the position of these interferences. This theory separates the carbon atoms of the molecule into two groups, each member of one group being connected to carbon atoms that belong to the other group. Thus it can be shown that an interference located at $E = 0$ is obtained only if the two electrodes are connected to two carbon atoms of the same group.

2.4 Inelastic Effects

Only the elastic electronic conduction through a molecule has been presented so far. In this framework, the incident electron does not transfer its energy to the molecule, and only scatters through its molecular orbitals. Inelastic effects in a molecular junction have been intensely studied both experimentally and theoretically [80, 81]. These effects imply a transfer of energy from the tunneling electrons to the molecule. This transfer of energy has been theoretically studied using, for example, the non-equilibrium Green's function [82, 83], leading to accurate simulation of inelastic tunneling electronic spectroscopy [84, 85]. These effects are also the source of current-induced dissociation of molecules under an STM tip [86–90]: from this one can predict STM-controlled chemical reactions, where the reaction pathway may be accurately controlled [91]. Inducing a rotational movement is another revolutionary possibility offered by the STM [92, 93]. The incident electrons, exciting vibrational modes in the molecule, are able to start a mechanical motion, opening the way to nano-mechanical devices that are the cornerstone of future molecular technology [94]. To excite a vibronic state, the incident electron must have a minimum energy corresponding to the transition between the first few vibronic levels. Therefore, inelastic effects occur only for a relatively large bias voltage that gives to the incident electrons enough energy to excite the vibronic states of the molecule. Under this threshold, inelastic effects can be safely neglected. Since our molecular logic gates are supposed to work close to equilibrium, the bias voltage is considered low enough not to consider inelastic process in the following. Our molecular logic gate will consequently be explored in the Landauer limit, and neither the modification of the conductance by inelastic effects nor heat generation [95, 96] will be taken into account.

3 Hybrid Devices

The basic structure of a hybrid molecular device is a single molecule ***B*** interacting with its two conducting ***AAA*** semi-infinite nano-pads in an |***AAABAAA***|-like electronic structure. Depending on the functionality of this molecule, its presence as a jumper between the two nano-pads should introduce a better tunneling path than a molecule-free |***AAA AAA***| junction. This is the basic premise already proposed in the seminal Aviram–Ratner rectifier paper [15]. The device consists of the molecule and of its two nano-pads. One can also add a third controlling lateral grid. Because of the finite mean free path of the electrons in the nano-pad ***AAA***, the electron phase will be lost only after a certain propagation length deep in the nano-pads. At room temperature, this can require at least 10–20 nm between the chemisorption site of the molecule on the nano-pad surfaces and the location on the nano-pad where the device is interconnected to the external world, or to another part of the circuit. The device functionality will be stabilized only after taking into account a certain portion of the conducting nano-pads: this suggests naming all the |***AAABAAA***|-like devices: "*hybrid molecular devices*." In practice, this distance corresponds very well to the asymptotic condition discussed in Sect. 2.1 to describe the transport phenomenon through a single electronic defect ***B*** embedded in a fully periodic |***AAAAAA***| structure. In principle, a complex electronic circuit can be built up by interconnecting such molecular devices, but respecting the standard electrical Kirchhoff mesh and node circuit laws. This means that, as soon as a hybrid molecular device is optimized, there is in principle no obstacle to designing complex circuitry. Many hybrid molecular devices have been proposed and calculated or simulated in a more or less well-described adsorption environment. Only a few have been experimentally worked out in their truly single-molecule version, that is, one single molecule interacting with a well-identified chemisorption site on the surface of each contacting nano-pad, and with its molecular conformation exactly determined in the junction. Their basic physical functionality can be easily classified using the valence-bond-like model of Sect. 2, as represented in Fig. 5, where one or two electronic states of the molecule are in interaction with two semi-infinite chains of electronic states describing the pads. Chemistry will put a lot of structural variability on the basic model of Fig. 5, together with the gating of the device when required. As presented in Fig. 5, the basic quantum physics behind all these hybrid molecular devices is the control of a tunneling current through an |***AAAABAAAA***| electronic structure by playing with some internal ***B*** parameter or with the ***A–B*** interactions parameters, since the device is hybrid by definition, i.e., involving the pads **AAA** and the molecule **B** to define its functionality. For a molecular switch (mainly Fig. 5a, b), electronic coupling (inside **B** or **A–B**) or electronic states manipulation in ***B*** have been proposed, and sometimes clearly observed experimentally. For molecular rectifiers or negative differential resistance (NDR) devices (Fig. 5b without gating or Fig. 5c), the ***B*** electronic states are designed on purpose to create an asymmetric *I–V* curve. For interference-like hybrid molecular devices

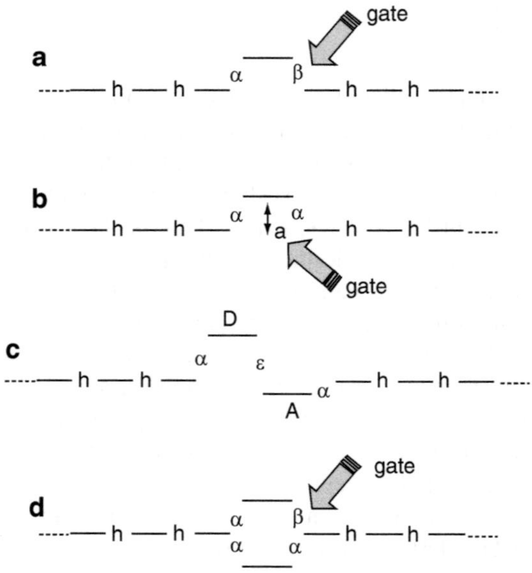

Fig. 5 The four generic electronic structures of a molecular device in a standard |*AAABAAA*| form: *h* is the coupling between two neighbor site *A* in the electrode, α and β are the coupling between the periodical electrodes and the impurity ***B*** and ε the coupling between the donor and the acceptor of a two-state impurity. (**a**) Molecular switch controlled by changing the electronic coupling α, β relative to *h* (**b**) Molecular switch controlled by changing the position of a molecular level as a function of an external gating. It is also the simplest model for a negative differential resistance molecular device. (**c**) Donor–acceptor (DA) molecular rectifier where ε indicates the compulsory small coupling between D and A in respect of the designed molecular electronic structure. (**d**) Simplest molecular interferometer structured with two equivalent electronic states with a coupling which can be positive or negative, as a function of the relative energy position of those two levels in reference with the Fermi level of the nano-pads ***AAA***. The interference can also be controlled by shifting the relative position in energy of the two electronic states

(Fig. 5d), the constructive (or destructive) interferences are controlled either by the change of a given ***B*** intramolecular electronic coupling, or by the shift of a ***B*** molecular electronic state. Those four generic types of molecular devices (or a combination of those in realistic molecular electronic structures) generate all the different hybrid molecular electronic devices proposed and studied so far. The "ON–OFF" ratio for a molecular switch or the gain for a molecular transistor can be improved by getting more ***B*** molecular electronic states involved in the switching or the gain process [97]. Note that in Fig. 5 classification, the gating electrode is not playing a role, because no leakage current was supposed between the gate and the ***AAA*** nano-pads of the molecular junction. If such a leak exists (or if it is required), the classification of Fig. 5 can be generalized, including a third ***AAAA*** chain converging towards the central ***B*** electronic structure using, for example, the NESQC calculation technique of Sect. 2.3 to simulate the functioning of the resulting molecular device.

3.1 Molecular Switches

From Fig. 5a, b, a molecular switch is a simple |***AAABAAA***| electronic structure, where the gating affects an electronic coupling in the ***ABA*** junction, or the energy position of some B electronic states. Both effects can be combined to increase the ON–OFF conductance ratio of the ***ABA*** switch. Those are well-known effects, back to the period where through-bond electron transfer processes were studied as a function of the conformation of the molecular ligand [98]. Any well-designed molecule ***B*** with a chemical group able to rotate or to undergo a *cis–trans* isomerization along the ***B*** electron transfer path between the two nano-pads will bring some electronic coupling variation along this path. Landauer was skeptical about such quantum-mechanical electronic transmission switching [99], because, whatever the molecule, one must still design the gating effect with a third electrode. From a more physical point of view, changing a ***B*** electronic coupling, or the energy position of some ***B*** electronic states involved in the transport process in the ***ABA*** junction, does not generally lead to a very abrupt change in the conductance of the ***ABA*** junction [59, 100]. Nevertheless, in an *AXeA*-like STM tunneling junction, Eigler nicely demonstrated that the STM manipulation of a single Xe atom in and out of the STM junction, using the electric field at the STM junction, can change the conductance of this junction by two orders of magnitude [101]. The Xe atom under the tip apex reduces the effective tip apex to surface tunnel barrier height compared to a vacuum. This reduction increases the penetration depth for tunneling electrons within the barrier, resulting in a larger tunneling current. This is a Fig. 5b type switch, where the intermediate Xe 6*s* electronic state is coupled or not to the junction [102]. Under the STM, the first example of a truly single molecular switch was found while looking at the adsorption of a legged Cu–porphyrin molecule on a Cu(211) surface [103]. At a Cu(211) step edge, one leg of this molecule is free to rotate between two positions: one perpendicular and one parallel to the surface. The conductance OFF (parallel) is more stable than the conductance ON (perpendicular) conformation on a Cu(211) surface. Therefore, positioning the tip apex at the location of this switchable leg leads to a nice switching effect, provided the tip remains at the same height for the two conformations of the switched leg (see Fig. 6). This molecular switch belongs to Fig. 5a type, where the central electronic states of the leg are almost not shifted during the switching. This leg is electronically coupled (decoupled) to the STM tip apex and to the surface in its perpendicular (parallel) surface conformation. The gating effect is very interesting. The switching leg of this molecule is like the moving mechanical part of a macroscopic switch but with a stable and a metastable conformation. While gently pushing on the leg to pass from the ON (perpendicular) to the OFF (planar) configuration, an energy barrier has to be overcome. This barrier was measured [104]. To be back to the ON state, one has to manipulate this leg a bit laterally to find a reaction path that overcomes the switching barrier. To optimize such a molecular switch, one has also to determine where the bistable potential energy path of the switching B molecule is coming from: from the nano-pad surface where the molecule is sitting or from the

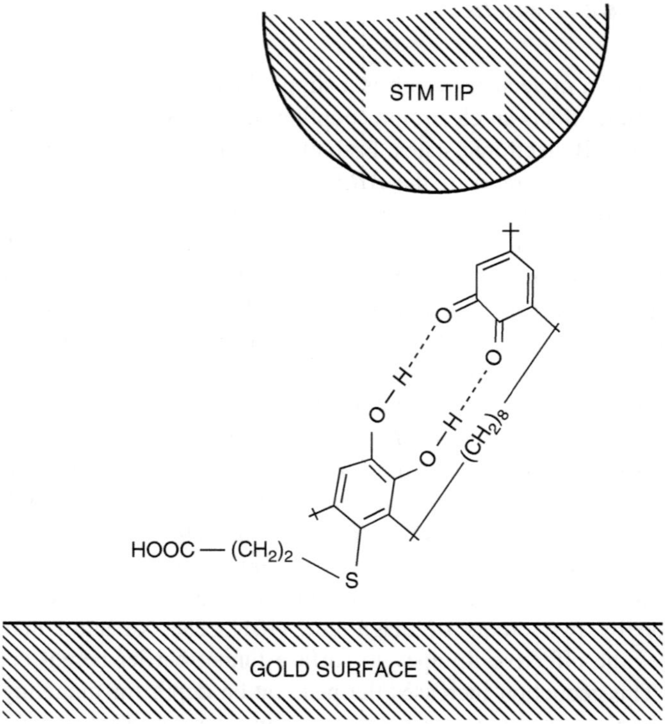

Fig. 6 An intrinsic molecular switch positioned in an STM tunnel junction. Here, the mechanism of intramolecular switching is independent of the surface. It is the intramolecular transfer of two protons which is supposed to shift the molecular orbitals of molecule since one phenyl is bonded to a sulfur atom. The protons transfer is triggered by the tip to surface electric field

molecule alone. After this legged Cu–porphyrin molecule on a Cu(211) surface example, many molecular switches have been proposed and a few experimentally studied, where again the nano-pad surface is determining the switching machinery: a biphenyl molecule on Si(100) [105] or a diazobenzene molecule on a metal surface [106]. All those switches belong to Fig. 5a type. The dependence of the switching ability of the molecule on its interaction with the electrode surface was truly demonstrated in the diazobenzene case where no switching occurs on Au(100) or Cu(111) surfaces as compared to an Au(111) surface [107]. To keep the origin of the switching mechanism in the molecule free from any surface influence, one can use, for example, an intramolecular proton transfer process. The molecule must be well designed for the proton not to escape from the molecule to the surface. Some belong to Fig. 1d interference type of switches [107] and others are of Fig. 1b type. They are based on a change of the ***B*** intramolecular electronic structure upon intramolecular proton transfer without changing (in first approximation) the ***B*** molecule electronic coupling to the contacting nano-pads. This is the case for the early intramolecular proton transfer switch, whose gating was also applied by the tip apex of an STM [20]. In this hemiquinone molecule (see Fig. 7), two protons

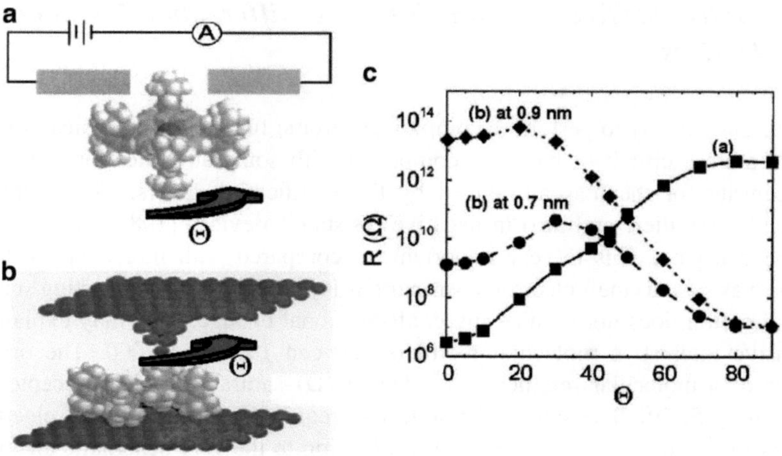

Fig. 7 A molecular switch made of the leg of a legged Cu–porphyrin adsorbed on a Cu(211) surface. (**a**) An idealized version of such a molecular switch where the switching leg is exactly interconnected to two atomic wires in a Fig. 1a like configuration. The device resistance is maximum for a perpendicular = 0 conformation. (**b**) The real experimental device where the STM tip apex can be maintained at 0.7 nm or 0.9 nm separation between the tip and the surface in between the leg switching. In this case, the resistance is minimum for perpendicular to the Cu(211) surface = 0 conformation. The energetic of the switching mechanism can be calculated and the switching barrier height was also determined experimentally

were supposed to be transferable up and down the molecule while an electric field is applied on the ***ABA*** junction. Such a proton transfer shifts the molecular orbitals of the molecule in a Fig. 5b type switching effect, resulting in a resonance appearing in the *I–V* characteristics of the ***ABA*** junction when the protons have been transferred.

There are many other ways to gate Fig. 5a or Fig. 5b type switches, like light induced conformation change or the application of an electric field on the junction. The previous diazobenzene molecular switches can be controlled by light or by an inelastic tunneling current. For a Fig. 5b switch, the objective upon the application of an electric field on an ***ABA*** junction is to shift largely in energy some of the ***B*** conjugated molecule electronic states [107]. The experimental problem with this approach is that an extremely large electric field is required to displace in energy the electronic states of a conjugated molecule for the result to be observable on the junction conductance. Since the electrical breakdown of an ***ABA*** tunnel junction is around a few volts per nanometer, the junction will break before any large "ON–OFF" conductance switching ratio is observed. With the molecular switches described above, the open question is how to design a miniature complex circuit by assembling a lot of them together. We are not speaking here about the available technology to do so. We are speaking about adapting the design of those switches to a planar atomic scale technology. For example, almost all the molecular switches proposed so far will not work on an insulating surface in a planar technology that lies in between well defined atomic scale nano-pads.

3.2 Molecular Rectifiers and Negative Differential Resistance Devices

A well-known way to perform a complex electronic function like Boolean logic is to assemble a circuit of rectifiers completed with some amplification devices to compensate for the loss introduced by the rectification layers. The interest in molecular rectifiers and also in negative resistance devices [108] is that they do not need a gate. This is very important, as compared with molecular switches, because as soon as their electronic structure is isolated from the supporting surface, their function does not require any conformational change. This may explain the extensive interest in molecular rectifiers between 1990 and 2000. The original design of a molecular rectifier was a Donor (D)–saturated bridge–Acceptor (A) molecule [15, 59]. The saturated bridge was required for the D and A electronic states of the molecule to be well-defined, leading to the very schematic electronic structure of an ***ABA*** molecular rectifier device (Fig. 5c). In this design [59], the alkyl bridge was very long, and the resulting current intensity through the molecule falls in the attoampere range. This was too low to use in a complex logic circuit. The same occurs with the molecular rectifiers calculated by the Mitre Corporation [27], whose output current is of the order of 100 fA. To get a larger output current, the bridge was reduced to its minimum length and the D and A groups acquired a very small lateral size. Very simple amino-phenyl (D) and nitro-phenyl (A) fragments were tested as the elementary rectifier chemical groups [109]. As presented in Fig. 8 and connected between two metallic nano-electrodes, such a simple molecular rectifier showed a very high monoelectronic tunneling transmission spectrum *T*(*E*). Even with a very short CH_2 bridge, the HOMO and LUMO electronic resonances of the D and A parts of the molecule can be very well identified in the

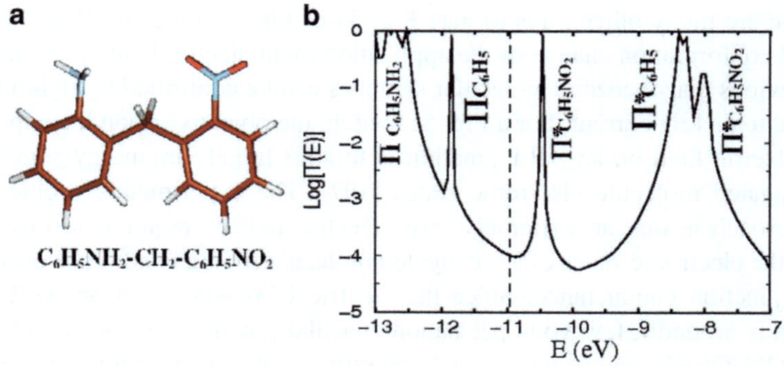

Fig. 8 (**a**) The chemical structure of a very simple D–A like molecule where the bridge had been restricted to a simple CH_2 chemical group for increasing the molecular rectifier running current intensity. (**b**) the calculated *T*(*E*) electronic transmission through this molecule while connected in a Au nano-pad tunnel junction. The frontier D–A, $\Pi C_6H_5NH_2$ and $\Pi^*C_6H_5NO_2$ molecular orbitals are identified. Note that in an elastic regime of transport through this molecule, the *I*–*V* asymmetry will come from the large width of the $\Pi C_6H_5NH_2$ electronic resonance, as compared to the very sharp $\Pi^*C_6H_5NO_2$ one. The width of the intercalated phenyl resonance is so small that it does not count in the *I*–*V* characteristic

$T(E)$ spectrum. When the Fermi level E_F is located between the D-HOMO and the A-LUMO resonances, a large rectification effect is observed where $T(EF)$ reaches almost 10^4. At a low 100 mV bias voltage and in a forward polarity, the tunnel current intensity reached around 1 nA. The $T(E)$ spectrum of Fig. 2b was calculated using the ESQC technique associated with a semiempirical description of the tunnel junction [110]. The full valence MO structure of the junction is taken into account in the calculation.

NDR was observed for certain slightly asymmetric molecules bonded to gold electrodes in restricted temperature ranges (around 60 K) [108]. This molecular analog to an Esaki tunnel diode would in principle be very useful, since in a device using it, the negative resistance from the "Au/molecule/Au" system, if put in series with an equal but opposite positive load resistance, would provide "infinite" power gain (division by zero). Alas, the effort to commercialize this NDR failed because the device was unreliable. Inspired by the Aviram–Ratner proposal [15], Metzger and coworkers synthesized and measured 11 different molecular rectifiers, mostly in monolayers (Al/molecule/Al, or Au/molecule/Au), with rectification ratios between 5 and 80 [111]. Other laboratories have studied about ten other rectifiers, mainly as single molecules, by STM [111]. However, at present the interest for molecular rectifiers has abated somewhat, probably because of their very low running current. This leads to extremely low conductance molecular rectifier devices, much smaller than G_0. Indeed, the Landauer–Büttiker formula predicts that the overall conductance for an "electrode/single-strand/electrode" (***ABA***) system cannot exceed G_0. Within that single strand (string of metal atoms, or single molecule) the conductance can be quite high, or even infinite for a perfect conductor, but the overall one-strand system conductance of *ABA* cannot exceed G_0. Together with an ***ABA*** stray capacitance of a few attofarads, the running frequency of one molecular rectifier will be lower than 100 MHz and the running frequency of a Boolean logic function lower than 10 MHz.

3.3 *Molecular Interferometer-Like Devices*

As presented in Sect. 2 Fig. 6, destructive electronic interferences in an ***ABA*** molecular junction can lead to a change in the conductance of the ***ABA*** junction as a function of an external perturbation by a few orders of magnitude, and hence to a large switching effect. This corresponds to a Fig. 5d type molecular device, where interference conditions occur when the effective coupling through the central two molecular electronic states is exactly zero at some energy in a few electron volts range around the Fermi levels of the nano-pad materials.

Note that through-molecule electronic interferences do not necessarily require a loop-like spatial topology of the molecule to be active. Molecular electronic states of different symmetry relative to the contacting nano-pads can do the job [112]. Through-molecule interferences can be controlled by changing the molecular conformation [113, 114], by applying a lateral electric field to the ***ABA*** junction [115], or by controlling the chemical composition of a pending chemical group bonded along the tunnel path [116]. Often, the gating effect used to reach an interference effect leads to a fantastic manipulation of the molecular states imaged by the STM [117] but with almost no difference in conductance between the ON and OFF states. But there is more required than a switch with a Fig. 1d-like ***ABA*** interference device. It was demonstrated that, for some molecule in the ***ABA*** junction like a C_{60} molecule, going out of a destructive interference regime is so easy with the external gating parameter that a small variation of this parameter leads to a very large variation of the ***ABA*** junction conductance [118]. This is the exact definition of gain. Among all four types of molecular devices presented in Fig. 5, only the interference in Fig. 5d can deliver a large gain. Of course (and as already discussed in the introduction of this section), one must have more than two active molecular electronic states in the ***ABA*** junction for the gain to be large and to get a molecular transistor. Starting from the simple schematic diagram of a discrete three-terminal device, a few molecular scale transistor-like devices have been proposed in the past [119]. The design of most of them is based on the topological resemblance between a well-designed molecule and a field-effect transistor (FET). But no gain was calculated for those molecules. In the case of the C_{60} electromechanical amplifier [120], the gate effect is produced by intramolecular interference effects controlling the extension through the molecule of the penetration depth of the tunneling electron coming from the electrodes of the nanojunction. With its fivefold HOMO and a threefold LUMO, a C_{60} molecule was a good candidate for a Fig. 5d-type molecular device benefiting from a large number of molecular electronic states in interaction with its nano-pads. With a working current intensity of 7.48 nA, the measured transconductance of the C_{60} amplifier is 3.974 S [120]. This trans-conductance cumulates the effect of reducing the tip-apex-to-surface distance and closing the C_{60} rapid gap, due to the molecular level repulsion effect, while pressing with the tip apex on the C_{60} molecular cage. A gain of 5 was observed experimentally while gently compressing the cage [120]. Calculations have indicated that a larger gain can be reached in an optimized ***ABA*** junction configuration [121]. This demonstrates that a single molecule can bring a gain in a circuit by itself and opens the quest for new high-gain molecules. This C_{60} amplifier (Fig. 9) can be made planar with a real separation between the source, the drain, and the electromechanical grid. In this planar design, a gain as high as 40 is expected [100]. It is a true molecular transistor with its source-drain channel resistance controlled by a third independent electrode. This grid can be qualified as "extrinsic" because its driving signal is a voltage applied on a small piezo-cantilever and not a current intensity. The association of such C_{60} transistors in series and in parallel to design logic gates was also simulated [100]. The results are disappointing because of the very high input impedance of those transistors. This high impedance problem was already underlined by Landauer, who noted that smaller devices often mean larger interconnection impedances [99]. A solution to this problem is to remain at the molecular scale by integrating a large number of circuits and devices in the same molecule, as discussed in the next section (Fig. 9).

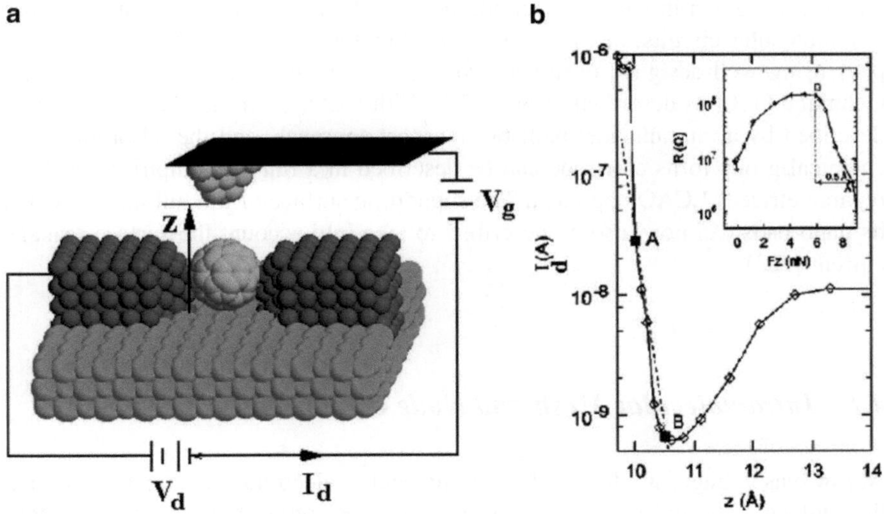

Fig. 9 (**a**) The atomic scale structure of a proposed planar C_{60} transistor where the source and drain Au nano-pads are supposed to be constructed on a surface with a large electronic gap. The gating effect is created by the tip apex of a miniaturized cantilever controlled by a miniature piezo electric crystal. The V_g gate voltage is controlling the z distance of the tip apex to the surface. (**b**) The variation of I_d source-drain current intensity as a function of z. The optimum working range of this electromechanical transistor is between the B high resistance and the A small resistance z values where the $I_d(z)$ variation is almost exponential. The resistance vs force curve is presented in the *inset*. It serves to evaluate the required switching energy from B to A. The current intensity was calculated for a 0.1 V source-drain bias voltage

4 Classical Monomolecular Devices

To push the electronic circuit miniaturization down to the nanoscale, Carter proposed in 1984 to integrate the circuitry and the molecular devices inside a very large single molecule [26]. This solution was also supposed to solve the difficult "tyranny of numbers" problem by synthesizing the full circuit at once, instead of assembling it molecule by molecule on a lattice of metallic nano-junctions. Some circuits embedded inside a single molecule have been proposed, culminating by the Mitre Corp. proposal of a single-molecule binary adder [27]. These intramolecular circuits were designed assuming the validity of the Kirchoff laws for the bonding in series and in parallel of molecular groups to form the intramolecular circuit. Unfortunately, the mesh Kirchoff circuit law does not apply inside a molecule, in either the ballistic or the tunnel-electron transport regime [60]. Therefore, the first step is to establish those laws. Then large molecule-circuits can be designed. In this section we have only selected a few simple monomolecular circuits, starting from the molecular OR gate [110] (with and without including rectifiers) and a molecular XOR gate [109] and then proposing the balancing of a standard (but now intramolecular) single-molecule Wheatstone bridge [122]. Aside from the intramolecular circuit rules given in the next section, the logic function fulfilled by any of those intramolecular circuits, i.e., a single molecule interconnected to N metallic nano-pads using well-designed chemical groups, can be calculated using the NESQC technique [110] as described in Sect. 2.3. With NESQC all the nano-pads can be described by a multichannel ballistic transport approach, and the adsorption of a chemical group to its electrode can be described in a full semiempirical (or DFT reparametrized) LCAO approach. The supporting surface of the full molecule and its nano-pads can now also be described to take into account the surface leakage current [123].

4.1 Intramolecular Mesh and Node Circuit Laws

At low bias voltage, and for any Fig. 10 molecules, electrons are tunneling through the molecular orbital tunnel paths offered by the conjugated branches of those molecular bridges. To a good approximation [23], the tunnel current intensity, measured by an external macroscopic circuit connected to any combination of two nano-pads *i* and *j* of the circuit, is proportional to $|V_{ij}|^2$. The electronic coupling V_{ij} is introduced between the chosen *i* and *j* nano-pads by the molecule itself. For the same interpads distance, V_{ij} is generally much larger than the through-space

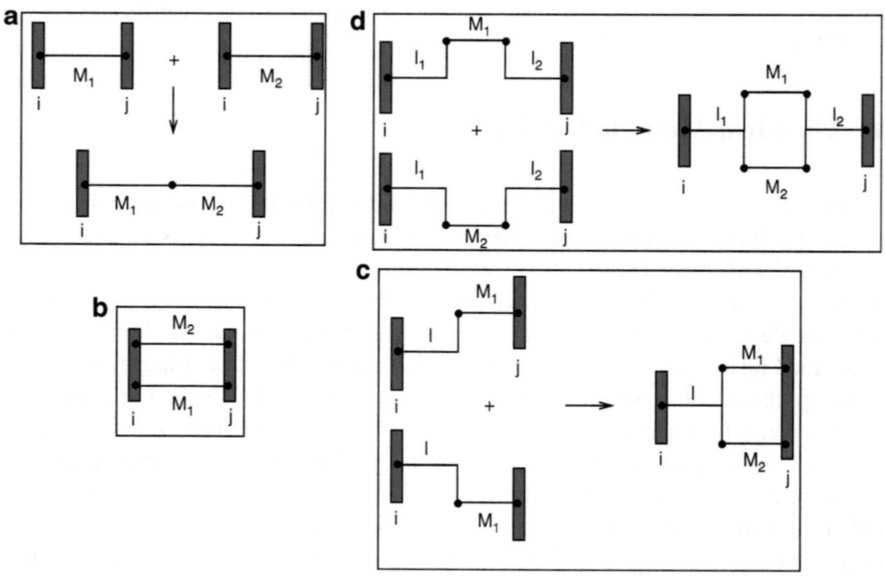

Fig. 10 Simple circuit diagrams of the different series and parallel association of molecular wires M_1 and M_2 discussed in the text. The two molecular wires are (**a**) bonded in series, (**b**) connected in parallel on the metallic pads, (**c**) forming a single molecule with one intramolecular node, and (**d**) forming a single molecule with two intramolecular nodes

coupling between the same i and j nano-pads, but with the molecule disconnected. In some cases, a molecule can introduce specific electronic interference effects by its specific topology or chemical composition, which reduces the low voltage conductance of the tunnel junction towards its vacuum level of conductance or below. As recalled in Sect. 2, V_{ij} is usually calculated using the effective Hamiltonian technique [124]. Let us choose two quantum states $|i\rangle$ and $|j\rangle$ of nano-pads i and j with an energy very close to the Fermi level of the pads. In this case, $V_{ij} = \langle i|P(U-1)HUP|j\rangle$ where P is the projector on the model subspace generated by the $|i\rangle$ and $|j\rangle$ states.

For a given electronic energy, U is a nonunitary transformation in charge of selecting the two H eigenstates which are the closest to the model subspace. H is the electronic Hamiltonian of the corresponding metal–molecule–metal tunnel junction. Let us now take two molecular wires M_1 and M_2. Each is able to introduce respectively an electronic coupling V_1 and V_2 when jumping over the tunnel nano-junction. At low coupling, i.e., for an electronic transparency of each individual molecular wire much lower than unity, the electronic coupling introduced between the i and j nano-pads by the serial association of the two molecular wires M_1–M_2 (see Fig. 10a) is $V_s = kV_1 \times V_2$ with k a factor. For two molecular wires bonded in parallel (see Fig. 10b), there are two different cases. When the nodes are on the metallic nano-pads, each molecular wire brings its independent V_i contribution to the electronic coupling between pads. When, as presented in Fig. 10d, the two nodes are integrated in the molecule, far from the pads and using two other auxiliary molecular groups l_1 and l_2 connected one side each to a nano-pad, the complete electronic coupling to consider is that introduced by the new molecule l_1–M_1//M_2–l_2. There is a complete mesh in the molecule, which creates two different intramolecular tunnel paths: the l_1–M_1–l_2 path accounting for the V_1 electronic coupling, and the l_1–M_2–l_2 path accounting for the V_2 electronic coupling between the two electrodes. In this case, the resulting electronic coupling between pads i and j introduced by the complete l_1–M_1//M_2–l_2 molecule is simply $V_p = V_1 + V_2$. Starting from these three ways of superposing electronic couplings through a molecule(s) between two nano-pads, the conductance G of a metal–molecule–metal tunnel junction is easily deduced from the proportionality, at low coupling, between the current intensity and the square of the electronic coupling [23]. In series, for the M_1–M_2 molecule, it becomes $G = G_1 \times G_2$, with G_1 and G_2, respectively, the conductance of M_1 and M_2 groups measured separately (as presented in Fig. 10a). In parallel, when two independent molecules M_1 and M_2 are both connected in parallel to the two electrodes of the tunnel junction, the standard $G = G_1 + G_2$ resumes (Fig. 10b), with some corrections depending on the adsorption distance between the two molecular wires M_1 and M_2 on the surface of the electrodes [125]. In parallel, with a single l_1–M_1//M_2–l_2 molecule (where the two nodes l_1 to M_1//M_2 and M_1//M_2 to l_2 belong to the molecule as presented in Fig. 10d),

$$G = G_1 + G_2 + 2(G_1 G_2)^{1/2} \tag{4}$$

with G_1 and G_2 the conductance of the l_1–M_1–l_2 and l_1–M_2–l_2 molecule taken separately [125]. Also in a parallel configuration, with an l–M_1//M_2 molecule, where only the l to M_1//M_2 node belongs to the molecule (see Fig. 10c), the standard $G = G_1 + G_2$ resumes, again with some corrections, depending on the adsorption

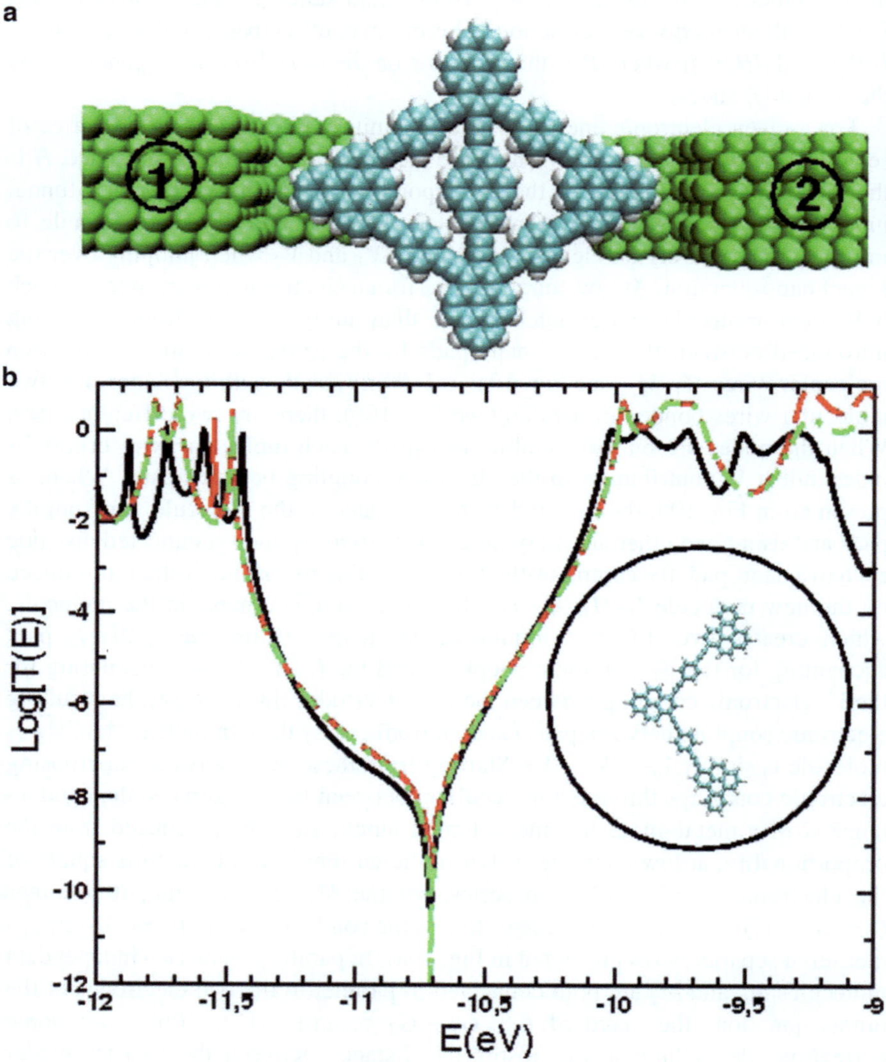

Fig. 11 The scattering properties of a five branches – four electrodes molecular bridge. (**a**) Detailed atomic structure of the molecule. A central perylene branch was included to mimic an internal measurement branch. (**b**) EHMO–ESQC calculated $T_{12}(E)$ transmission coefficient (*plain*) and predicted $T_{12}(E)$ transmission coefficient (*dashed*), applying the intramolecular circuit rules discussed for the four molecular fragments given in Fig. 12. The *dashed* (*dotted*) *line* is the $T_{12}(E)$ variation for the single molecular branch, as presented in the *inset*, to show the origin of the destructive interference

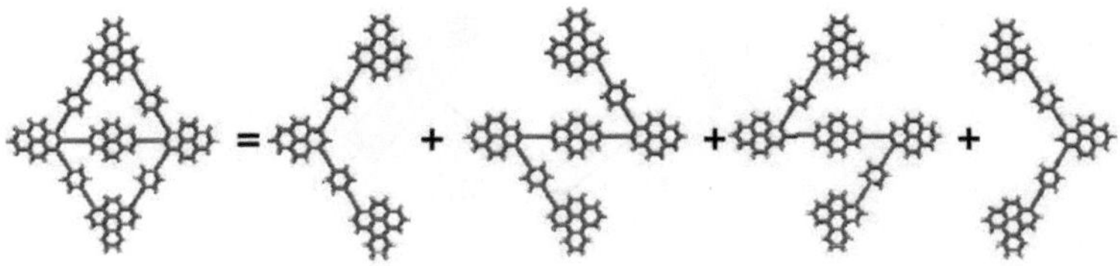

Fig. 12 The Fig. 11 intramolecular circuit can be decomposed into four tunneling paths, to apply the parallel superposition rule, and predict the transmission coefficient through Fig. 11 molecule. Molecules 1 and 2 are for the contribution of two short tunnel paths and molecules 3 and 4 for the contribution of the two longer paths through the central perylene wire

distance between the two molecular wires M_1 and M_2 on the surface of the electrodes [60]. One interesting and simple example is the intramolecular circuit presented in Fig. 11: it is made of five branches and is connected to only two electrodes. Its two-terminal behavior can be calculated using the $N = 2$ version of EHMO–NESQC [126]. As expected, there is a pronounced interference effect in the middle of the HOMO–LUMO gap of this five-branch molecule (Fig. 11b). The overall electronic conductance of this molecule, can be calculated by decomposing it in four specific molecular wires M_1, M_2, M_3, and M_4 (see Fig. 12) and by applying the circuit rules discussed above to this decomposition.

There are two left M_1 and right M_2 lateral molecular branches going from nano-pad 1 to nano-pad 2, with a conductance G_1 and G_2 with $G_1 = G_2$ for a planar conformation. There are also two zigzag left M_4 and right M_3 tunneling paths via the horizontal perylene molecular wires. They have a conductance G_4 and G_3 with $G_3 = G_4$ for a planar conformation. The conductance of each path is defined by interconnecting the corresponding molecular wire given in Fig. 14 to two metallic electrodes, using benzoperylene end-groups for the interconnection, and by calculating the transparency of each of them. According to our circuit rules, $G = G_1 \times \mathrm{G}_3$ and $G = G_2 \times G_4$, because there is one molecular group more in series in the molecular wire M_3 and M_4 than in M_1 and M_2. The application of this product superposition rule gives a fast decrease of the conductance of a molecular wire with an increase of its length. The conductance of the total molecule is given by our parallel superposition rule, specific to the case where the two nodes of the circuit are inside the molecule, that is $G = 4(G_1 + G_3) + 8(G_1 G_3)^{1/2}$ because $G_1 = G_2$ and $G_3 = G_4$. In Fig. 11b, the energy-dependent transmission coefficient $T_{12}(E)$ deduced from this rule is compared to that obtained by a complete molecule transparency calculation using the EHMO–ESQC technique. There is almost no difference in the HOMO–LUMO gap of the molecule and the destructive interference is still there, confirming that they are introduced by the M_1 and M_2 branches independently. As discussed in the next section, those rules are applied to the balancing rules of a single-molecule Wheatstone bridge. The next step in intramolecular circuit theory is to connect a single molecule to three nano-pads with a single node in the circuit, where the three molecular branches M_1, M_2, and M_3 are converging (see Fig. 13). The simple case corresponds to $M_1 = M_2 = M_3$ with

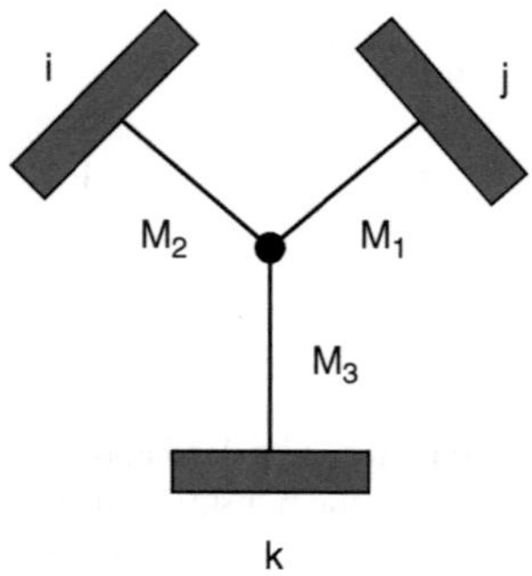

Fig. 13 The simple circuit diagram of a central molecular node connected to three nano-pads *i*, *j* and *k*. The three M_1, M_2 and M_3 branches form a single molecule with the central node

a symmetrical central node. In this case charge conservation occurs, resulting in an identical conductance for all the two-terminal molecular tunnel junctions that can be defined in Fig. 13 three-terminal circuit [60]. Of course, each elementary conductance depends on the chemical composition of the third branch.

A very good example is the conductance of a dianthra[*a*,*c*]naphtacene starphene-like molecule presented in Fig. 20, interacting with three metallic nano-pads. The EHMO–NESQC *T*(*E*) transmission spectrum per tunnel junction looks like a standard conjugated molecule *T*(*E*) with well-identified molecular orbitals and their resonances. For the Fig. 20 case all the *T*(*E*) are the same. One can note a small deviation after the LUMO resonance, due to a little asymmetry in the adsorption site between the three branches on the nano-pads [127]. A lot of asymmetric star-like three-molecular-branches system can be constructed, in particular in reference to chemical composition of the central node. This had been analyzed in detail [60]. But in this case, each molecule becomes a peculiar case. The next section presents one application of this central-node case to construct molecule OR and molecule XOR logic gates.

4.2 *Intramolecular Logic Circuits with Only One Node*

The simplest way to design a logic gate (or any electronic circuit) in a molecule is to map the chemical structure of the molecule with the corresponding electrical circuit diagram. It was first practiced by Carter [26] and then by the Mitre Corporation researchers [27], but without any simulation of the electrical conductance, either of their chemical drawing or with respect to the new circuit rules discussed above. Using the above intramolecular circuit laws and the EHMO–NESQC technique to trouble-check in detail the multiple *T*(*E*) tunneling paths through the molecule as a function of its detailed chemical structure, it is now possible to go a step further, and propose simple logic gates based on the same architecture as the diode logic of

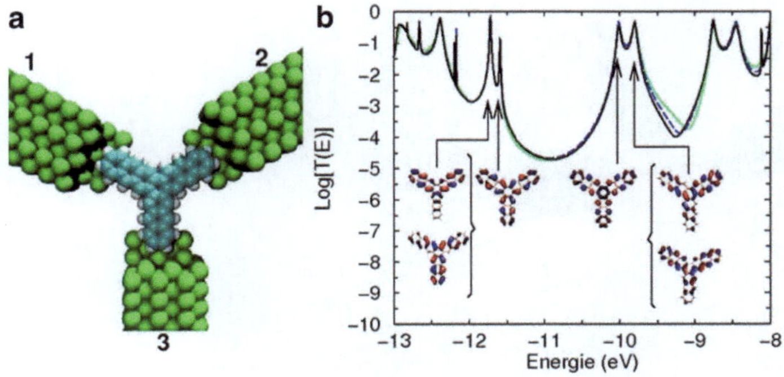

Fig. 14 The three-branches dianthra[*a,c*]naphtacene molecule circuit of symmetry formed by three anthracene fragments equivalently bonded to a central phenyl group. The molecule is adsorbed by the three branch ending phenyls onto the Au nano-pads. A semilogarithmic plot of the $T_{ij}(E)$ EHMO–NESQC electron transmission spectra (in valence energy range) per pair of branches. The presented frontier MOs show how the valence π electrons are delocalized on the molecule. At resonance, this provides a good electronic conductance through each pair of molecular branches, almost one quantum of conductance

the 1950s. Presented below are simple examples of a large variety of chemical structures, which can be calculated to trouble-check the Boolean truth tables they should obey. No supporting surface was added, but can be added now in the new NESQC version, as noted above, to provide a realistic description of the molecule circuit, its interconnection nano-pads, and the electronic structure of its supporting surface. Let us start with a simple OR gate. As recalled in Fig. 14 and in simple diode logic, an OR is simply two rectifiers connected to a common node.

From this node, the output current intensity must respect an OR Boolean truth table. The rectifiers are there to ensure no back current when one of the two input voltages is set to zero. Since the Kirchoff current superposition rule holds for a molecular node, as discussed above (see Figs. 11 and 12), two Aviram–Ratner rectifier groups can be chemically bonded in series with a central phenyl group to form the two input branches of an OR gate (see Fig. 15), mapping Fig. 14 diode logic circuit diagram. The central phenyl is the intramolecular node of the molecule-OR circuit. The output current is collected directly on this node by bonding a short acetylinic chain to the phenyl node. The complete atomic structure of the full circuit, including the central OR-molecule and its interconnection to the metallic electrodes, is presented in Fig. 15. The output molecular wire group is connected to the current collecting electrode A via a thiol bond, while the acceptor parts of each rectifier group are bonded in a planar conformation to the input metallic electrodes B and C. From the A, B, and C nano-pads, the full scattering matrix of this three-branch molecule was calculated, using the EHMO–NESQC technique. All the valence molecular orbitals of the molecule-OR have been taken into account in the calculation. This molecule-OR is characterized by three transmission coefficients: $T_{AB}(E)$ and $T_{AC}(E)$ for the electronic transmission through the

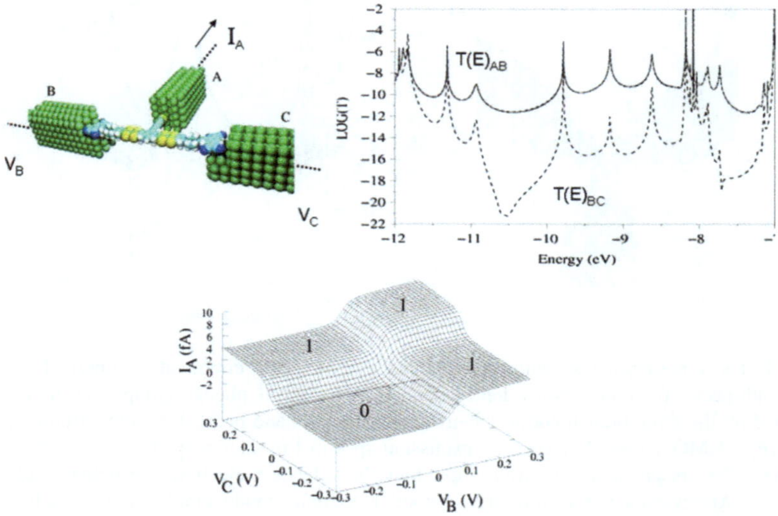

Fig. 15 A molecular OR gate, whose chemical structure maps the electrical circuit diagram shown in Fig. 20a. Two Aviram–Ratner molecular rectifier chemical groups have been bonded to a central chemical node. This intramolecular circuit with one simple node can be easily designed, because the node Kirchoff node law is valid here. Note that the molecular orbital of each partner can be still identified on the 2 $T(E)$ because of their weak interactions through the CH_2 bridge. This is not always the case. The obtained logic surface demonstrates an OR function for well-selected values of the input voltage, but with two logical level "1" outputs which would have to be corrected using an additional output circuit

molecule from one input electrode to the output electrode A and the leakage transmission $T_{BC}(E)$ between the inputs B and C. $T_{AB}(E)$, $T_{AC}(E)$, and $T_{BC}(E)$ are presented in Fig. 15. All the molecular orbitals of the OR molecular logical gate can be identified on those tunneling spectra, like the π and π^* doublets coming from the two rectifiers in $T_{AB}(E)$ and $T_{AC}(E)$. From B to C, through the series of the two rectifiers, the leakage current is ten orders of magnitude lower than the direct current, showing the good insulation of our design. From the B and C electrodes, the central phenyl is a bad tunnel splitter, compared for example to a three-ways splitter [60]. In the HOMO–LUMO gap, the electronic transparency of the molecule-OR is $T_{AB}(E_f) = 10^{-12}$. This small value is due to the methyl CH_2 groups introduced between the phenyl node and each rectifier group. They are needed to preserve the electronic integrity of the donor part of each molecular rectifier. They are at the origin of a large electronic reflection coefficient on the molecule-OR at the B and C input electrodes. This molecular OR, designed using Aviram–Ratner molecular rectifiers, delivers a current intensity in the 10 fA range. Its detailed logic surface is presented in Fig. 15. Aside from the very low output current, this molecule gate delivers two output logic levels "1," which would need to be compensated by an external circuit. Figure 15 molecule logic gate running current is too low to build up realistic complex logic gates. The same occurs for example using the molecular rectifiers proposed by the Mitre Corporation, whose output current is of the order of 100 fA, when calculated using the NESQC technique. In both cases, this low output intensity comes from the long bridge separating the electronic structure of the donor (D) and acceptor (A) chemical groups of those molecular rectifiers. To get larger output currents, the bridge has to be reduced to its minimum length and the D and A groups must be of a very small lateral extension. In Fig. 16, the chemical structure of a very simple D–σ–A molecular structure is described, where for a 100 mV bias voltage forward polarity, the tunnel current intensity can reach around 1 nA [109]. Following the Fig. 15 molecular circuit structure, two such molecular rectifiers were bonded via a pyrrolyl group to a central pyrenyl node, to get a molecule OR logic gate with better output performance (see Fig. 16). Its logic surface is presented in Fig. 16. For a planar pyrene conformation, a tunneling current intensity as high as 50 nA is expected for the (1,1) input status using a 400 mV input voltage to encode "1" on inputs 2 and 3.

For the same logical output "1," the expected different values of the output current intensity is recovered, because there is no voltage drop along the molecule. The tunneling current of the 2–1 and 3–1 branches simply adds up through the pyrene part of the OR-molecule. A molecule AND can also be designed, mapping again a standard AND rectifier logic circuit on the chemical structure of the molecule [110]. As presented in Fig. 17, a molecule-AND is essentially structured around a central node, where four molecular branches are chemically bonded instead of three for an OR gate. As in Fig. 15, standard Aviram and Ratner rectifier chemical groups were still used in Fig. 17 design, leading to a very low running current intensity for this gate. As presented in Fig. 17, its logic surface is very close to the truth table of an AND gate.

Another very good example of the mapping procedure, which can be practiced to design a semiclassical intramolecular logic gate molecular circuit, is the nontrivial

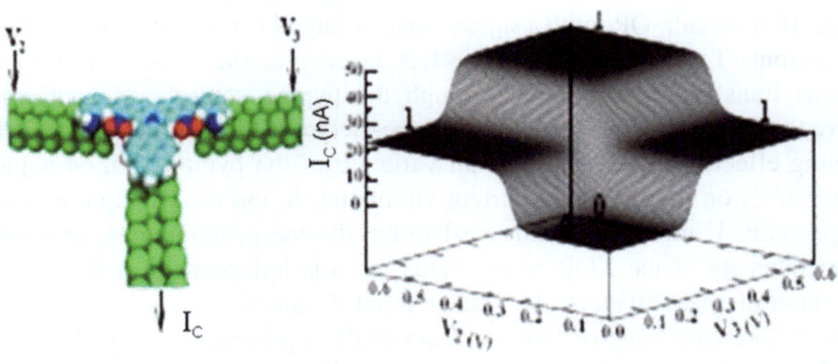

Fig. 16 An optimized molecular OR, built up using two molecular rectifiers bound to a central pyrrolyl, whose output current is collected via a pyrene wire. The chemical structure of a given molecular rectifier chemical group is made of an amino-phenyl donor (D) and nitro-phenyl acceptor (A) bound together via a single CH_2 fragment. The corresponding surface logic is calculated with a supposed planar conformation of the molecule. Since there is no voltage drop along the molecule, there are two output current intensities for the same "1" logic output. Note the large (1,1) 50 nA output current, as compared with Fig. 15 molecule OR

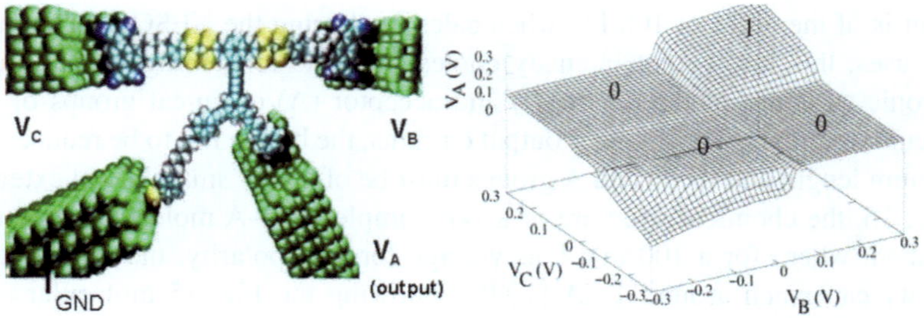

Fig. 17 A molecular AND gate, whose chemical structure maps the standard AND diode logic circuit diagram (not shown). The molecule is also made of two Aviram–Ratner molecular rectifier chemical groups bonded to a central chemical node. A fourth molecular branch had been added for the reference ground (GND) following the classical circuit design. The obtained logic surface demonstrates a good AND function for well-selected values of the input voltage but with two logical level "1" outputs. Note that this gate output must be measured in voltage and not in current

case of an XOR gate. In one circuit, an XOR gate combines an OR function, followed by an AND, which forces the output to zero when the two inputs have a "1" logical status. In standard electronics, there are many ways of designing such more complex circuits, like combining OR, NAND, and AND gates in a logical circuit, or adding a zener diode on the output branch of an OR gate, to force the current intensity to zero when both inputs are "1." As an example, we have chosen to map in a molecule Fig. 18 electrical circuit, which is also an OR gate whose output current intensity is detected by a relay which switches the output current of the full gate whenever the output current is too high in intensity, that is when both logical inputs are in a "1" status.

The chemical structure and its interconnection configuration to the nano-pads of this molecule-XOR is presented in Fig. 19. Following Fig. 18 design principle, it is a Fig. 16 molecule-OR, with a supplementary lateral nano-electrode to access the XOR output. To get the nonlinear effect leading to the XOR, the number of electrons transferred per second through the pyrene controls its conformation relative to the planar axis of the (1, 2, 3) three-electrode tunnel junction via inelastic tunneling effects [110]. Then, a small variation of the pyrene rotation angle has a large effect on the tunneling current circulating in the (1–4) output mesh. The voltage source V introduced in this mesh brings the energy required to set up the I14 current intensity of the XOR output. This mesh is independent of the (3–1) and (2–1) input meshes driven by the voltage input V_2 and V_3.

The full surface logic of this molecule-XOR is presented in Fig. 19. The "0" logical output corresponds to an I_{14} current, stabilized around 100 pA for $V = 100$ mV. The "1" logical plateau is large enough to stabilize an I_{14} current intensity around 220 pA for $V = 100$ mV. The difference between "0" and "1" is large enough to be detectable, even if (as presented in Fig. 19 inset), our design deforms the ideal XOR response logic, therefore reducing the immunity of the molecule-XOR gate to input voltage noise, especially for the "0" inputs).

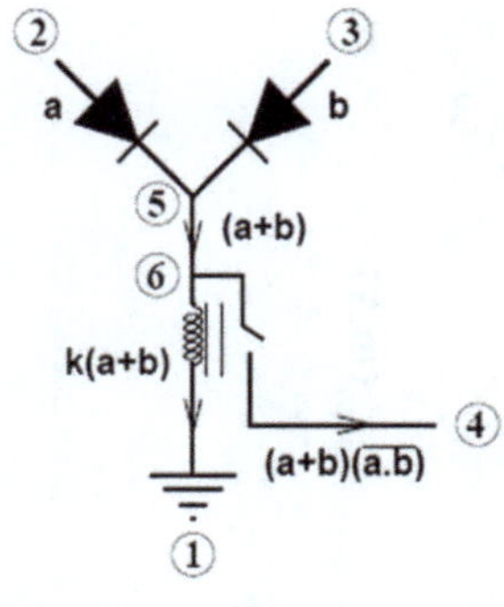

Fig. 18 For reference, the classical electrical circuit diagram of an XOR gate in diode logic with its two top rectifiers and the relay to detect the logical complete (1,1) input configuration. For a (1,1) input configuration the output current is forced to zero by the relay

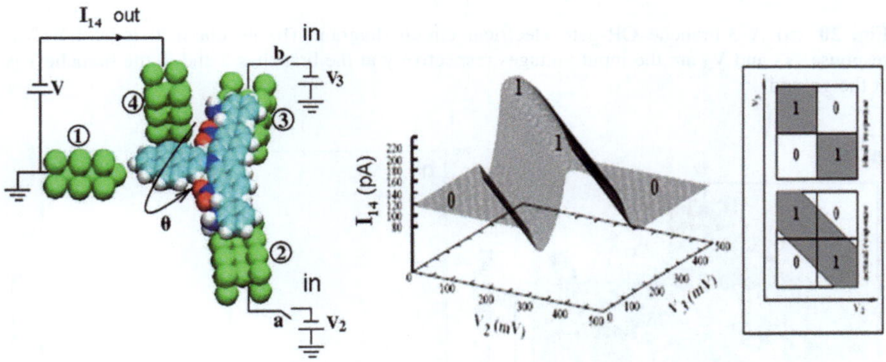

Fig. 19 Set-up and surface logic of a molecular XOR logic gate embedded in a single molecule. Nano-pad 1 is shifted down in space, to optimize the central I_c current and nano-pad 4 is laterally shifted on the central pyrene, to optimize both the electronic and inelastic mechanical response of the molecule. With its bias voltage V, the (1–4) output mesh is independent of the logical inputs. The logic surface was calculated with $V = 100$ mV in the 1–4 output mesh. The "a" and "b" logical input status are encoded in V_2 and V_3 with $V_i < 300$ mV for a "0" logical input and $V_i > 400$ mV for a "1" logical input. *Inset*: the logic surface top view of an ideal XOR gate and of that obtained with the present molecule-XOR intramolecular circuit

4.3 Balancing a Four Branches Monomolecular Wheatstone Bridge

One of the first applications of the new mesh and node intramolecular circuit rules discussed above is the well-known problem in electrical circuit theory of the balancing of a Wheatstone bridge. In Fig. 21, a molecular Wheatstone bridge is presented, made of loop-like 4 tolane molecular wires bonded via benzopyrene end-groups for nano-pads 1 and 3, and via pyrene end-groups for nano-pads 2 and 4. This four-electrode and four-branch molecule is connected to a battery and an ammeter.

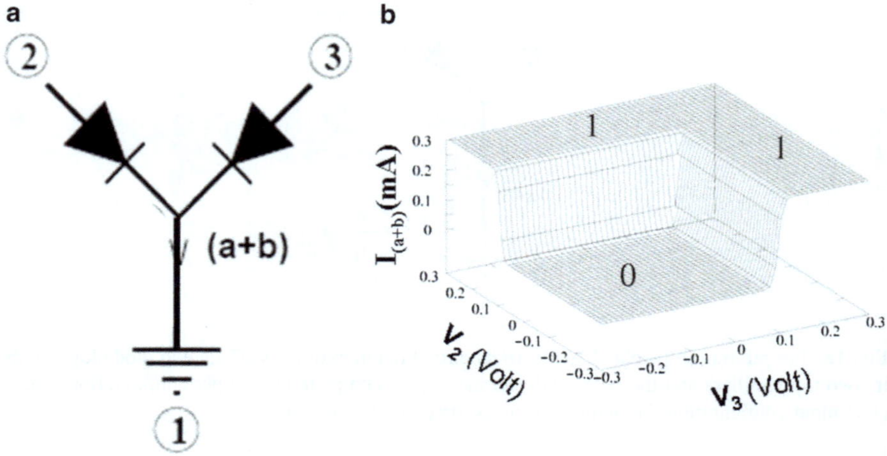

Fig. 20 (**a**) A 3-branche OR-gate electrical circuit diagram, (**b**) its classical logic $I(V_2,V_3)$ response. V_2 and V_3 are the input voltages respectively at the branches 2 and 3, the branche 1 is at the ground

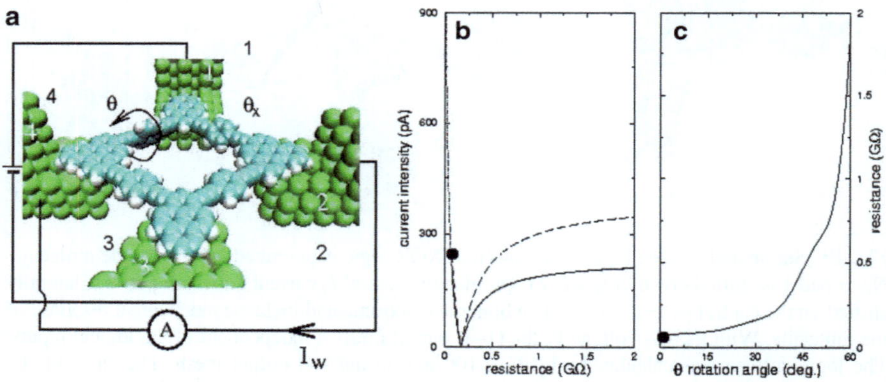

Fig. 21 The variation of the balancing tunneling current of the four branches four electrodes monomolecular Wheatstone bridge connected as presented in (**a**). In (**b**), the *dashed line* is for the current intensity I_w (in absolute value) measured by the ammeter A and deduced from the standard Kirchoff laws calculating each molecular wire tunneling junction resistance of the bridge one after the other from the EHMO–ESQC technique. In (**b**), the *full line* is the same tunnel current intensity but obtained with the new intramolecular circuit rules discussed in Sect. 2. (**c**) The resistance of the branch used to balance the bridge as a function of its rotation angle. The minimum accessible resistance by rotation is 78 MΩ for the short tolane molecular wire used here

One tolane of the molecular bridge is rotated by an angle $\theta_x = 30°$. It will play the role of the unknown resistance to be determined by a good balancing of the bridge. The tolane rotation corresponds to a resistance of $R_x = 175$ MΩ [122]. The molecular Wheatstone bridge is balanced when the tunneling current intensity I_W measured by the ammeter A in Fig. 21 is zero. Keeping this θ_x value, the balancing is obtained by rotating another tolane of the bridge by an angle. According to the standard Wheatstone mesh and node laws, the tuning would be reached for [128]

$$R_\theta R_{23} = R_x R_{34}. \tag{5}$$

Note that all the resistances are defined here by the resistance value of the corresponding junctions: the (1–2) molecular wire tunnel junction for R_x, (2–3) for R_{23}, (3–4) for R_{34}, and (4–1) for R_θ. Since the minimum value of R_{23} and R_{34} is 78 MΩ by keeping their corresponding tolane conformation angles to zero, the variations of I_w as a function of R_θ are readily calculated using the standard expression of the central branch current intensity of a Wheatstone bridge with no resistance in this branch [128]. The result is presented in Fig. 21, and the balancing of the bridge would be obtained when $R_\theta = R_x$. The basic principles behind (5) and behind this standard calculation for I_w is that, for example, two resistances in series add. But in the tunneling regime, this is not the case, because two resistances in series multiply, as discussed above. Therefore, since I_w is measured outside the molecular bridge by the classical ammeter A, I_w is simply the superposition of two tunnel currents I_{wl} and I_{wr} with $I_w = I_{wl} - I_{wr}$. The current intensity I_{wl} results from electrons tunneling through the tuning tolane molecular branch of resistance R_θ, then flowing via electrode 4 towards the ammeter A and then tunneling again through the (2–3) molecular tunneling junction. The molecular wire current from electrode 1 to electrode 3 is much too low, compared to that of electrode 1 to electrode 4. Therefore, $I_{wl} = V/(R_\theta + R_{23})$, since the molecular junctions (1–4) and (2–3) are connected in series via a metallic wire under a Boltzmann regime of transport. Following the same argument, $I_{wr} = V/(R_x + R_{34})$. Therefore, the balancing condition of the bridge becomes

$$R_\theta + R_{23} = R_x + R_{34}. \tag{6}$$

The molecular bridge is still balanced for $R_\theta = Rx$. But, as presented in Fig. 21, the variations of I_w as a function of R_θ are different with our new circuit rules, as compared to that given by the Kirchoff laws. Furthermore, the conformational variations used to balance the bridge do not permit to explore values of R_θ below the planar conformation. This restricts the exploration of the possible I_w intensity as presented in Fig. 21.

5 Quantum Monomolecular Devices

The monomolecular approach presented in the previous section provides very interesting devices that go way beyond the minimum size possible for solid-state circuits [129]. On top of this technological aspect, it proposes to use the unique characteristics of molecular electronic conduction to our advantage. However it suffers from an important drawback: the exponential decay of the tunneling current intensity with the size of the molecule [130].

To implement a complex logic function, a large molecule is required, and only a very small current can go through this large molecular device. The monomolecular approach is based on a powerful but classical idea: stacking simple elementary blocs together, like molecular diodes, to construct an electronic circuit. This classical point of view is the reason why a complex molecule is required to realize a complex logic function. When dealing with quantum systems, this powerful stacking approach seems not to be the best one. The unique resources offered by a quantum system can be used to implement complex functions in a very small system. In order to do this, the classical rules used to construct electronic circuits have to be replaced by new ones adapted to the quantum world. In Sect. 3.2 we have seen how a conformational change of the molecule perturbs the transmission of the junction and how to design a molecular switch accordingly. Enlarging this idea, any modification on the Hamiltonian of the molecule, denoted H_{m}, rearranges its molecular orbitals and consequently modifies its electronic conduction. Therefore a specific controlled modification on H_{m} can be used to encode one bit of information carried by one logical input. This modification of H_{m} can be induced by a conformational change, the displacement of a surface atom in the vicinity of the molecule, etc. If one can modify H_{m} at two different points, then two logical inputs can be encoded and so on. Designing correctly the molecule, the variations of its electronic transmission, induced by a change of the logical inputs, can respect a given truth table and lead to the implementation of a Boolean function. This simple idea is the basis of the so-called Quantum Hamiltonian Computing (QHC) approach discussed in this section.

5.1 Design of QHC circuits

The design of transistor-based electronic circuits was revolutionized by the symbolic analysis developed by Shannon [4]. This seminal work provides general rules to connect switches together to design an electronic circuit that performs the desired Boolean function. A similar approach is highly sought after to design molecular circuits following the QHC approach. However, the cumbersome expression of the electronic transmission, even through a simple quantum system, is not amenable for a symbolic analysis. We have seen in Sect. 2.1 the relationship linking the oscillation frequency, Ω, between the scattering states of the electrodes and the transmission coefficient. The rather simple analytical expression of Ω is the starting point of a symbolic analysis of QHC circuits. Its complete demonstration can be found in [131], and only the general idea is presented here. Since the scattering states are weakly connected to the molecule, the Löwdin partitioning shows that the oscillation frequency can be accurately approximated by a series of Dirac functions, each one corresponding to the high oscillation frequency obtained when the scattering states are at the resonance with one eigenstate of the molecule. Then, decomposing this oscillation frequency over the different values that the logical inputs take, a pseudo-Boolean equation, where Boolean operators and Dirac distribution are associated, is obtained. In the simple case, where only two logical inputs labeled α and β control the Hamiltonian, this decomposition leads to

$$\begin{aligned}\Omega(\alpha,\beta) = \bar{\alpha}\cdot\bar{\beta} \quad \delta(\mathscr{F}_{00}) + \bar{\alpha}\cdot\beta \quad \delta(\mathscr{F}_{01}) \\ + \alpha\cdot\bar{\beta} \quad \delta(\mathscr{F}_{10}) + \alpha\cdot\beta \quad \delta(\mathscr{F}_{11}),\end{aligned}$$

where $\bar{x}$ is the logical complement of x defined by $\bar{0} = 1$ and $\bar{1} = 0$ function in argument of the Dirac distribution are defined by all the parameters of the molecule and can have rather complex expressions. This general expression is given in [131] and specific examples are given in the next section. Controlling the zeros of these functions by tuning the values of the structural parameters, one or several Boolean operators can be selected, leading to the implementation of a given logic function. For example, it has been shown that the model system represented in Fig. 22 can perform six different logic functions, depending on the values of its structural parameters, e_i and k. The parameters $e_i = k = 0$ lead to the implementation of an XOR logic gate, whose output can be read either in the value of $T(E)$ at $E = 1$ eV or in the tunneling current intensity integrating the $T(E)$ from $E = 0$ eV to $E = -1$ eV. This current intensity is represented in the inset of Fig. 22 and for varying α and β continuously from 0 to 1 eV. Stable plateaux at the corner of this map naturally correct small deviations in the inputs that lead to even smaller deviations in the output. This fundamental property of any logical device is due here to the sharp resonances that are pushed in or out of the integration region by α and β. However, this efficient programmable system is a very abstract model. More realistic systems have to be studied in order to bring the QHC approach into the molecular electronics family.

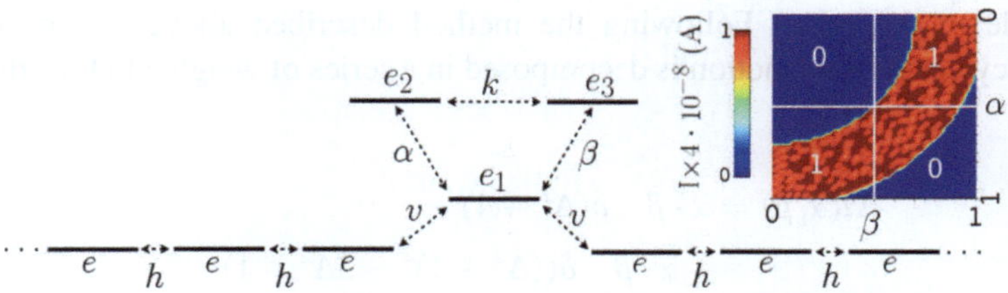

Fig. 22 (**a**) Model system able to perform six different logic functions depending on its structural parameters e_i and k. Current intensity passing through this system for $e_i = k = 0$ eV, $v = 5$ meV, and α and β going from 0 to 1 eV. The variation of the current respects the XOR truth table: a strong current is obtained for = 0 = 1 and = 1 = 0 and a weak one for = 0 = 0 and = 1 = 1. Due to the stable plateaux at the corners of the map, this device naturally corrects small deviations in the inputs that lead to even smaller deviations in the output

5.2 Molecular Implementation

Accounting for the topology of a given molecule during the symbolic analysis is much harder than playing with a model system whose elements can be tuned at will. A simple molecule like cyclobutadiene is a reasonable starting point to see whether this method can be applied to more realistic models. In a tight binding model, similar to those studied in Sect. 2, the four molecular orbitals of the network of this molecule introduce resonant tunneling channels in the junction. The two electrodes are connected to two neighboring carbon atoms but are supposed not to interact directly since it would blur the conduction of the molecule. This critical point is discussed in the following. Several solutions can be investigated to encode the logical inputs, for example different rotating functional groups can be covalently bound to the two remaining carbon atoms of the molecule. Introducing resonances in the HOMO–LUMO gap of the molecule, the NO_2 group has been found to be the best candidate to control the overall conduction of the molecule in this energetic region. Thus, the molecular junction represented in Fig. 23a is studied in the frame of the QHC approach to implement logical functions. Each logical input, α and β, controls the rotation of one of the nitro groups that can either be perpendicular to the board if the input is 0 or parallel to the board if the input is 1. To study this molecule in the frame of the symbolic analysis presented in the previous section, the model represented in the inset of Fig. 23a is used. Aside from the four p_z orbitals that form the network of the molecule, two supplementary states modeling the nitro groups are introduced. If the NO_2 is perpendicular to the molecule, the oxygen atoms screens the p_z orbital of the nitrogen atom and the NO_2 group does not modify the conduction of the naked cyclobutadiene. In this case the supplementary state is not connected to the skeleton of the molecule. On the other hand, if the NO_2 is parallel to the board the p_z orbital of the nitrogen atom introduces a new electronic pathway that consequently modifies the electronic conduction of the molecule. The supplementary state is connected to the network of the molecule. As for the model system presented in the previous section, an effective Hamiltonian, only defined in the scattering states of the electrode, can be derived using Lowdin partitioning. This effective Hamiltonian grasps the main characteristic of the evolution from one electrode to the other. Following the method described above, the oscillation frequency across the junction is decomposed in a series of weighted Dirac distribution as

$$
\begin{aligned}
\Omega(\alpha,\beta) = \bar{\alpha}\cdot\bar{\beta} \quad & \delta(\Delta^2-4) \\
+\,\bar{\alpha}\cdot\beta \quad & \delta((\Delta^2-1)^2-3\Delta^2+1) \\
+\,\alpha\cdot\bar{\beta} \quad & \delta((\Delta^2-1)^2-3\Delta^2+1) \\
+\,\alpha\cdot\beta \quad & \delta((\Delta^2-1)^3-\Delta^2(3\Delta^2-2))
\end{aligned}
$$

with $\Delta = E - e$ where E is the scattering energy and e the energy of the p_z orbitals. Tuning values allows the selection of one or several Dirac distributions

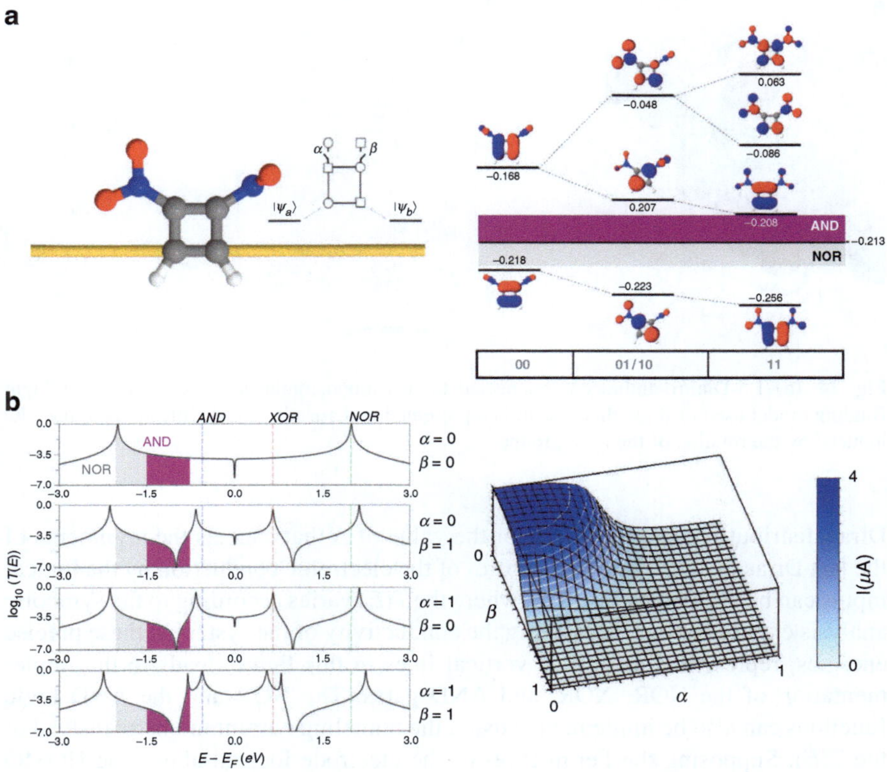

Fig. 23 (**a**) Dinitro-cyclobutadiene when connected to electrodes. Each nitro group is used to encode one bit of information each. When perpendicular to the anthracene plane, the NO2 encodes for a logical 0 whereas when in a planar configuration it encodes for a logical 1. The *inset* shows the model system used to simulate this device. (**b**) Electronic conduction of the model stem depending on the values of α and β. The *vertical dashed lines* represent the energetic values where the output status can be measured using the conductivity of the molecule. The *filled area* represents the integration limit to encode a NOR (*gray*) or an AND (*purple*) logic gate, whose output status is measured in the tunneling current intensity. (**c**) The tunneling current intensity map in the NOR configuration. The current is much higher when $\alpha = \beta = 0$. (**d**) Electronic calculation of the device using the B3LYP exchange functional and the 6-31G* basis set. The NOR and the AND gates are still implemented in this device when represented by a much more sophisticated model

and implementation of a specific logic function. The value $\Delta = \pm\sqrt{2}$ cancels out the argument of the first Dirac distribution that is weighted by the Boolean operator $\overline{\alpha} \cdot \overline{\beta}$. Therefore for this value of Δ, Ω and consequently the transmission coefficient are much larger when $\alpha = \beta = 0$ than for any other values of the logical input. The molecule is therefore a NOR logic gate whose Boolean expression is of course $\overline{\alpha} \cdot \overline{\beta}$. By the same token, one can find the values of Δ that lead to the implementation of the XOR and the AND logic functions. The XOR function is obtained for the value of Δ that cancels the arguments of the second and third

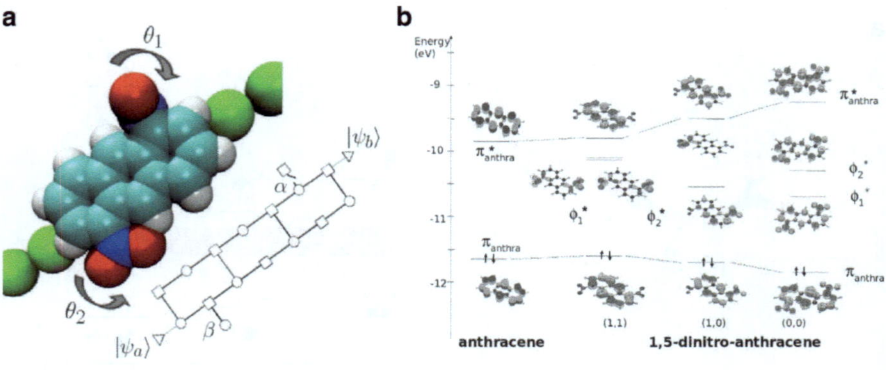

Fig. 24 (**a**) 1,5-Dinitro-anthracene connected to two monoatomic gold electrodes. The Tight Binding model used to study this system is represented underneath. (**b**) Modifications of the MO induced by the rotation of the nitro groups

Dirac distribution, and the AND for the value of Δ that cancels the arguments of the last Dirac distribution. This control of the electronic conduction by the logical inputs can be verified by Fig. 23b, where the $T(E)$ varies according to the symbolic analysis exposed above. Measuring the conductivity of the system at these precise energies, represented as dashed vertical lines in this figure, leads to the implementation of the NOR, XOR, and AND gates. The NOR and the AND logic functions can also be implemented using the tunneling current intensity instead of the $T(E)$. Supposing the Fermi level of the electrode located above the HOMO with $\alpha = \beta = 0$ configuration, and below the LUMO with $\alpha = \beta = 1$ configuration, to be around -1.5 eV, the output status of the NOR logic gate can be measured by integrating the $T(E)$ from this energy to -2 eV (gray area in Fig. 23b). This corresponds to a negative bias voltage applied on one of the two electrodes. The AND logic gate is obtained by integrating the $T(E)$ from -1.5 to -0.8 eV, which corresponds to a positive voltage applied on one of the electrode (purple area in Fig. 23b). The current intensity map obtained for the NOR logic gate is represented in Fig. 23c. A high intensity current is measured for $\alpha = \beta = 0$, whereas a very low current intensity is measured for different values. One could think that the extremely simple model used here is misleading for studying dinitrocyclobutadiene. The MOs of this molecule, using DFT with B3LYP exchange function and the 6-31G* basis set, are represented in Fig. 23c. Despite its simplicity, the model used above captures the main features of the molecule, since the AND and NOR logic functions are still implemented in the molecule, using this much more elaborate electronic structure calculation. The main problem of this design is consequently the direct coupling between the two electrodes that will screen the molecule in the junction. The next step in our design is therefore to extend the molecule to avoid this direct through-space electronic coupling between the two electrodes. Using Dewar's rules on alternating hydrocarbons is a convenient way to achieve this extension. In Fig. 23 one of the

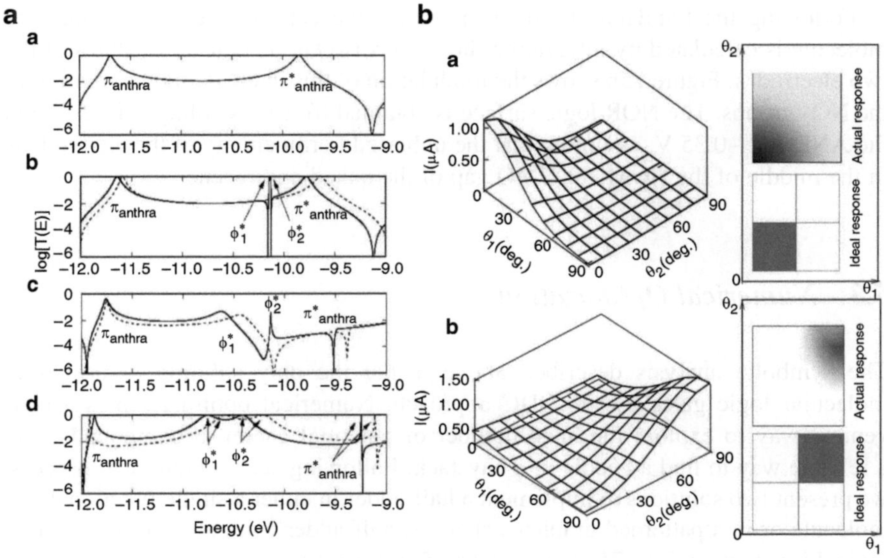

Fig. 25 (**A**) Tight-binding calculation (*dashed*) and EHMO (*plain*) of the electronic conduction, for the naked anthracene (**a**) when the two NO_2 groups are perpendicular to the molecule (**b**), when one NO_2 is rotated (**c**) and when the two NO_2 are rotated (**d**). (**B**) Tunneling current intensity for the NOR gate (**a**) and the AND gate (**b**) depending on the orientation angle of the two NO_2 groups

two electrodes is connected to a ○ site, and the other electrode to a □ site. The design can be extended by symmetry, respecting the alternation between the □ and ○. Increasing the number of states in the molecule leads to a more complicated function F, whose zeros can be calculated numerically. However, due to its symmetry, the system will perform the same functions as the simple dinitro-cyclobutadiene. We have chosen to use the [1–5]-dinitro-anthracene to embody this molecular logic gate as represented in Fig. 24a. The two electrodes, as well as the two NO_2 groups, are connected one to a ○ atom, and one to a ○ atom. The electronic conduction of this system, calculated in an EHMO model, is represented in Fig. 25b. The $T(E)$ of the naked anthracene shows the broad π_{anthra} and $\pi_{\text{anthra}}{}^*$ resonances that define the HOMO–LUMO gap of the molecule. In the perpendicular conformation, the nitro groups introduce two extremely narrow resonances labeled ϕ_1 and ϕ_2 between the HOMO and the LUMO of the molecule. The width of these resonances is due to their very weak interactions with the electrodes. Interacting with the molecular orbitals of the anthracene, they nonetheless shift slightly the HOMO and LUMO of the naked anthracene. When an NO_2 is parallel to the molecule, it introduces a new electronic pathway and therefore a supplementary resonance in the $T(E)$. It also creates interference just below the sharp π_{anthra} resonance. Finally, when they are both in the plane of the molecule, the two NO_2 groups introduce two supplementary resonances that are split by their mutual interaction through the anthracene board.

Following the Landauer–Büttiker approach, the current intensity through the molecule is calculated by integrating the *T*(*E*) between the density of states of the two electrodes. Figure 25b shows the modulation of this intensity by the rotation of the NO_2 groups. The NOR logic surface is obtained for a bias voltage of 0.5 V and the AND for -0.35 V, assuming that the unbiased Fermi energy of the electrode is in the middle of the HOMO–LUMO gap of the naked anthracene.

5.3 Numerical Optimization

The symbolic analysis described above is not the only solution to designing molecular logic gates by the QHC approach. Numerical optimization is a convenient way to explore the huge number of potential chemical compounds, and a reliable way to find a needle in a haystack. Following such a numerical process, we present two solutions to implement a half-adder in either a single functionalized molecule or in a patterned graphene sheet. A half-adder computes the sum of two logical inputs α_1 and α_2. Therefore, the half-adder has two output statuses: the sum $s = \alpha_1 \oplus \alpha_2$ and the carry $c = \alpha_1 \cdot \alpha_2$. The half-adder is consequently the superposition of an XOR and an AND logic gate. In order to implement this function in a single molecule, this latter has to be connected to three electrodes: one to deliver the current and two to measure the logical output status. The molecule also needs two switchable groups to encode the value of α_1 and α_2.

5.3.1 Building Up a Molecule

The first solution to use numerical optimization in the frame of the QHC is to optimize the chemical structure of a molecule. Defining simple chemical rules to assemble several fragments, a systematic search among millions of chemical components can be numerically achieved to find those whose electronic conduction respects the truth table of the half-adder. As in the precedent examples, the logical inputs are encoded in the rotation of two NO_2 groups. A “0” encodes for a planar conformation, and a “1” for a perpendicular configuration. The algorithm automatically generates stable chemical compounds incorporating two nitro groups. Their electronic conductions are computed between all the possible points where the three electrodes can be connected and for the four different conformations of the NO_2 groups. If the response of one compound corresponds to the half-addition, the algorithm saves it and discards it otherwise. After searching among millions of molecules, the dibenzo[*a*,*e*]fluoranthene presented in Fig. 26 has been found to be the best candidate to perform the half-addition. As shown in Fig. 26b, the resistance of the molecule is much lower when the corresponding logical output equals one than when it is zero, ensuring a lower current intensity in the latter case.

a

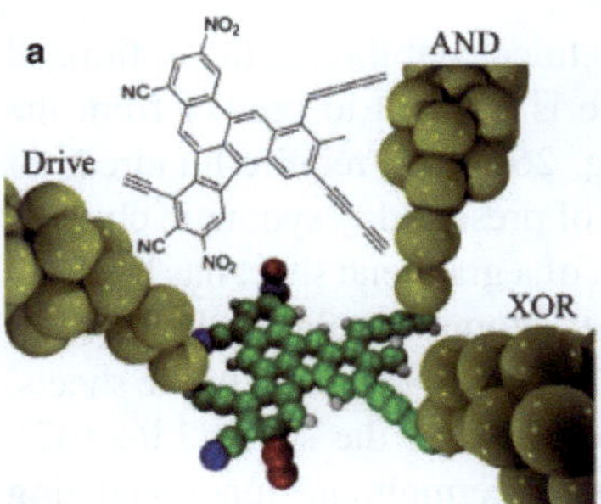

b

input	R_{XOR}		R_{AND}	
00	79.7	$G\Omega$	516.0	$G\Omega$
01	4.79	$G\Omega$	3.46	$G\Omega$
10	1.86	$G\Omega$	1.13	$G\Omega$
11	147.0	$G\Omega$	6.22	$M\Omega$

b

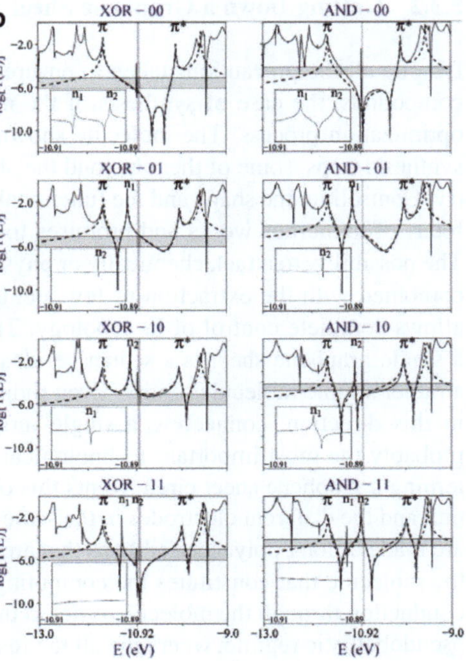

Fig. 26 (**a**) The chemical structure of the molecular half-adder. The conformation of each NO_2 group encodes the logic input while the output status is encoded in the resistance between the drive and the output nano-electrodes. The complete truth table for the XOR and the AND outputs. Note the difference in magnitude between the XOR "1" and the AND "1". (**b**) The $T(E)$ spectra of the junction represented in Fig. 26 for all the logic inputs (*solid line*). Each *inset* emphasizes the modification of the conductance near the Fermi energy of the molecule. Each $T(E)$ spectrum had been fitted in the active area to determine the minimum number of quantum levels required to reproduce it (*dashed line*)

In order to reveal the mechanism of this molecular half-adder, the $T(E)$ spectra of the molecule are presented in Fig. 26b. When perpendicular to the plane of the molecule, each NO_2 contributes a very sharp resonance which does not participate in the overall conductance. When rotated by 90°, an NO_2 introduces a supplementary resonance in the gap of the molecule. Due to its asymmetrical delocalization over the atomic orbitals, this resonance increase the conduction between the drive and the XOR electrode, but not between the drive and the AND electrode. This insures a "1" output for the former and a "0" for the latter. When the two NO_2s are rotated, the two resonances they introduce create a deep interference between the drive and the XOR electrode. Located on the Fermi energy of the molecule, this interference leads to a low conductance state and a "0" logical output for the XOR gate. In contrast, the two resonances do not interfere destructively between the drive and the AND electrode, leading to a high conductance state and a "1" logical output.

5.3.2 Cutting Down a Graphene Sheet

Despite all the precautions taken to ensure the chemical stability of the optimized compounds, the ease of synthesis of the molecule is difficult to predict from the optimization process. The molecule shown in Fig. 26 would require hundreds of synthesis steps, some of them beyond the abilities of present-day synthetic chemistry. Controlling the shape and the functionalization of a graphene sheet has been the focus of numerous works and promises to reach an atomic resolution [132, 133]. The possibility to attack chemically or physically the edges [134, 135] of the sheets, combined with the extraction of few atoms in the center of the sheet [136, 137], allows complete control of its topology. Therefore patterning and functionalizing a single graphene sheet is a seductive idea to implement a complex logic gate as a rather simple molecular entity. Even though much effort is currently being made in this direction, connecting a single molecule to more than two electrodes is probably the most important technological bottleneck of the QHC approach. Patterning a graphene sheet circumvents this obstacle by incorporating the computing unit and the different electrodes in the same sheet. In this framework, the electrodes are made of long polyacene ribbons that are covalently bonded to the central part of the molecule that constitutes the computing unit. In this framework the electronic conduction through the molecule is not in the tunneling regime any more, but in the pseudoballistic regime, where the all the resonances are broadened. Controlling the rotation of the NO_2 groups to encode the logical input is another experimental challenge. Other solutions can be explored to encode the classical logical input in the Hamiltonian of the molecule. Since they do not require supplementary electrical contacts on the molecule, photochromic groups seem to be interesting candidates to encode such logical inputs. The *cis–trans*-isomerization of stilbene is one of the well-known optically-triggered conformational transformations [138–140]. When illuminated with the right wavelength, this photochromic compound can switch from an open to a closed configuration. If such a compound is used to functionalize a graphene sheet, changing its conformation allows modification of the overall conductance of the entire graphene sheet. We consequently choose to encode each logical input in the photoisomerization state of a single stilbene molecule that is supposed to be attached to the edge of the sheet. Using two slightly different stilbene groups allows encoding of the two logical inputs that are then controlled by different wavelengths. An optimization process very similar to that used to build up the molecule represented in Fig. 26 has been used to shape a small graphene sheet. For each position of the electrode, the shape of the sheet is modified by removing atoms, while preserving the aromaticity of the resulting sheet. If the electronic conduction of the resulting system respects the truth table of the half-adder, the system is kept and is discarded otherwise. This optimization process leads to the graphene sheet shown in Fig. 27a. A single carbon atom has been removed from a small hexagonal sheet connected to three electrodes. The two stilbene groups are positioned at the end of small polyacene ribbons. Switching these systems from a closed to an open configuration perturbs sufficiently the

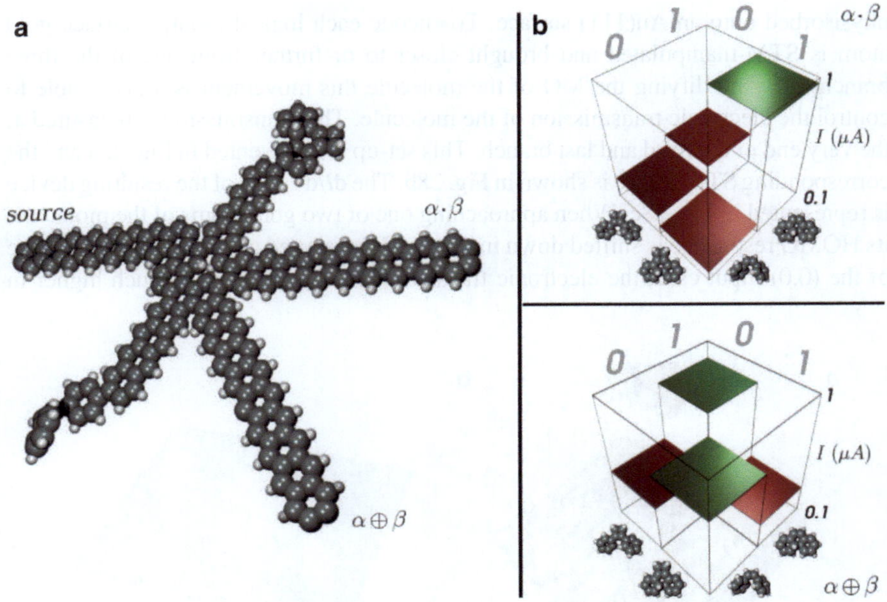

Fig. 27 (**a**) Optimized graphene sheet for the realization of a half-adder. Each logical input noted α and β controls the photoisomerization state of one of the two stilbene groups. Depending on this isomerization state, the overall conductance of the molecule between the three electrodes is modified. (**b**) Current intensity calculated in the two output electrodes depending on the conformation of the stilbene groups

transmission through the molecule, and the system will perform the half addition. The current intensity going from the drive electrode to the two output electrodes is represented for the four different conformations of the stilbene in Fig. 27b. Due to the strong coupling between the electrodes and the molecule, a high current intensity but a small margin between a 0 and a 1 output status are obtained. This top-down-like approach of a QHC circuit allows foreseeing of the implementation of small computing units in periodical systems patterned with atomic precision. The same approach is currently applied to SiH (111) surfaces, where a single hydrogen atom can be selectively removed from the surface [141–143]. Using this extraordinary technique one can imagine directly drawing a QHC circuit!

5.4 *Experimental Realization*

The QHC logic gate presented in the previous section performs complex logic functions but is still inaccessible experimentally. To ensure the validity of the QHC approach, a simple logic gate should be implemented in a realistic system and tested experimentally. To embody such a realistic QHC logic gate, the starphene molecule represented in Fig. 27 has been chosen [144, 145]. The three branches of this molecule allow encoding the input and reading the output easily when this molecule is physisorbed onto an Au(111) surface. To encode each logical input, a surface gold atom is STM-manipulated and brought closer to or further from one of the three branches. By modifying the MO of the molecule this movement is a priori able to control the electronic transmission of the molecule. This transmission is measured at the very end of the third and last branch. This set-up is represented in Fig. 28a and the corresponding STM image is shown in Fig. 28b. The d*I*/d*V* map of the resulting device is represented in Fig. 28c. When approaching one or two gold atoms of the molecule, its HOMO resonance is shifted down in energy. If measured at the HOMO resonance of the (0,0) input case, the electronic transmission of the system is much higher in

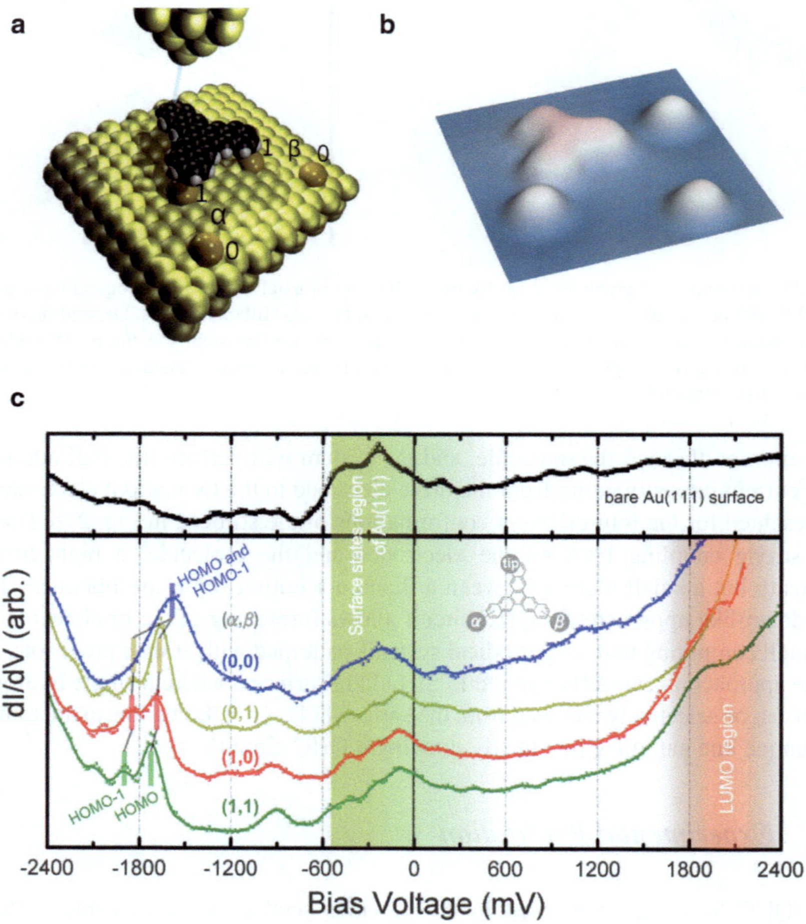

Fig. 28 (**a**) Model of the starphene molecule physisorbed on an Au(111) surface with an STM tip positioned to read the output of the logic gate. The position of the two surface atoms encodes the value of the logical input. (**b**) The corresponding experimental image with three Au surface atoms in the vicinity of the molecule. (**c**) The experimental d*I*/d*V* of the molecule showing how the energetic position of the HOMO is shifted down when the Au atoms are brought closer from the molecule

the (0,0) configuration than in the other three: (0,1), (1,0), and (1,1). This leads to the experimental realization of a NOR logic gate following the QHC approach.

6 Conclusion

Single-molecule logic devices can be designed in many different architectures. The simplest is very classical, one molecule plus the number of required metallic nano-pads. This forms a device if those nano-pads are long enough for the conductance of the "nano-pads–molecule–nano-pads" molecular device to be stabilized. In this case, a complex logic circuit can be simply assembled by respecting the classical Kirchoff nodes and meshes laws of electrical circuits, and also the Shannon Boolean logic symbolic analysis. The drawback of this classical design is that all the three terminal devices designed or experimented with so far have a very small gain and a very large on-resistance [121]. Therefore, the number of logic layers is very limited even in a crossbar-like architecture. The second type of architecture is to integrate all the circuitry inside a single molecule, forcing the molecule to assume the shape of an electronic logic circuit. This semiclassical mapping leads to a simple molecule logic gate, like an OR, an AND, or an XOR gate as presented here. The running current is very low, which forbids the extension of the lateral molecular size further than simple logic gates. The third type of architecture is to play with intramolecular quantum phenomena, either by inputting the data to be manipulated on the initial state vector or on the Hamiltonian [146]. The first case is the well-known quantum computer (QC) qubit approach. The second is the more recent QHC approach [131, 146–148]. In both cases, the results of the quantum computation can be measured on the state vector reached after the required computing period of time. QHC can be called a hemiquantum approach. It also opens up the possibility to get the output measured as a current [144, 145]. As compared to the semiclassical approach, it is not necessary for the output current to pass through the whole molecule. Quantum information is distributed among the different computing units of the molecule. The current is measured only at some specific well-designed output ports of the molecule. Both QC and QHC are opening the possibility to design very complex logic gates, without increasing too much the width of the molecule. Whatever the architecture solution, an important condition for such logic gates to work properly is that the molecular structure and its nano-pads (if necessary also the interconnecting nano-wires between the internal parts) be supported by a surface. It turns out that few molecular devices or logic gates have been designed or experimented with in a surface planar technology. One difficulty is the surface leakage current between the devices or directly between the interconnecting nano-pads. The second difficulty is the possible deformation of the desired molecular electronic structure of the device due to surface adsorption. The third difficulty is that, to avoid any random behavior of the device, the nano-pads must be fabricated with atomic-scale precision. The fourth difficulty is that the end molecular orbital overlaps between the molecular devices and its contacting nano-pads must be very precisely defined [149]. Chemisorption is not favorable because it deforms the electronic density of states of the surface nano-pads, and the junction conductance is generally very small in this case. Physisorption may introduce an electronic coupling between the molecule end and the surface nano-pads that is too small. One solution to those problems is to abandon the use of molecules, to construct complex nano-device, and to use surface atomic-scale circuits. It may seem strange to dig out surface atomic circuits in a book about molecular devices. But after all, such circuits are constructed, atom by atom, on the passivated surface of a semiconductor [150]. Therefore, surface atomic orbitals can be shaped in such a way as to form specific surface molecular orbitals with a quantum behavior similar to that of molecular orbitals. Semiclassical, hemiquantum (QHC), and qubit circuits can, in principle, be designed, playing with surface dangling bond states, as one can play with molecular states to design a molecular logic gate. The difference is that atomic-scale wires can also ensure parts of the interconnections. The conductance of surface atomic wires and of surface atomic scale tunnel junctions has already been calculated [150]. It may turn out that a hybrid combination of surface atomic wires and specific molecular switches will be the grand solution for constructing atomic-scale surface logic gates. Whatever the material (single molecule or surface atomic-scale circuit), it remains that a new technology must now be developed to construct atomically precise planar devices and logic gates, and, in the near future, more complex circuits. This is now underway with the very new atomic scale interconnection machines [151] that can provide at the same time the required atomic-scale precision of multiaccess single molecule (or surface atomic scale) circuit interconnects and the nano-to-micro interconnects, without perturbing the surface atomic-scale order of the constructed circuit. A specific nano-packaging [111] will also have to be worked out for molecular logic gate to work at ambient atmosphere and temperature.

References

1. Tinder RF (2000) Engineering digital design, revised 2nd edn. Elsevier Academic, Amsterdam
2. Horowitz P, Hill W (1989) The art of electronics. Cambridge University Press, Cambridge
3. Riordan M (2005) How Europe missed the transistor. IEEE Spectrum 42:52
4. Shanon CE (1936) A symbolic analysis of relay and switching circuits. Master's thesis, MIT 55. MIT, Cambridge
5. Saxena AN (2009) Invention of integrated circuits: untold important facts. World Scientific, Singapore
6. Landauer R (1990) Advanced technology and truth in advertising. Physica A 168:75
7. Intel Corporation (2009) Fact Sheet, Fun facts : exactly how small (and cool) is 32 nanometers?
8. Moore GE (1965) Cramming more components onto integrated circuits. Electronics 38:114
9. Tri-gate transistor architechture with high-k gate dielectrics: metal gates and strain engineering. Digest of Technical Papers, IEEE Trans VLSI Tech Symp 62
10. Landauer R (1996) Need for critical assessment. IEEE Trans Electron Devices 43:1637
11. Wigington RL (1959) A new concept in computing. Proc IRE 47:516
12. Goto E (1959) The parametron, a digital computing element which utilizes parametric oscillation. IEEE Trans Proc IRE 47:1304
13. Reimann OA, Kosonocky WF (1965) Progress in optical computer research. IEEE Spectrum 2:181

14. Pei-li L, De-xui H, Xin-Liang Z, Guang-xi Z (2006) Ultrahigh-speed all-optical half adder on four-wave mixing in semiconductor optical amplifier. Opt Express 14:11840
15. Aviram A, Ratner M (1974) Molecular rectifiers. Chem Phys Lett 29:277
16. Rectifying characteristics of Mg[(C16H33Q3CNQ LB film)]Pt structures. J Chem Soc Chem Commun 1374
17. Metzger RM (2003) Unimolecular electrical rectifiers. Chem Rev 103:3803
18. Binnig G, Rohrer H (1982) Scanning tunneling microscopy. Helv Phys Acta 55:726
19. Binnig G, Rohrer H (2000) Scanning tunneling microscopy. IBM J Res Dev 44:279
20. Aviram A, Joachim C, Pomerantz M (1988) Evidence of switching and rectification by a single molecule effected by a scanning tunneling microscope. Chem Phys Lett 146:490
21. Aviram A, Joachim C, Pomerantz M (1989) Chem Phys Lett 162:416
22. Strikov DB, Snider GS, Stewart DR, Williams RS (2008) The missing Memristor found Nature 453:80
23. Joachim C, Gimzewski JK, Aviram A (2000) Electronics using hybrid-molecular and mono-molecular devices. Nature 408:541
24. Cuevas JC (2010) Molecular electronics: an introduction to theory and experiment. World Scientific Series in Nanotechnology and Nanoscience. World Scientific, Singapore
25. Aviram A (1988) Molecules for memory, logic, and amplification. J Am Chem Soc 110:5687
26. Carter FL (1984) The molecular device computer: point of departure for large scale cellular automata. Physica D 10:175
27. Ellenbogen JC, Love JC (2000) Architectures for molecular electronic computers. Proc IEEE 88:386
28. Duchemin I, Renaud N, Joachim C (2008) A intramolecular digital 1/2-adder. Chem Phys Lett 452:269
29. Tersoff J, Hamann DR (1983) Theory and application for the scanning tunneling microscope. Phys Rev Lett 50:1998
30. Landauer R (1957) Spatial variation of currents and fields due to localized scatterers. IBM J Res Dev 1:223
31. Lippmann BA, Schwinger J (1950) Variation principle for scattering processes I. Phys Rev 79:469
32. Mujica V, Kemp M, Ratner MA (1994) Electron conduction in molecular wires. I. A scattering formalism. J Chem Phys 101:6849
33. Mujica V, Kemp M, Ratner M (1994) Electron conduction in molecular wires. II. Application to scanning tunneling microscopy. J Chem Phys 101:6856
34. Pendry JB, Pretre AB, Krutzen BCH (1991) Theory of the scanning tunneling microscope. J Phys Condens Matter 3:4313
35. Todorov TN, Briggs GAD, Sutton AP (1993) Elastic quantum transport through small structures. J Phys Condens Matter 5:2389
36. Nieminen JA, Niemi E, Rieder KH (2004) Interference between competing tunneling channels and chemical resolution of STM. Surf Sci Lett 552:L47
37. Samanta MP, Tian W, Datta S, Henderson JI, Kubiak CP (1996) Electronic conduction through organic molecules. Phys Rev B 53:R7626
38. Tian W, Datta S, Hong S, Reifenberger R, Henderson JI, Kubiac CP (1998) Conductance spectra of molecular wires. J Chem Phys 109:2874
39. Heurich J, Cuevas JC, Wenze W, Schön G (2002) Electrical transport through single-molecule junctions: from molecular orbitals to conduction channels. Phys Rev Lett 88:256803
40. Cuevas JC, Heurich J, Pauly F, Wenzel W, Schön G (2003) Theoretical description of the electrical conduction in atomic and molecular junctions. Nanotechnology 14:R29
41. Nieminem J, Lahti S, Paavilainen S (2002) Contrast changes in STM images and relations between different tunneling models. Phys Rev B 66:165421
42. Cerda J, van Hove MA (1997) Efficient method for the simulation of STM images. I. Generalized green-function formalism. Phys Rev B 56:15885

43. Pickup BT, Fowler PW (2008) An analytical model for steady-state currents in conjugated systems. Chem Phys Lett 459:198
44. Ernzerhof M, Bahmann H, Goyer F, Zhuang M, Rocheleau P (2006) Electron transmission through aromatic molecules. J Chem Theor Comput 2:1291
45. Solomon GC, Andrews DQ, Hansen T, Goldsmith RH, Wasielezski MR, van Duyne RP, Ratner MA (2008) Understanding quantum interferences in coherent molecular conduction. J Chem Phys 129:054701
46. Yeganeh S, Ratner MA, Mujica V (2007) Dynamics of charge transfer: rate processes formulated with nonequilibrium Green's function. J Chem Phys 126:161103
47. Nitzan A (2001) A relationship between electron-transfer rates and molecular conduction. J Phys Chem A 105:2677
48. Hansen T, Solomon GC, Andrews DQ, Ratner MA (2009) Interfering pathway in benzene: an analytical treatment. J Chem Phys 131:194704
49. Solomon GC, Hermann C, Hansen T, Mujica V, Ratner MA (2010) Exploring local currents in molecular junctions. Nat Chem 2:223
50. Bar-Joseph I, Gurvitz SA (1991) Time-dependent approach to resonant tunneling and inelastic scattering. Phys Rev B 44:3332
51. Ness H, Fisher AJ (1997) Nonperturbative evaluation of STM tunneling probability from ab initio calculations. Phys Rev B56:12469
52. Sanchez CG, Stamenova M, Sanvito S, Bowler DR, Horsfield AP, Todorov N (2006) Molecular conduction: does time-dependent simulation tell you more than the Landauer approach? J Chem Phys 124:214708
53. Subotnik JE, Hansen T, Ratner MA, Nitzan A (2009) Nonequilibrium steady-state transport via the reduced density-matrix operator. J Chem Phys 130:144105
54. Stratford K, Beeby JL (1993) A time-dependent approach to conductance in narrow channel. J Phys Condens Matter 5:L289
55. Doyen G (1993) Tunnel current and generalized Ehrenfest theorem. J Phys Condens Matter 5:3305
56. Joachim C, Ratner MA (2005) Molecular electronics: some views on transport junctions and beyond. Proc Natl Acad Sci USA 102:8801
57. Remacle F, Levine RD (2006) Electrical transport in saturated and conjugated molecular wires. Faraday Discuss 131:45
58. Maciejko J, Wang J, Guo H (2006) Time-dependent quantum transport far from equilibriam: an exact nonlinear response theory. Phys Rev B 74:085324
59. Sautet P, Joachim C (1988) Electronic transmission coefficient for the single-impurity problem in the scattering-matrix approach. Phys Rev B 38:12238
60. Ami S, Joachim C (2002) Intramolecular circuits connected to N electrodes using a scattering matrix approach. Phys Rev B 65:155419
61. English RA, Davison SG (1994) Transmission properties of molecular switches in semiconducting polymers. Phys Rev B 49:8718
62. Villagomez CJ, Zambelli T, Gauthier S, Gourdon A, Barthes C, Stojkovic S, Joachim C (2007) A local view on hyperconjugation. Chem Phys Lett 450:107
63. Villagomez CJ, Zambelli T, Gauthier S, Gourdon A, Stojkovic S, Joachim C (2009) STM images of a large organic molecule adsorbed on a bare metal substrate or on a thin insulating layer: visualization of HOMO and LUMO. Surf Sci 603:1526
64. Bellec A, Ample F, Riedel D, Dujardin G, Joachim C (2009) Imaging molecular orbitals by scanning tunneling microscopy on a passivated semiconductor. Nano Lett 9:144
65. Fano U (1961) Effects of interaction configuration on intensities and phase shifts. Phys Rev 124:1866
66. Fano U, Rau ARP (1986) Atomic collisions and spectra. Academic, Orlando, FL, USA, p 57
67. Mies FH (1968) Configuration interaction theory. effects of overlapping resonances. Phys Rev 175:164
68. Durand P, Paidarová I, Gadéa FX (2001) Theory of Fano profiles. J Phys B 34:1953

69. Durand P, Paidarová I (2002) Theory of generalized Fano profiles. J Phys B 35:469
70. Gurvitz SA, Kalbermann G (1987) Decay width and the shift of a quasistationary state. Phys Rev Lett 59:262
71. Reuter MG, Hansen T, Seideman T, Ratner MA (2009) Molecular transport junction with semiconductor electrodes: analytical forms for one-dimensional self-energies J. Phys Chem A 113:4665
72. Renaud N, Joachim C (2008) Design and stability of NOR and NAND logic gates constructed with three quantum states. Phys Rev A 78:062316
73. Joachim C, Launay JP (1986) Bloch effective Hamiltonian for the possibility of molecular switching in the ruthenium-bipyridylbutadiene-ruthenium system. Chem Phys 109:93
74. Soe W, Manzano C, Renaud N, de Mandoza P, de Sakrar A, Ample F, Echavarren AM, Chandrasekhar N, Hliwa M, Joachim C (2011) Manipulating molecular quantum states with classical metal atom inputs: demonstration of a single molecule NOR logic gate. ACS Nano 5:1436
75. Bloch C (1958) Sur la Théorie des Perturbations des États Liés. Nucl Phys 6:329
76. Durand P (1983) Direct determination of effective Hamiltonians by wave-operator methods. I. General formalism. Phys Rev A 28:3184
77. Sautet P, Joachim C (1988) Electronic interferences produced by benzene embedded in a polyacetylene chain. Chem Phys Lett 153:511
78. Solomon GC, Andrews DQ, Van Duyne RP, Ratner MA (2008) When things are not as they seem: quantum interference turns molecular electron transfer "rules" upside down. J Am Chem Soc 130:7788
79. Cacciolati O, Joachim C, Martinez JP, Carcenac F (2004) Fabrication of N-electrodes nanojunction for monomolecular electronic interconnects. Int J Nanosci 3:233
80. Galperin M, Ratner MA, Nitzan A (2007) Molecular transport junctions: vibrational effects. J Phys Condens Matter 19:103201
81. Fu Q, Luo Y, Yang J, Hou J (2010) Understanding the concept of randomness in inelastic electron tunneling excitations. Phys Chem Chem Phys 12:12012
82. Sergueev N, Demkov AA, Guo H (2007) Inelastic resonant tunneling in C_{60} molecular junction. Phys Rev B75:233418
83. Seideman T, Guo H (2003) Quantum transport and current-triggered dynamics in molecular junctions. J Theor Comput Chem 2:439
84. Solomon GC, Gagliardi A, Pecchia A, Frauenheim T, Di Carlo A, Reimers JR, Hush NS (2006) Understanding the inelastic-tunneling spectra of alkanedithiols on gold. J Chem Phys 124:094704
85. Galperin M, Ratner MA, Nitzan A, Troisi A (2008) Nuclear coupling and polarization in molecular transport junction: beyond tunneling to function. Science 319:1056
86. Wintterlin J, Schuster R, Ertl G (1996) Existence of a 'hot' atoms mechanism for the dissociation of O_2 on Pt(111). Phys Rev Lett 77:123
87. Stipe BC, Rezaei MA, Ho W, Gao S, Persson M, Lundquist BI (1997) Single molecule dissociations by tunneling electrons. Phys Rev Lett 78:4410
88. Hahn JR, Lee HJ, Ho W (2000) Electronic resonance and symmetry in single-molecule inelastic electron tunneling. Phys Rev Lett 85:1914
89. Zambelli T, Barth JV, Wintterlin J (2002) Thermal dissociation of chemisorbed oxygen molecules on Ag(110): an investigation by scanning tunneling microscopy. J Phys Condens Matter 14:4241
90. Sloan PA, Palmer RE (2005) Two-electron dissociation of single molecules by atomic manipulation at room temperature. Nature 434:367
91. Pascual JI, Lorente N, Song Z, Conrad H, Rust HP (2003) Selectivity in vibrationally mediated single-molecule chemistry. Nature 423:525
92. Stipe BC, Rezaei MA, Ho W (1998) Inducing and viewing the rotational motion of a single molecule. Science 279:1907

93. Gimzewski JK, Joachim C (1999) Nanoscale science of single molecule using local probes. Science 283:1683
94. Manzano C, Soe WH, Wong HS, Ample F, Gourdon A, Chandrasekhar N, Joachim C (2009) Step-by-step rotation of a molecule-gear mounted on an atomic-scale axis. Nat Mater 8:576
95. Schwarzer D, Kutne P, Schroder C, Troe J (2004) Intramolecular vibrational energy redistribution in bridged azulene-anthracene compounds: ballistic energy transport through molecular chains. J Chem Phys 121:1754
96. Galperin M, Nitzan A, Ratner MA (2007) Heat conduction in molecular transport junction. Phys Rev B 75:155312
97. Joachim C (1987) Control of the quantum path-target state distance: bistable-like characteristic in a small tight-binding system. J Phys A 20:L1149
98. Joachim C (1987) Ligand-length dependence of the intramolecular electron transfer through-bond coupling parameter. Chem Phys 116:339
99. Landauer R (1989) Can we switch by control of quantum mechanical transmission? Phys Today 42:119
100. Ami S, Joachim C (2001) Logic gates and memory cells based on single C_{60} electromechanical transistor. Nanotechnology 12:44
101. Eigler D, Lutz CP, Rudge WE (1991) An atomic switch realized with the scanning tunneling microscope. Nature 352:600
102. Joachim C, Sautet P, Lagier P (1992) The tip apex structure of the Eigler atomic switch. EuroPhys Lett 20:697
103. Moresco F, Meyer G, Rieder KH, Tang H, Gourdon A, Joachim C (2001) Recording intramolecular mechanics during the manipulation of a large molecule. Phys Rev Lett 87:088302
104. Loppacher C, Guggisberg M, Pfeiffer O, Meyer E, Bammerlin M, Lüthi R, Gimzewski JK, Joachim C (2003) Direct determination of the energy required to operate a single molecule switch. Phys Rev Lett 90:066107
105. Lastapis M, Martin M, Riedel D, Hellner L, Comtet G, Dujardin G (2005) Picometer-scale electronic control of molecular dynamics inside a single molecule. Science 308:1000
106. Hugel T, Holland NB, Cattani A, Moroder L, Seitz M, Gaub H (2002) Single-molecule optomechanical cycle. Science 296:1102
107. Sautet P, Joachim C (1989) The salicylideneaniline molecular switch revisited. Chem Phys 135:99
108. Chen J, Reed MA, Rawlett AM, Tour JM (1999) Large on-off ratios negative differential resistance in molecular electronic devices. Science 286:1550
109. Jlidat N, Hliwa M, Joachim C (2008) A semi-classical XOR logic gate integrated in a single molecule. Chem Phys Lett 451:270
110. Ami S, Hliwa M, Joachim C (2003) Molecular OR and AND logic gates integrated in a single molecule. Chem Phys Lett 367:662
111. Metzger RM (2008) Unimolecular electronics. J Mater Chem 18:4364
112. Sautet P, Joachim C (1992) Are electronic interference effects important for stm imaging of substrates and adsorbates? A theoretical analysis. Ultramicroscopy 42:115
113. Higelin D, Sixl H (1983) Spectroscopic studies of the photochromism on N-salicylideneaniline mixed crystals and glasses. Chem Phys 77:391
114. Solomon GC, Andrews DQ, Van Duyne RP, Ratner MA (2009) Electron transport through conjugated molecules: when the π system only tells part of the story. ChemPhysChem 10:257
115. Bauer R, Neuhauser D (2002) Phase coherent electronics: a molecular switch based on quantum interference. J Am Chem Soc 124:4200
116. Ricks AB, Solomon GC, Colvin MT, Scott AM, Chen K, Ratner MA, Wasielewski MR (2010) Controlling electron transfer in donor-bridge-acceptor molecules using cross-conjugated bridges. J Am Chem Soc 132:15427

117. Liljeroth P, Repp J, Meyer G (2007) Current-induced hydrogen tautomerization and conductance switching of naphthalocyanine molecules. Science 317:1203
118. Joachim C, Gimzewski JK, Tang H (1998) Physical principles of the single-C_{60} transistor effect. Phys Rev B 58:16407
119. Tour JM, Kozaki M, Seminario JM (1998) Molecular scale electronics: a synthetic/computational approach to digital computing. J Am Chem Soc 120:8486
120. Joachim C, Gimzewski JK (1997) Electromechanical amplifier using a single molecule. Chem Phys Lett 265:353
121. Joachim C, Gimzewski JK (1998) A nanoscale single-molecule amplifier and its consequences. Proc IEEE 86:184
122. Ami S, Hliwa M, Joachim C (2003) Balancing a four-branch-single-molecule nanoscale Wheatstone bridge. Nanotechnology 14:283
123. Ample F, Duchemin I, Hliwa M, Joachim C (2011) Single OR-molecule and OR atomic circuit logic gates interconnected on Si(100)H surface. J Phys Condens Matter 23:125303
124. Joachim C, Launay JP, Woitellier S (1990) Distance dependence of the effective coupling parameters through conjugated ligands of the polyene type. Chem Phys 147:131
125. Magoga M, Joachim C (1991) Conductance of molecular wires connected or bonded in parallel. Phys Rev B 59:16011
126. Sautet P, Joachim C (1991) Calculation of the benzene on rhodium STM image. Chem Phys Lett 185:23
127. Jlidat N, Hliwa M, Joachim C (2009) A molecule OR logic gate with no molecular rectifier. Chem Phys Lett 470:275
128. Terman FE, Pettit JM (1952) Electronic measurements, 2nd edn. Radio Engineers Handbook, McGraw-Hill, London
129. Arden W, Brillouët M, Cogez P, Graef M, Huizing B and Mahnkopf R (2009) More-than-Moore, ITRS white paper.
130. Lafferentz L, Ample F, Yu H, Hecht S, Joachim C, Grill L (2009) Conductance of a single conjugated polymer as continuous function of its length. Science 323:1193
131. Renaud N, Ratner MA, Joachim C (2011) A time-dependent approach to electronic transmission in model molecular junctions. J Phys Chem B. doi:dx.doi.org/10.1021/jp111384d
132. Geim AK, Novoselov KS (2007) The rise of graphene. Nat Mater 6:183
133. Wu J, Pisula W, Mullen K (2007) Graphene molecules as potential material for electronics. Chem Rev 107:718
134. Li X, Wang X, Shang L, Lee S, Dai H (2008) Chemical derived, ultrasmooth graphene nanoribbon semiconductors. Science 319:1229
135. Dayen JF, Mahmood A, Golubev DS, Roch-Jeune I, Salles P, Dujardin E (2008) Side-gated transport in focused-ion-beam-fabricated multilayered graphene nanoribbons. Small 4:716
136. Park JG, Zhang C, Liang R, Wang B (2007) Nano-machining of highly oriented pyrolytic graphite using conductive atomic force microscope tips and carbon nanotubes. Nanotechnology 18:405306
137. Ewels CP, Heggie MI, Briddon PR (2002) Adatoms and nanoengineering of carbon. Chem Phys Lett 351:178
138. Frederick JH, Fujiwara Y, Penn JH, Yoshihara K, Petek H (1991) Models for stilbene photoisomerisation: experimental and theoretical studies of the excited-states dynamics of 1,2-diphenylcycloalkenes. J Phys Chem 95:2845
139. Laarhoven WH (1984) Aspects of the photochemistry of aryl ethylenes. Pure Appl Chem 56:1225
140. Irie M (2000) Photochromism: memories and switches - introduction. Chem Rev 100:1685
141. Mayne AJ, Riedel D, Comtet G, Dujardin G (2006) Atomic-scale studies of hydrogenated semiconductor surfaces. Prog Surf Sci 81:1
142. Bellec A, Riedel D, Dujardin G (2008) Dihydride dimer structures on the Si(100):H surface studied by low-temperature scanning tunneling microscopy. Phys Rev B 78:165302

143. Riedel D, Cranney M, Martin M, Guillory R, Dujardin G, Dubois M, Sautet P (2009) Surface-isomerization dynamics of trans-stilbene molecule adsorbed on Si(100)-2x1. J Am Chem Soc 131:5414
144. Soe W, Manzano C, Renaud N, de Mandoza P, de Sakrar A, Ample F, Echavarren AM, Chandrasekhar N, Hliwa M, Joachim C (2011) Demonstration of a NOR logic gate using a single molecule and two surface gold atoms to encode the logical input. Phys. Rev B 83:155443
145. Stojkovic S, Joachim C, Grill L, Moresco F (2005) The contact conductance on a molecular wire. Chem Phys Lett 408:134
146. Renaud N, Joachim C (2011) Classical Boolean logic gates with quantum systems. J Phys A 44:155302
147. Duchemin I, Renaud N, Joachim C (2005) A quantum digital half-adder inside a single molecule. Chem Phys Lett 406:167
148. Renaud N, Ito M, Shangguan W, Saeys N, Hliwa M, Joachim C (2009) A NOR-AND quantum running gate molecule. Chem Phys Lett 472:74
149. Kawai H, Yeo YK, Saeys M, Joachim C (2010) Conductance decay of a surface hydrogen tunneling junction fabricated along a Si(001)-(2x1)-H atomic wire. Phys Rev B 81:195316
150. Joachim C, Martrou D, Rezeq M, Troadec C, Jie D, Chandrasekhar N, Gauthier S (2010) Multiple atomic-scale solid surface interconnects for atom circuits and molecule logic gates. J Phys Condens Matter 22:084025
151. Lwin MHT, Tun TN, Kim HH, Kajen RS, Chandrasekhar N, Joachim C (2010) Silicon on insulator nanoscale backside interconnects for atomic and molecular scale circuits. J Vac Sci Tech B 28:978

Index

MIX
Papier aus verantwortungsvollen Quellen
Paper from responsible sources
FSC® C105338

If you have any concerns about our products,
you can contact us on
ProductSafety@springernature.com

In case Publisher is established outside the EU,
the EU authorized representative is:
Springer Nature Customer Service Center GmbH
Europaplatz 3, 69115 Heidelberg, Germany

Printed by Libri Plureos GmbH
in Hamburg, Germany